3rd International Conference on Advanced Materials Research and Applications (AMRA 2016)

IOP Conference Series: Materials Science and Engineering
Volume 170

Guangzhou, China
18 – 21 December 2016

Editor:

N.M. Ravindra

ISBN: 978-1-5108-3674-7
ISSN: 1757-8981

Printed from e-media with permission by:

Curran Associates, Inc.
57 Morehouse Lane
Red Hook, NY 12571

CURRAN ASSOCIATES INC.
proceedings
.com

Some format issues inherent in the e-media version may also appear in this print version.

Printed by Curran Associates, Inc. (2017)

For permission requests, please contact the Institute of Physics
at the address below.

Institute of Physics
Dirac House, Temple Back
Bristol BS1 6BE UK

Phone: 44 1 17 929 7481
Fax: 44 1 17 920 0979

techtracking@iop.org

Additional copies of this publication are available from:

Curran Associates, Inc.
57 Morehouse Lane
Red Hook, NY 12571 USA
Phone: 845-758-0400
Fax: 845-758-2633
Email: curran@proceedings.com
Web: www.proceedings.com

Table of contents

Volume 170

3rd International Conference on Advanced Materials Research and Applications (AMRA 2016)

18–21 December 2016, Guangzhou, China

Accepted papers received: 19 January 2017
Published online: 17 February 2017

Preface

Papers

IOP Conf. Series: Materials Science and Engineering **170** (2017) 011001 doi:10.1088/1757-899X/170/1/011001

Preface

The 3rd International Conference on Advanced Materials Research and Applications (AMRA 2016) was successfully held from December 18-21, 2016 in Guangzhou, China. The Conference addressed recent advancements in Materials Science & Engineering and the relevant applications. The Technical Program consisted of 2 international keynote speakers, 25 oral presentations, and a poster session with 13 posters.

This volume of IOP Conference Series: Materials Science & Engineering (MSE) is a compilation of the accepted papers in AMRA2016 and represents important results that were presented in the Conference. The electronic submission and handling of the manuscripts via the Conference Website, including the selection of reviewers and evaluation of manuscripts, were identical to the procedures applied to manuscripts submitted as regular contributions for publication in peer reviewed journals.

The organization of this conference and the preparation of proceedings volume would have been impossible without the tremendous efforts and dedication of many individuals. Prof. Nuggehalli M. Ravindra oversaw the organization of the Conference. We would like to acknowledge a large team of reviewers for their timely submission of quality reports. Without this support, the Conference could not have been successful. We sincerely hope that AMRA2016 will serve as an ideal forum for academia, industrial practitioners and government agents.

Prof. N.M. RAVINDRA (RAVI)

General Chairman of AMRA 2016

Peer review statement

All papers published in this volume of *IOP Conference Series: Materials Science and Engineering* have been peer reviewed through processes administered by the proceedings Editors. Reviews were conducted by expert referees to the professional and scientific standards expected of a proceedings journal published by IOP Publishing.

IOP Conf. Series: Materials Science and Engineering **170** (2017) 012001 doi:10.1088/1757-899X/170/1/012001

Synthesis and characterization of a novel bio-based resin from maleated soybean oil polyols

Y T Li, L T Yang, H Zhang

School of Chemistry and Environment, South China Normal University, Guangzhou P.R. China

E-mail: yanglt@scnu.edu.cn

Abstract. In this paper, a novel bio-based resin was prepared by the radical copolymerization of maleated soybean oil polyols (MSBOP) and styrene (ST). Structure of the product was studied by Fourier transformation infrared spectrometer (FT-IR), and the result was found to be consistent with that of theoretical structure. Swelling experiments indicated that the crosslinking degree increased with the increase of hydroxyl value. Thermal analysis by differential scanning calorimetry (DSC) and thermo-gravimetric analysis (TG) revealed that glass transition temperature (T_g) of the polymer increased with increasing hydroxyl values, and that its thermal stability showed a good correlation with the hydroxyl value. The tensile strength and impact strength were significantly affected by the hydroxyl value of soybean oil polyols. With increasing hydroxyl value, the tensile strength presented an increasing trend, while the impact strength showed a decreasing one. Moreover, the property of the polymer from elastomer to plastic character also depended on the functionality of the hydroxyl value of soybean oil polyols.

1. Introduction

In recent years, environmentally friendly substitutes of raw materials for preparing polymers have become crying needs with the development of materials science. Many scientific investigations were focused on polymer materials based on renewable resources. The main achievement is devoted to the preparation of polyols from plant oil due to its widely available, economical and eco-friendly characteristics [1]. Currently, polymer materials derived from renewable plant resources are becoming more and more important as the replacement of petroleum based polymers owing to various advantages, and have been widely used in many engineering and commodity applications [2-4]. Plant oil is one of the most interesting environmental resources for the synthesis of monomers or polymers, because its aliphatic chains and functional groups contribute a lot to the backbone of polymer as the primary element [5]. Consequently, plant oil has attracted much attention as a renewable resource for preparing chemicals and polymers [6].

Soybean oil (SBO) is an unsaturated triglyceride, which mainly consists of linolenic acid, linoleic acid, oleic acid, palmitic acid, and stearic acid. Double bonds in SBO molecules must be functionalized to form polymers, since it is difficult for them to participate in radical-initiated polymerization reaction [7, 8]. Many groups such as amines, hydroxyls, epoxies, thiols and halides have been introduced to plant oil triglycerides. Specially, by introducing reactive groups to soybean oil

[9], epoxidized soybean oil (ESO) and acrylated epoxidized soybean oil (AESO) can be prepared, and they have been widely used as monomers based on plant oil.

In previous work, we have successfully functionalized SBO, corn oil (CO) and other plant oils to introduce polymerizable functional groups, such as conjugation [10], epoxidation [11], hydroxylation [12], acrylation [13], and esterification [14]. Hydroxyl groups, ester groups and double bonds in these plant oil molecules make it available for performance modification and polymerization reaction [15]. The functional plant oils can be used to prepare poly urethane (PU), epoxy acrylate, alkyd resin and epoxy resin, and the applications of the product involve casting plastic, foamed plastic, coating, adhesive, paint and many other fields. It is noticed that SBOP is mainly used to prepare PU, whereas few studies use SBOP to prepare polymer materials through reaction with maleic anhydride. Moreover, SBOP with different hydroxyl values has seldom been used to prepare polymers and study the relationship between hydroxyl value and material properties.

In this work, a new monomer based on radically polymerizable triglyceride was synthesized by the reaction of SBOP and maleic anhydride (MA). Firstly, SBOP with different hydroxyl values has been synthetized. Then the prepared polyols were used to prepare MSBOP through the mono-esterification reaction with MA. Finally, MSBOP was polymerized with vinyl monomers to form the polymer. On this basis, the structure of the network, and the relationship between hydroxyl value and the properties of final materials were studied by several suitable test methods.

2. Experimental

2.1. Materials

Soybean oil was obtained from Nanhai Oil Co. Ltd., Guangzhou, China. Glacial acetic acid (99.5%), sulfuric acid (98%), sodium carbonate, hydrogen peroxide (30%), aqueous ammonia (30% in water), fluoroboric acid (48% in water), maleic anhydride, methanol, methylbenzene, styrene, and benzoyl peroxide (BPO) were purchased from Damao Chemical Reagent Co., Tianjin, China. All the reagents were analytically pure (AR).

2.2. Methods

Epoxy value was determined by the ASTM D1652 testing standard. Hydroxyl value was determined by the ASTM D1957 testing standard. Degree of crosslinking of the polymer was determined by the swelling equilibrium method.

FTIR spectra of the samples were recorded using a Prestige-21 spectrometer (Shimadzu, Japan). DSC curves were recorded on a 200PC instrument (Netzsch, Germany). In each measurement, 8-10 mg of sample was used, and the sample was heated from -50 to 200℃ at a heating rate of 20℃/min with nitrogen as the purge gas. Thermo-gravimetric analysis was carried out on an STA 409PC instrument (Netzsch, Germany). The sample was heated from 30 to 600℃ at a heating rate of 10℃ /min, with air as the purge gas. Tensile strength was determined on a WSM-100 tensile testing machine (Changchun, China) with an extension rate of 10 mm/min. Impact strength was determined on a JJ-20 impact testing machine (Changchun, China).

2.3. Preparation of ESO and SBOP

ESO with varying epoxy values was prepared through the reaction of SBO and peroxyacetic acid at 65℃ for different reaction times. Peroxyacetic acid was prepared in situ by the reaction of various mixtures of glacial acetic acid and H_2O_2, with sulphuric acid as the catalyst. The extents of epoxidation were 0.197 mol/100g, 0.220 mol/100g, 0.264 mol/100g, 0.331 mol/100g and 0.375 mol/100g, respectively.

SBOP with different hydroxyl values was synthesized from the reaction of methanol with ESO. The molar ratio of epoxy group to hydroxyl group is 1:11. Fluoroboric acid was used as the catalyst at a concentration of 0.2% in weight. The reaction mixtures were kept in water bath at 70℃ for 3 h. After complete reaction, the mixture was cooled to room temperature, and then washed by ammonia sodium. Finally, after removing the water and solvent, the samples were obtained. The hydroxyl values of the polyols were 86 mgKOH/g, 117 mgKOH/g, 126 mgKOH/g, 141 mgKOH/g and 156 mgKOH/g, respectively.

2.4. Preparation of MSBOP

MSBOP with different saturation degrees was obtained by the reaction of MA with the prepared SBOP. The molar ratio of hydroxy group to maleic anhydride is 1: 2, in order to ensure the complete reaction of hydroxy groups. SBOP and methylbenzene were placed into a 500 mL round bottom flask and kept at 110℃ for 1 h to ensure that there was no water in the mixture. After the reaction, the mixture was cooled down, and then washed by deionized water. After removing the water and solvent, the samples were obtained. The reaction extent was over 90% and the yield of the prepared MSBOP was over 95%.

2.5. Synthesis of the cast resin

The polymer was prepared by mixing the prepared MSBOP with styrene (ST), with BPO at a concentration of 3% in weight as the initiator. The mass fractions of ST and MSBOP were 40% and 60% in weight, respectively. BPO was dissolved in ST by moderate stirring, and then blended with MSBOP. The mixture was put into vacuum oven to eliminate air bubbles, and then poured into a mould. Afterwards, the mixture in the mould was heated to 60 ℃ for one-hour heat preservation, and then to 100℃ for another three-hour heat preservation to complete the polymerization reaction.

3. Results and discussion

3.1. Characterization of MSBOP

Scheme 1 presents the reactions for preparing MSBOP and polymer, through which a series of ESO with different epoxy values and SBOP with different hydroxyl values can be synthesized. Table 1 indicates that the hydroxyl value increases with the epoxy value. The crosslinking degree of the polymer was characterized by the swelling experiments. It is found that the gel fraction increases with the hydroxyl value obviously. The polymer mainly consists of linear polymer or low polymer at low hydroxyl values, while cross-linked polymer is dominated at high hydroxyl values. The polymer with low gel fraction behaves as an elastomer, whereas that with higher gel fraction displays rigid plastic properties.

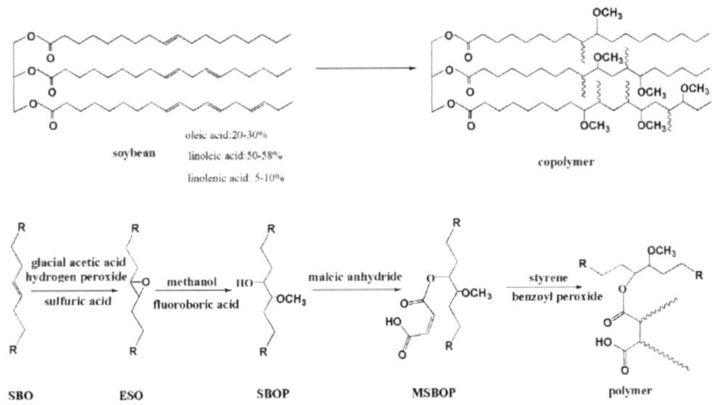

Scheme 1. Synthesis routes of MSBOP and polymer

Table 1. Gel fraction of polymer derived from MSBOP

Sample	a	b	c	d	e
Epoxy value of the epoxidized soybean oil/(mol/100g)	0.197	0.220	0.264	0.331	0.370
Hydroxyl value of the polyols / (mgKOH/g)	86	117	126	141	156
Properties of the polyols	liquid	liquid	grease	grease	grease
gel fraction of the cast resin/%	34.84	62.59	71.12	85.49	93.41
Properties of the cast resins	elastomer	elastomer	elastomer	elastomer	plastic

3.2. FTIR spectra analysis

Figure 1 presents the FTIR spectra. The peak at 3009 cm^{-1} in the spectrum of SBO is attributed to the carbon-hydrogen stretching vibration in the carbon-carbon double bond, and it disappears in the spectrum of ESO. The peak at 825 cm^{-1} in the spectrum of ESO corresponds to the epoxy groups, and its appearance indicates the occurrence of epoxidation reaction. Compared with the spectrum of ESO, the peak at 3464 cm^{-1} is attributed to the hydroxy group stretching vibration of SBOP, while that at 825 cm^{-1} corresponding to the epoxy groups disappears. For MSBOP, the peak at 1645 cm^{-1} corresponding to the stretching vibration of carbon-carbon double bond and that at 3464 cm^{-1} corresponding to the hydroxy groups both widen and present dispersive characteristics, indicating that single-esterification reaction occurs for SBOP and MA. This means that alcoholic hydroxyl groups have transformed into carboxyl groups. In the spectrum of the polymer, the peaks at 3064 cm^{-1} and 3034 cm^{-1} correspond to the carbon-hydrogen stretching vibration of the benzene ring, and those at 1584 cm^{-1}, 1494 cm^{-1}, 1453 cm^{-1} and 1402 cm^{-1} are attributed to the bending vibration of carbon-carbon double bond in the benzene ring. However, the stretching vibration peak of the carbon-carbon double bond at 1653 cm^{-1} almost disappears, implying that the polymerization reaction has occurred.

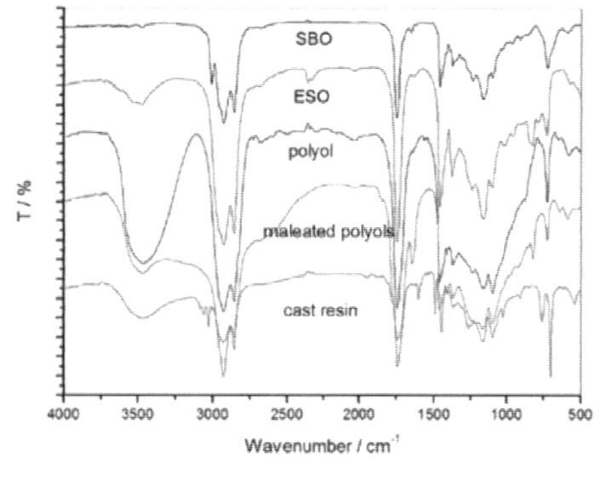

Figure 1. Infrared spectra of the samples

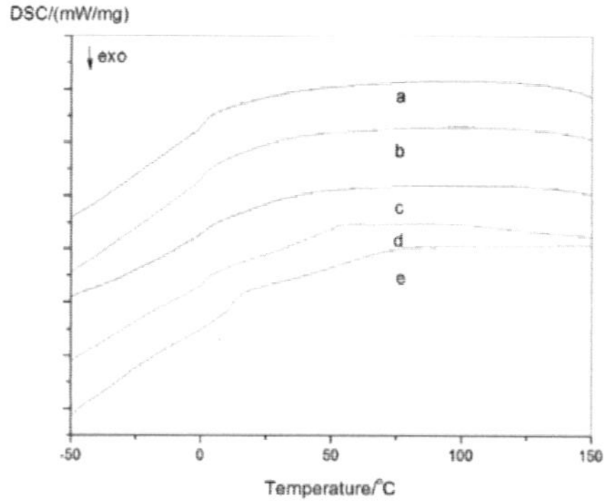

Figure 2. DSC curves of polymers derived from MSBOP

3.3. Analysis of thermal properties

Properties of polymer are significantly affected by the cross linking degree, which depends on the functionality degree of monomer molecules, such as hydroxyl groups and epoxy groups. Figure 2 shows the DSC curves of the polymers, and the glass transition temperatures (T_g) of the polymers are provided in Table 2. As can be seen, T_g of the polymers increases with the hydroxyl value and the epoxy value. Since the number of active carbon-carbon double bonds increases with the reaction of MSBOP, the degree of functionality also increases accordingly. In addition, carbon-carbon double bonds in the triglyceride chain have an obvious effect on the flexibility of macromolecule chain. T_g of the polymers with hydroxyl values ranging from 86 to 126 mg KOH/g is lower than room temperature and displays only one glass transition temperature because of the variable crosslinking degree and the low gel fraction. Consequently, these samples have a large amount of linear polymer, low polymer or unreactive species. Thus, the resulting materials are soft, weak, and rubbery. In contrast, polymers made from polyols with hydroxyl values of 141 to 156 mg KOH/g have high T_g and display two glass transition temperatures. This can be attributed to the phase separation in the polymers caused by the high gel fraction. To be specific, the gel park with high crosslinking density has higher T_g, while the sol park shows lower T_g. Hence, the two samples present rigid plastic properties, which can be proved by the analysis of mechanical properties.

Table 2. Thermal properties of polymers derived from MSBOP

Sample	T_g	$T_{5\%}$/℃	$T_{10\%}$/℃	$T_{50\%}$/℃	T_{max}/℃
a	0.4	314.6	349.9	401.8	405.4
b	0.7	324.3	351.8	399.6	402.1
c	19.1	311.4	345.6	399.2	403.3
d	50.4	247.0	325.2	395.0	400.3

e	58.6	243.5	312.0	386.3	391.8

Figure 3 shows the TG and DTG thermograms of the polymers in air atmosphere. The temperatures corresponding to 5%, 10% and 50% mass loss, and the temperature at which mass loss occurs with the maximum rate (T_{max}) are listed in Table 2. Thermal stability of the polymers shows a good correlation with the hydroxyl value. With the increase of hydroxyl value of SBOP, the decomposition temperature of the polymers increases at lower hydroxyl values, while decreases at higher hydroxyl values. Meanwhile, as the hydroxyl value increases, the number of active carbon-carbon double bonds introduced from MA also increases, thus increasing the crosslinking degree of the polymers. Besides, the number of carboxyl groups in the resulting materials increases with increasing hydroxyl values. Because of the low thermal stability of the carboxyl groups, the thermal stability of the polymers shows a decreasing trend within limits.

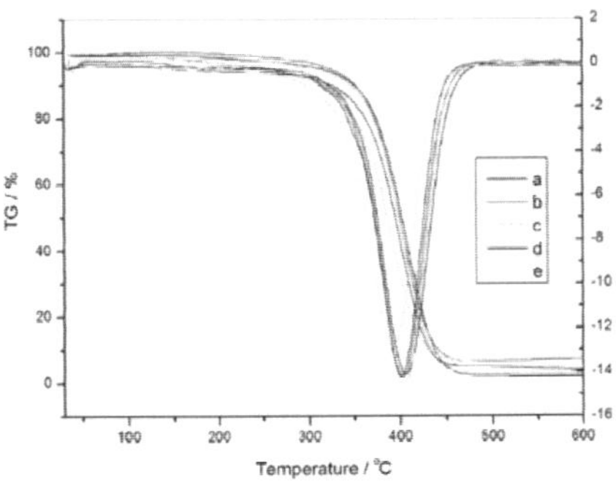

Figure 3. TG and DTG curves of polymers derived from MSBOP

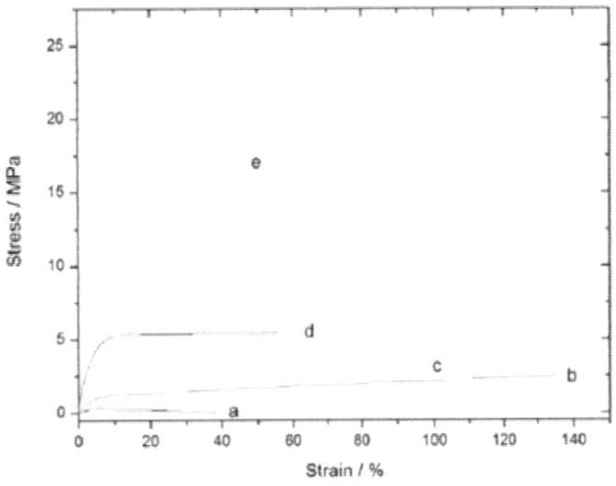

Figure 4. Stress-strain curves of polymers derived from MSBOP

3.4. Mechanical properties of the cast resin

Figure 4 presents the stress-strain curves of the polymers, and the tensile strengths are listed in Table 3. The polymer derived from SBOP with a hydroxyl value of 86 mgKOH/g has a very low tensile strength of 0.37 MPa and a low elongation of 40.47%, which displays soft and weak characteristics. In this case, due to the low average functionality degree of MSOP, the crosslinking degree is low and the network is imperfect. Higher hydroxyl values contribute to higher crosslinking degree and the formation of relatively complete network, and can improve the tensile strength of glassy polymers. With a hydroxyl value of 156 mgKOH/g, the polymer presents a high tensile strength of 23.63 MPa and a moderate elongation of 43.08%, which displays rigid plastic properties.

Table 3. Mechanical properties of polymers derived from MSBOP

Sample	Hydroxyl value of the polyols / (mgKOH/g)	Tensile strength /MPa	Elongation at break/%	Impact strength /KJ/m^2
a	86	0.37±0.02	40.47±4.24	5.28±0.63
b	117	2.11±0.32	136.63±2.19	25.28±5.99
c	126	2.37±0.15	119.78±0.16	15.29±3.56
d	141	5.50±0.06	55.15±1.20	4.38±0.62
e	156	23.63±1.39	43.08±2.38	4.11±0.35

The impact strengths of the polymers with different hydroxyl values are also given in Table 3. As it reveals, the impact strength of the polymers decreases with increasing hydroxyl values, due to the decrease of the flexibility of polymer chains. Especially, the polymer derived from SBOP with a hydroxyl value of 86 mgKOH/g is distinguished from others, and shows a lower impact strength of 5.28 KJ/m^2 due to its lower crosslinking degree and imperfect networks. At higher hydroxyl values, since the strengths of the polymer are mainly controlled by the strong cross-linked network, the effect of the flexibility of the polymer chain is not obvious, so that the strength decrease of the materials is inconspicuous. Meanwhile, the soybean oil molecules also make certain contributions to the flexibility of the polymer chain.

4. Conclusions

In summary, polymers derived from SBOP with different hydroxyl values were obtained, and the effects of hydroxyl value on the properties of the polymers were studied. Mechanical and thermal properties of the polymers depended on the functionality degree of SBOP. The wide distribution of hydroxyl values determined different crosslinking degrees of the polymers. Thermal analysis revealed that T_g increased with increasing hydroxyl values. Mechanical analysis illustrated that the tensile strength increased while the impact strength decreased with increasing hydroxyl values. Polymers derived from SBOP with lower hydroxyl values behaved as elastomers, whereas those from SBOP with higher hydroxyl values displayed rigid plastic properties. The bio-based polymers could be prepared from renewable environmentally friendly resources, and showed attractive properties and wide application prospects in many fields.

References

[1] Bakhshi et al. 2013 Synthesis and characterization of antibacterial polyurethane coatings from quaternary ammonium salts functionalized soybean oil based polyols *Materials science and engineering: C* **33** 153
[2] Nagarjuna R P, Smita M, Nayak S K 2015 Mechanical and thermal properties of sisal fiber reinforced acrylated epoxidized castor oil toughened diglycidyl ether of bisphenol A epoxy nanocomposites *J. Reinf. Plast. Comp.* **34** 1476
[3] Reddy P N, Mohanty S,Nayak S K 2014 Synthesis and modifications of epoxy resins and their composites: a review *Polym. Plast. Technol. Eng.* **16:** 1723
[4] Sapuan F M, AL-Oqla S M 2014 Natural fiber reinforced polymer composites in industrial applications: feasibility of date palm fibers for sustainable automotive industry *J Cle. Prod.* **66** 347
[5] Audrey L, Patrick-Kurt D, Marc von C 2016 Renewability is not enough: recent advances in the sustainable synthesis of biomass-derived monomers and polymers *Chem. Eur. J.* **22** 11510
[6] Muhammad Y A, Muhammad I 2015 Synthesis and properties of cross-linked polymers from epoxidized rubber seed oil and triethylenetetramine *J. Appl. Polym. Sci.* **10** 42591
[7] Tan S G, Z. Ahmad, W S, Chow 2013 Relationships of cure kinetics and processing for epoxidized soybean oil bio-thermoset *Ind. Crop. Prod.,* **43** 378
[8] Salih N, Salimon J, Yousif E 2011 The physicochemical and tribological properties of oleic acid based trimester biolubricants *Ind. Crop. Prod.* **34** 1089-1096
[9] Huseyin E, Gokhan S 2016 *Eur. J. Lipid Sci. Technol.* **118** 959
[10] Yang L T, Dai C L, Ma LJ, Lin SQ 2011 Conjugation of soybean oil and it's free-radical copolymerization with acrylonitrile *J. Polym. Environ.* **19** 189
[11] Yang L T, Dai H H, Yi A H, Lin B 2008 Structure and properties of partially epoxidized soybean oil *J. Therm. Anan. Cal.* **93** 875
[12] Dai H H, Yang L T, Lin B, Wang C S, Shi G 2009 Synthesis and characterization of the different soy-based polyols by ring opening of epoxidized soybean oil with methanol, 1,2-ethanediol and 1,2-propanediol *J. Am. Oil Chem. Soc.* **86** 261
[13] Fu L Y, Yang L T, Zhao C S, Dai C L 2010 Thermal and mechanical properties of acrylated expoxidized soybean oil -based thermosets *J. Appl. Polym. Sci.* **117** 2220
[14] Zheng XQ, Yang L T, Wang K P 2011 Synthesis and characterization of aikyd resin based on soybean oil polyols *Adv. Mater. Res.* **239** 1721
[15] Wang H J, Rong M Z, Zhang M. Q 2008 Biodegradable foam plastics based on castor oil *Biomacromolecules* **9** 615

IOP Conf. Series: Materials Science and Engineering **170** (2017) 012002 doi:10.1088/1757-899X/170/1/012002

The alcohol-sensing behaviour of SnO_2 nanorods prepared by a facile solid state reaction

F Gao[1], X P Ren[2], W J Wan[1], Y P Zhao [1], Y H Li[3], H Y Zhao[1,4]

[1] School of Materials Science and Engineering, Yunnan University, Kunming, 650091, PR China.
[2] Yunnan Security and Technology Co. Ltd., Kunming, 650033, PR China.
[3] Advanced Measurement and Analysis Center of Dali University, Dali, 671200, PR China
[4] Yunnan Key Laboratory for Micro/Nano Materials and Technology, Kunming 650091, PR China

E-mail: hyzhao@ynu.edu.cn ; rxp89@126.com

Abstract. SnO_2 nanorods with the range of 12-85 nm in diameter were fabricated by a facile solid state reaction in the medium of NaCl-KCl mixture at room temperature and calcined at 600, 680, 760 and 840 °C, respectively. The XRD, TEM and XPS were employed to characterize the structure and morphology of the SnO_2 nanorods. The influence of the calcination temperature on the gas sensing behaviour of the SnO_2 nanorods with different diameter was investigated. The result showed that all the sensors had good response to alcohol. The response of the gracile nanorods prepared at a low calcined temperature demonstrated significantly better than the thick nanorods prepared at a high calcined temperature. The mechanism was attributed to the nonstoichiometric ratio of Sn/O and larger surface area of the gracile nanorods to enhance the oxygen surface adsorption.

1. Introduction

Gas monitoring and detection is significant for the applications in environmental protection, air quality control and healthcare. Gas sensor based on semiconductor metal oxides, including SnO_2, ZnO, In_2O_3 and TiO_2, play an important role to be used in the detection of volatile organic compound vapors (VOCs). As a typical gas sensing material, n-type semiconductor SnO_2 has gained prominent attention for the excellent electron conduction and good chemical and thermal stability [1]. In recent years, nanostructured semiconductor oxides are attracting a great deal of attention to improve the gas-sensing performances. Among them, one-dimensional (1D) nanostructure materials stood out as a gas-sensing materials due to their high surface–volume ratio, abundant surface states and enhanced surface reactivity [2]. Therefore, many novel one-dimensional nanostructures SnO_2 such as nanorods [3], nanobelts [4], nanowires [5], nanotubes [6] and nanofibers [7] have been found to show excellent response to VOCs. Although various chemical and physical methods were used for synthesis SnO_2 one-dimensional nanocrystals with different shapes, it is still challenging to fabricate SnO_2 nanostructures with controlled shape and crystallinity for obtaining excellent sensing performances. In this paper, SnO_2 nanorods with various diameters were fabricated by an adjusted solid state reaction in the presence of NaCl-KCl mixture annealed at different calcined temperature. The influence of calcination temperature on the alcohol-sensing behaviour of the SnO_2 nanorods was investigated. The mechanism of the enhanced response of gracile SnO_2 nanorod was discussed.

2. Experimental

All utilized analytical grade chemical reagents were purchased from the Sinopharm Chemical Reagent Co. without further purification. SnO_2 nanorods were prepared by calcining the precursor powders, using a route of solid state reaction between $SnCl_4 \cdot 5H_2O$ and KBH_4 mixed with surfactants (NP5+NP9) at room temperature, in which was participated by KCl-NaCl mixture. In order to obtain different diameters of the nanorods, the precursor mixture was annealed at 600, 680, 760 and 840 °C for 2h in the atmosphere (denoted as SA, SB, SC and SD, respectively). In this case, the 1:1 molar rate of the NaCl and KCl is the key condition due to they form flux at lower calcined temperature, which is conducive to adjust the calcination temperature in a wide range of 600-900 °C.

The structure and morphology of the SnO_2 nanorods were determined by powder X-ray diffraction (XRD, Siemens D5005 diffractometer with Cu Ka radiation) and transmission electron microscope (TEM, JEOL 2010, 200 kV). Meanwhile, the surface analysis of the nanorods was characterized by X-ray photoelectron spectroscopy (XPS) and the specific surface area was assessed by surface area analyzer (CHEMBET 3000) and the BET method. The indirect-heating sensor was elected to investigate the response of the SnO_2 nanorods [8]. The measurement of the gas-sensing properties was performed on a JF02F measurement system to obtain the electrical resistance of the sensor in atmospheric air and in the target gas in a relative humidity range of 40–70%. The response was defined as the ratio of R_a/R_g, where R_a and R_g represents the resistance in atmospheric air and in target gas, respectively.

3. Results and discussion

3.1 Characterization of SnO_2 nanorods

Fig. 1 shows a typical XRD pattern of SnO_2 products. The distinct and sharp reflection peaks observed from Fig.1 suggest the highly crystalline of all samples. All samples could be identified the rutile structured SnO_2 crystal, according to the peaks conformed to the standard data file of the JCPDS 41-1445. The curve (a) shown in Fig. 1 demonstrates the wide diffraction peaks of SnO_2 nanorods synthesized at 600 °C, indicating the small crystallite size of SnO_2 at lower calcination temperature. As the curves shown in Fig. 1b, 1c and 1d, the diffraction peaks of SnO_2 turn sharper gradually with the increase of the temperature from 680 ◦C to 840 ◦C, attributing to the growth up of the SnO_2 nanorods. The Scherer's formula was used to estimate the crystalline size of the SnO_2 nanorods employing XRD data of the (110) plane diffraction peak, and the result was about 11.8, 29.4, 53.2 and 79.1 nm, respectively. It betokens that the crystalline size of the SnO_2 samples gradually increases with the calcination temperature, which will be confirmed by the results of the TEM.

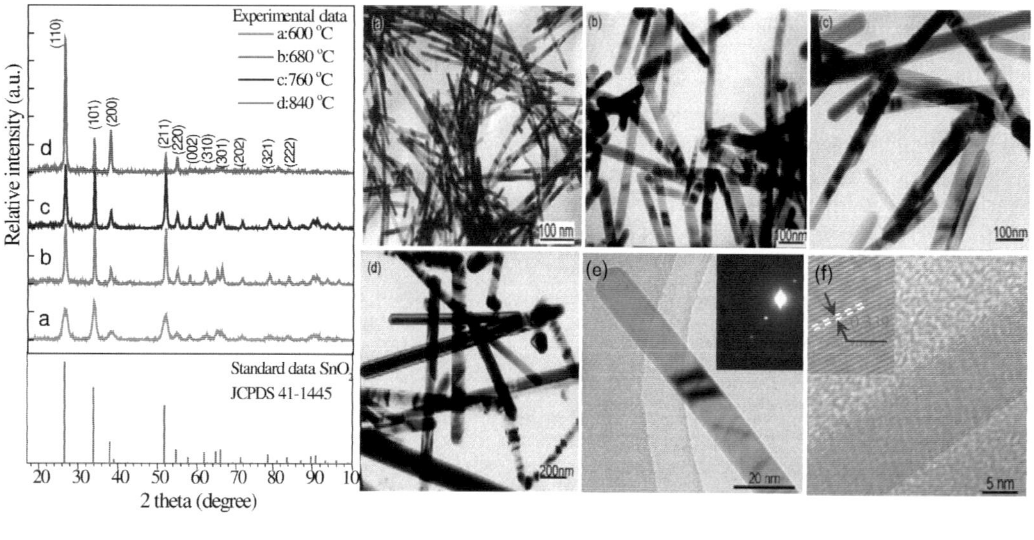

Fig.1. XRD patterns of SnO_2 nanorods prepared at different calcination temperatures.

Fig. 2. TEM images of SnO_2 nanorods prepared at different calcination temperatures. (a) 600 °C, (b) 680 °C, (c) 760 °C, (d) 840 °C, (e) A single SnO_2 nanorods prepared at 600 °C, (f) HRTEM image.

As shown in Fig.2, the morphology of the SnO_2 nanorods prepared at different calcination temperature depicts that the as-prepared samples mainly consist of solid rod-like structure SnO_2. The straight nanorod has a diameter of 10–100 nm and up to several micrometers long. As seen from the Fig. 2(a)-(d), the average diameter of the nanorods prepared at different calcination temperature was statistically estimated to be approximately of 12.2, 30.5, 52.4 and 81.6 nm, respectively. Evidently, as a function of calcination temperature, the diameter of the nanorod grew thick with the increase of the calcination temperature, which is in accordance with the XRD results. An enlarge TEM image individual SnO_2 nanorods shown in Fig. 2e demonstrates a smooth surface and an obtuse angle dome top of the nanorods. The corresponding SAED pattern (inset) indicates single-crystalline of the SnO_2 nanorods. Fig. 2f shows a high-resolution transmission electron micrograph (HRTEM) of an SnO_2 nanorod from SA sample. The uniform structure and clear lattice fringes revealing 0.339 nm space between two adjacent lattice fringe corresponds to the (110) lattice planes of SnO_2 in tetragonal cassiterite structure [9].

Fig. 3 shows the XPS spectra of Sn and O regions of the SnO_2 nanorods prepared at 600 °C. As shown in Fig. 3b, the high-resolution XPS spectra of Sn 3d demonstrates that Sn 3d5/2 peak and Sn 3d3/2 peak occur at 486.0 eV and 494.5 eV, respectively. The energy difference of 8.5eV between of them coincide with the energy splitting reported for SnO_2 [10]. The asymmetric O1s XPS spectrum survey shown in Fig. 3(b) indicates various states of oxygen nearby the nanorods surface region. The O1s peak was deconvoluted into two components located at 529.9 eV and 531.0 eV. The binding energy located at 529.9 eV is related to the crystal lattice O^{2-} anions corresponding to Sn-O bonds within the matrix of SnO_2 nanorods, while the binding energy sited at 531.0 eV causes for the adsorbed O^- or O^{2-} ions in the oxygen deficient regions within the matrix of SnO_2, which will react with the tested gas to enhance the response of SnO_2 nanorods for alcohol. In addition, the atomic concentration ratio of Sn and O was evaluated to be about 1:1.42 (≈0.704) using sensitivity factor, indicating serious deviation from the stoichiometry of SnO_2 composition (1:2). The XPS survey of other three samples calcined at 680, 760 and 840 °C was also measured (not shown). The atomic ratio of Sn/O was evaluated to be 1:1.57 (≈0.64), 1:1.71 (≈0.585) and 1:1.89 (≈0.53), respectively. It shows that the deviation from stoichiometry (0.5) of Sn/O atomic ratio changed from 0.7 to 0.53 caused by oxygen vacancies. The oxygen vacancy is conducive to improving the adsorption oxygen ions, which could contribute to the gas response. Thus, it can be speculated that the nanorods sample prepared at 600 °C would exhibit better gas-sensing response than other three samples.

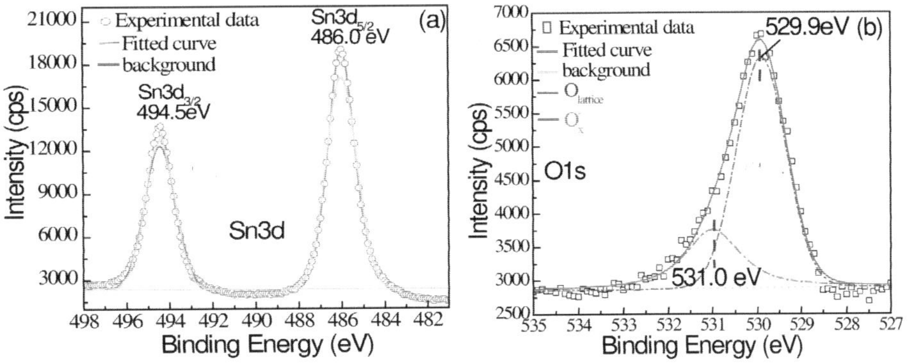

Fig. 3. XPS spectra of SnO_2 nanorods prepared at 600 °C: (a) Sn3d, (b) O 1s.

The specific surface area of SnO_2 nanorods was determined by nitrogen adsorption–desorption measurements and Brunauer–Emmett–Teller (BET) analyzer method. The results show that the surface area of the SnO_2 nanorods was calculated to be 69.6, 53.14, 38.56 and 23.84 m^2g^{-1}, respectively, indicating the decrease of the surface area with the increase of the SnO_2 nanorods diameter. It implies that the SnO_2 nanorods sample calcined at low temperature has a larger surface area than the sample calcined at higher calcination temperature. Generally, larger specific surface area may cause a better response due to more active sites to adsorb gas molecules.

3.2. Gas-sensing characteristics

The sensors were exposed to 1000 ppm ethanol at different operating temperature (175–425 °C) to determine the optimum working temperature and the result was shown in Fig.4(a). It can be seen from Fig. 4(a) that the response increase with the operating temperature and achieve a maximum value of 275 °C, and then the response rapidly descend with the increasing temperature again, which indicates the operable optimum temperature of the 275 °C. Fig. 4(b) depicted the response and recovery characteristics of the sensors based on SnO_2 nanorods prepared at 600, 680, 760 and 840 °C to 1000 ppm ethanol at 275 °C. For observing clearly, only one cycle was laid out. An overview from Fig. 4(b) is clear that the response amplitude of the sensor gradually increases with the diameter of SnO_2 nanorod decrease, indicating the sensing performance of sensor shows good dependence on the diameter of SnO_2 nanorods. Not surprisingly, the sensing performance of the SnO_2 nanorods with a diameter of about 12 nm exhibits an obviously better than the other sensors. It can be observed that the response and recovery time of SnO_2 nanorods calcined at 600 °C are about 6 s and 14 s, respectively, which is short enough for practical application.

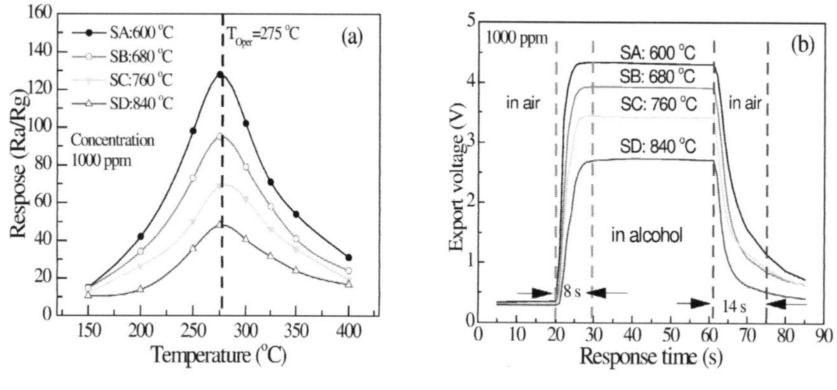

Fig. 4. (a) Dependence on the operating temperature of gas response. (b) Dynamic response and recovery behaviour of SnO_2 nanorods. (Only one cycle laid out).

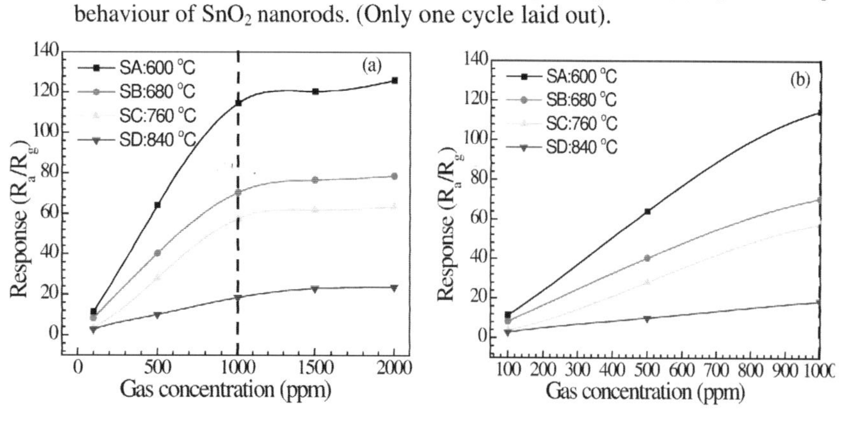

Fig. 5 (a) Gas response of sensors to ethanol concentrations at 275 °C. (b) Linear dependence of the response on concentration in the range of 100-1000 ppm.

The dependence of the response on the alcohol concentration at the optimum operating temperature of 275 °C was shown in Fig. 5. It can be seen from Fig. 5 that the response of all sensors increase with the alcohol concentration. Fig. 5(b) shows a good linear relationship between the response and alcohol concentration in the wide variation range of 100-1000 ppm. Above 1000 ppm, the response increase slowly with the increasing gas concentration, indicating the sensors achieve a saturated state. In particular, Fig. 5 presents that sensor based on the SnO_2 nanorods calcined at 600 °C exhibits much higher response to alcohol than other three SnO_2 nanorod sensors. For alcohol concentration of 100, 500, 1000, 1500 and 2000 ppm, the most sensitive SnO_2 sensor response is 11.38, 64.13, 114.46, 120.43 and 126.07, respectively. The higher response is probably ascribed to the absorption of more oxygen molecules to react with alcohol molecules.

As for the mechanism, the theory built on the change of the resistance caused by adsorption and desorption process of oxygen molecules on the SnO_2 nanorods surface is the widely accepted [11]. In atmospheric environment, oxygen molecules are chemisorbed on the surface and capture electrons from the conduction band of the SnO_2, causing an electron depletion layer in the SnO_2 nanorods surface layer and resulting in the resistance increase. When the alcohol molecules overstep the surface of the SnO_2 nanorods sensor, these gas molecules are oxidized by the adsorbed oxygen species to reduce the coverage of oxygen ions and the depleted electrons are fed back into the SnO_2 conductance band. It could result in a narrowed depletion layer and the decrease of the sensor resistance. Thus, an electrical signal is generated and output to the test system. The reaction between alcohol molecules and ionic oxygen species can be described in detail as Eq. (1) and (2) [12].

$$C_2H_5OH + 6O^-(ads) \rightarrow 2CO_2 + 3H_2O + 6e^- \quad (1)$$

$$C_2H_5OH + 6O_2^-(ads) \rightarrow 2CO_2 + 3H_2O + 12e^- \quad (2)$$

According to the above reactions, it is clear that a large specific surface area can provide more active sites for the gas adsorption to improve the adsorption capacity of oxygen molecules. In order to elucidate the influence of calcination temperature on the gas sensing behaviour of the SnO_2 nanorods with different diameter, we plot the specific surface area as a function of the average nanorods diameter (Fig. 6a), and the gas response measured at optimum temperature (275 °C) as a function of the atomic ratio of Sn/O (Fig. 6b), respectively. Fig. 6(a) implies that the SnO_2 nanorods prepared at low temperature has a larger surface area than the sample prepared at a higher calcination temperature. Larger specific surface areas have a more adsorption capacity of oxygen molecules, which lead to the enhancement of surface properties such as catalytic activity or surface adsorption. Moreover,

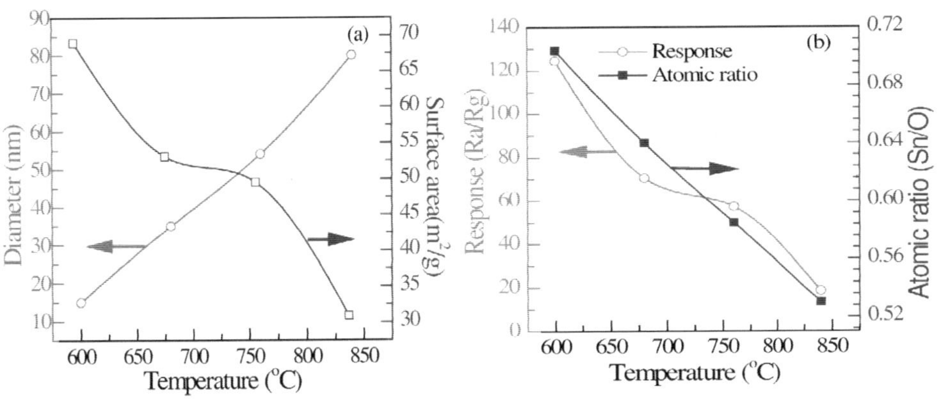

Fig. 6. (a) Correlation between the specific surface area (average diameter) and the calcinated temperature. (b) Correlation between the responses (or Sn/O atomic ratio) and the calcinated temperature.

depending upon the principle of crystallography, the calcination temperature affects the crystallinity of SnO_2 confirmed by the XRD results shown in Fig. 1. The low calcination temperature generally causes poor crystallinity of the SnO_2 crystal, resulting in a large number of oxygen vacancies in the surface layer of nanorods. It could enhance the absorption of oxygen molecules from the atmosphere. The results of the XPS shown in Fig. 3 have been confirmed that large amounts of oxygen molecules was adsorbed on the surface of gracile SnO_2 nanorods prepared at a low calcination temperature of 600 ^oC, which cause a positive effect on the gas response. The correlation between the response as a function of the Sn/O ratio of the SnO_2 nanorods with different diameter was shown in Fig. 6b. It can be seen that the gas sensitivity increases with the decrease of the Sn/O ratio deviation from stoichiometric ratio. Therefore, the gracile SnO_2 nanorods fabricated at lower calcination temperature shows a significantly better sensing performance, attributing to the larger effective surface area and the strongly stoichiometry deviation of the Sn/O ratio.

4. Conclusions

In summary, SnO_2 nanorods with different diameters were synthesized by an adjusted solid state reaction in the presence of mixed KCl-NaCl mixture and surfactants at room temperature and calcined at different temperature. The diameter of nanorods grew thick with the increase of the calcination temperature. The influence of the calcination temperature on the gas sensing behaviour of SnO_2 nanorods was investigated. All sensors showed good responses to ethanol gas. The response of the gracile nanorods prepared at low calcination temperature showed significantly better than the thick nanorods prepared at high calcination temperature due to the larger effective surface areas and the strongly deviated stoichiometric ratio of Sn/O. The experimental results indicate the great potential applications of utilizing gracile SnO_2 nanorods for fabricating gas sensor to detect ethanol.

Acknowledgements

The authors gratefully acknowledge the financial supported by the National Natural Science Foundation of China (no. 61161008, 61564009), the Natural Science Foundation of Yunan Province (2009CD015) the Graduate Student Scientific Innovation Research Projectof Yunnan University (Grant no.YNUY2016103)

References

[1] Sun G, Qi F X, Zhang S S, Li Y W, Wang Y, Cao J L, Bala H, Wang X D, Jia T K, Zhang Z Y 2014 *J. Alloy. Comp.* **617** 192
[2] Comini E, Baratto C, Faglia G, Ferroni M, Vomiero A, Sberveglieri G 2009 *Prog. Mater. Sci.* **54** 1
[3] Huang H, Lee Y C, Tan O K, Zhou W, Peng N, Zhang Q 2009 *Nanotech.* **20** 115501
[4] Fields L L, Zheng J P, Cheng Y, Xiong P 2006 *Appl. Phys. Lett.* **88** 263102
[5] Mohamed S H 2012 *J. Alloy. Comp.* **510** 119–124
[6] Shi L, Lin H L 2011 *Langmuir* **27** 3977
[7] Xu S, Kan K, Yang Y, Jiang C, Gao J, Jing L Q, Shen P K, Li L, Shi K Y 2015 *J. Alloy. Comp.* **618** 240
[8] Luo L, Jiang Q P, Qin G H, Zhao K, Du G F, Wang H, Zhao H Y 2015 *Sens. Actuators* B **218** 205
[9] Oviedo J, Gillan M J 2000 *Surf. Sci.* **463** 93
[10] Dai Z F, Jia L C, Duan G T, Li Y, Zhang H W, Wang J J, Hu J L, Cai W P 2013 *Chem. Euro. J.* **19** 13387
[11] Batzill M T, Diebold U R 2007 *Phys. Chem. Chem. Phys.* **9** 2307
[12] Wang T T, Ma S Y, Cheng L, Luo J, Jiang X H, Jin W X 2015 *Sens. Actuators* B **216** 212

Evaluation of Fatigue Performance of Asphalt Based on Constant Strain DSR Test

H Z Zhu[1,2, a], E H Yan[1, b] and Z T Lu[2, c]

[1]The Key Laboratory of Road Structure & Material Ministry of Transport, Beijing, PRC
[2]Chongqing Jiaotong University, Chongqing, PRC

E-mail:[a]zhuhongzhouchina@126.com, [b]eh.yan@rioh.cn, [c]502858541@qq.com

Abstract. Asphalt performance has important effect on the fatigue resistance performance of asphalt mixture. This research based on the DSR time scanning mode, investigated the constant strain performance of 70 # matrix asphalt and SBS modified asphalt. Based on 50% G* 0 to simulate the fatigue performance of two kinds of the asphalt.

1. Introduction

Fatigue cracking is one of the major failure mode of asphalt pavement and considered as the main factors of asphalt pavement structure design. Fatigue resistance of asphalt mixture is one of the important factors affecting the service life of the asphalt pavement. In recent years, researchers at home and abroad adopt the phenomenological method to study the factors affecting the fatigue performance of asphalt mixture. In the same time, different fatigue model of asphalt mixture is put forward, and some scholars also explore the fatigue mechanism of asphalt mixture in terms of damage perspective.

In asphalt mixture, asphalt binder bonds the aggregate together to form the structure of pavement. Under the loads, asphalt and asphalt mortar is the principal part to endure tensile stress and tensile strain of asphalt pavement, so the fatigue performance of asphalt directly affect the fatigue performance of asphalt mixture. In recent years Fatigue performance of asphalt were studied at home and abroad. Tan with the aid of dynamic shear rheometer, study the normal fatigue characteristic of asphalt through the stress mode of continuous loading fatigue test [1]. By the continuous load fatigue test of strain mode, Li Xiaomin analyzed the mechanical response of asphalt mortar under repeated load [2]. Shenoy A studied the influence of temperature on the fatigue test performance of the asphalt [3]. Yuan Yan study the fatigue properties of modified asphalt by stress and the strain mode of continuous load fatigue test [4]. K.S. Bonnetti chose two kinds of common asphalt and five kinds of modified asphalt as the research object, discussed the load mode, loading frequency, test temperature and initial dissipation can affect the performance of asphalt fatigue [5].

This paper used phenomenology method, based on the DSR time scanning mode, carried out strain fatigue test on 70 # ordinary asphalt and SBS modified asphalt, got the fatigue equation of different asphalt. Using based on 50% G_0^* to evaluate the two kinds of the fatigue performance of asphalt, which can provide reference for the design of the asphalt mixture.

2. Experiment materials and method

IOP Conf. Series: Materials Science and Engineering **170** (2017) 012003 doi:10.1088/1757-899X/170/1/012003

2.1. Materials

In our country 70 # matrix asphalt and SBS modified asphalt used mostly in asphalt pavement, so this paper used this two kinds of bitumen as the research object. In accordance with the Highway Engineering Asphalt and Asphalt Mixture Test Procedures (JTG E20-2011), the provisions of the related technical indexes were tested. 70 # matrix asphalt technical index test results are shown in table 1, meet the heavy traffic road asphalt quality requirements; SBS modified asphalt technology index test results are shown in table 2, conform to the requirements of the quality of modified asphalt.

Table 1. 70 # matrix asphalt technical indicators.

Test project	Technical requirements	Test results	Test method
Penetration (100g 5s 25°C) (0.1mm)	40~60	68	T0604
Softening point (°C)	≮60	48.5	T0604
Ductility (5cm/min,5°C) (cm)	≮20	28	T0605
Flash point (°C)	≮230	—	T0611
Solubility (%)	≮99	—	T0607
elastic recovery 25°C (%)	≮75	81	T0662
separation	≯2.5	0.4	T0661
kinematic viscosity (135°C) (Pa·s)	≯3	2.3	T0625

Table 2. SBS modified asphalt technical indicators.

Test project	Technical requirements	Test results	Test method
penetration (100g 5s 25°C) (0.1mm)	40~60	57	T0604
Softening point (°C)	≮60	67	T0604
Ductility (5cm/min,5°C) (cm)	≮20	28	T0605
Flash point (°C)	≮230	—	T0611
Solubility (%)	≮99	—	T0607
elastic recovery 25°C (%)	≮75	81	T0662
separation	≯2.5	0.4	T0661
kinematic viscosity (135°C) (Pa·s)	≯3	2.3	T0625

2.2. Experiment theory and method

2.2.1. DSR theory. In the late 1980s, the SHRP project introduced Dynamic Shear Rheometer (DSR) from material field into civil engineering, solving formlessness issue during asphalt fatigue experiment, which made high-speed development in asphalt fatigue property realized. Until now, the main study on asphalt fatigue property is implemented by DSR parallel plate. In this paper, asphalt samples are time swept via DSR, and are forced by sinusoidal dynamic shear stress or strain at the same time. The load effect is shown in Figure 1. Before the test, asphalt samples are places on the lower plate, which is fixed, and the distance between parallel plates is adjusted by shifting upper plate. After loading, one cycle begins with upper plate moving from point A to B, and then back to point C before finally reaching A.

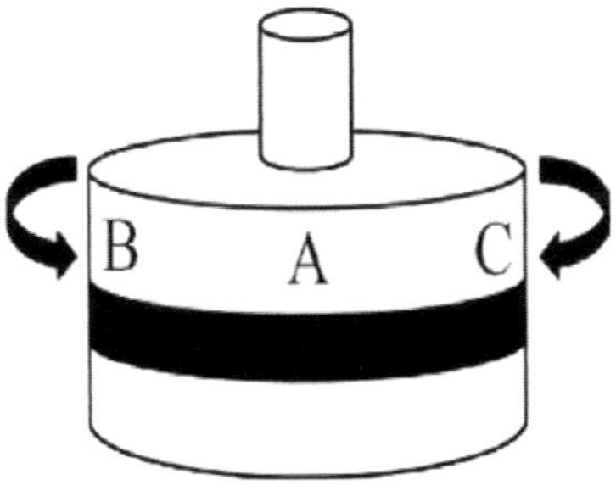

Figure 1. Illustrates the DSR loads.

Dynamic mechanical response rules of material associated with its own viscoelastic properties. Assumptions on the material according to the sine law of change of dynamic shear strain, angular frequency ω, as shown in type 1:

$$\gamma(t) = \gamma_0 \sin(\omega \cdot t) \tag{1}$$

Type, γ_0 — Strain amplitude;
ω — The angular frequency, rad/s.

For line elastomer, stress and strain is in building a balanced instantaneous, so

$$\tau(t) = G\gamma(t) = \gamma_0 G \sin(\omega \cdot t) \tag{2}$$

That is to say, the stress and strain with the same phase and frequency, amplitude $\sigma_0 = \gamma_0 G$, as shown in figure 2 (a). For the linear viscous fluid, according to the Newton's law:

$$\tau(t) = \eta \dot{\gamma} = \eta d\gamma(t)/dt = \gamma_0 \eta\omega \cos(\omega \cdot t) = \gamma_0 \eta\omega \sin(\omega \cdot t + \frac{\pi}{2}) \tag{3}$$

The stress and strain of the linear viscous fluid with the same frequency, the phase difference $\pi/2$, stress and strain lagging behind amplitude for $\pi/2$, and the size of the frequency, as shown in figure 2 (b).

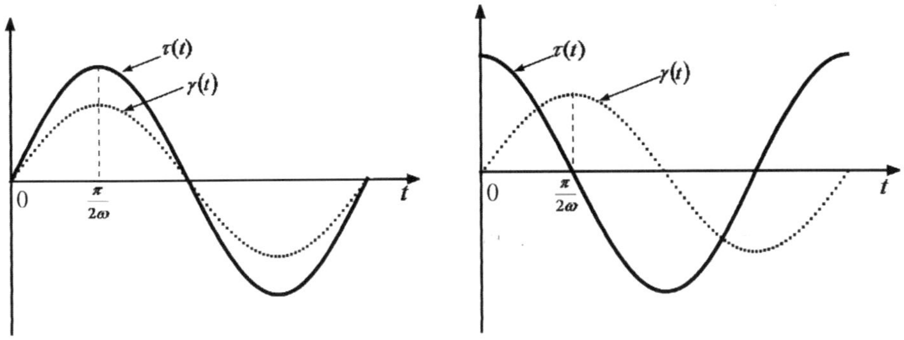

(a) Elastic material stress strain relationship (b) Viscous material stress-strain relationship

Figure 2. Material stress-strain diagram.

Under the effect of a certain amount of stress or strain, the mechanical behavior of materials with the environment temperature and load on the role of time, and the effect of loading history is closely linked.Under usually pavement temperature and traffic load, asphalt for viscoelastic body material can be sold and turned into money.When the asphalt applied by sinusoidal variation of strain and the stress is also changes according to the sine function.The same frequency, but the phase lag in the stress strain δ $(0 < \delta < \pi / 2)$, as shown in figure 3, the stress function can be expressed as:

$$\tau(t) = \tau_0 \sin(\omega \cdot t + \delta) \tag{4}$$

Type τ_0 — Stress amplitude (Pa);

ω — The angular frequency ; rad/s

δ — Phase Angle , °

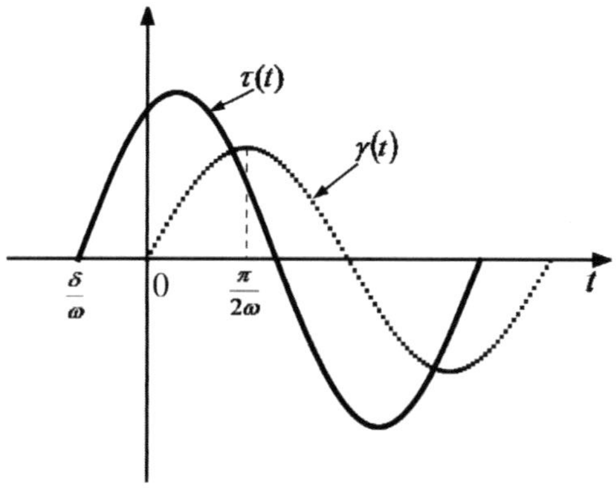

Figure 3. Asphalt stress-strain diagram.

In general, use the plural form description of alternating to physical quantities can simplify the calculation, so the alternating stress and cyclic strain can be represented as:

$$\gamma^* = \gamma_0 e^{i \cdot \omega \cdot t} \tag{5}$$

$$\sigma^* = \sigma_0 e^{i(\omega \cdot t + \delta)} \tag{6}$$

Define the complex shear modulus as:

$$G^*(\omega) = \frac{\sigma^*}{\gamma^*} = \frac{\sigma_0}{\gamma_0} e^{i\delta} = \frac{\sigma_0}{\gamma_0}(\cos\delta + \sin\delta) = G^{'}(\omega) + iG^{''}(\omega) \tag{7}$$

Type, $G^{'}(\omega)$— storage modulus, said the elastic and viscoelastic body shows that the size of the linear viscoelastic body stored energy (Pa);

$G^{''}(\omega)$— loss modulus and viscoelastic body of viscous, shows that the size of the loss of linear viscoelastic body energy (Pa).

According to the type 7, $G^{'}(\omega)$ and $G^{''}(\omega)$ can be expressed as:

$$G'(\omega) = \frac{\sigma_0}{\gamma_0}\cos\delta \tag{8}$$

$$G''(\omega) = \frac{\sigma_0}{\gamma_0}\sin\delta \tag{9}$$

The mode of complex shear modulus $G^*(\omega)$ is the dynamic modulus:

$$\left|G^*(\omega)\right| = G(\omega) = \left\{\left[G'(\omega)\right]^2 + \left[G''(\omega)\right]^2\right\}^{\frac{1}{2}} \tag{10}$$

Similarly follow the general definition of viscosity, define the plural shear viscosity for:

$$\eta^*(\omega) = \frac{\sigma^*}{\dot{\gamma}^*} = \frac{\sigma_0}{\omega\gamma_0}e^{i(\delta-\frac{\pi}{2})} = -\frac{i\sigma_0}{\omega\gamma_0}(\sin\delta - i\cos\delta) = \eta'(\omega) - i\eta''(\omega) \tag{11}$$

2.2.2. Testing scheme. Using DSR time scanning mode, selection of diameter of 8 mm parallel powder, control test temperature of 30°C, load frequency of 10 Hz, at 5%, 8%, 10%, 12% strain condition, the two often repeated strain loading fatigue test of asphalt for continuous scanning test (time).

3. Test results and analysis

3.1. fatigue damage standard

Complex shear modulus attenuation for the initial value of 50% is the most widely used definition of asphalt fatigue damage one of the methods and the definition of asphalt mixture fatigue damage of the most commonly used method for stiffness modulus attenuation to the initial value is less than 50% of the (S). As a result, the definition of asphalt fatigue damage way makes the research of asphalt and mixture. Studies have shown that under the same strain level and the test temperature, according to determine the fatigue life of asphalt and 50% G* and 50% S for determine the fatigue life of asphalt mixture in good correlation, the correlation coefficient is 0.7 ~ 0.9.

3.2. Different asphalt damage process

Figures 4 and 5 are respectively the DSR time scanning experiment, under the condition of different strain, 70 # asphalt and SBS modified asphalt complex shear modulus changing with the load times.

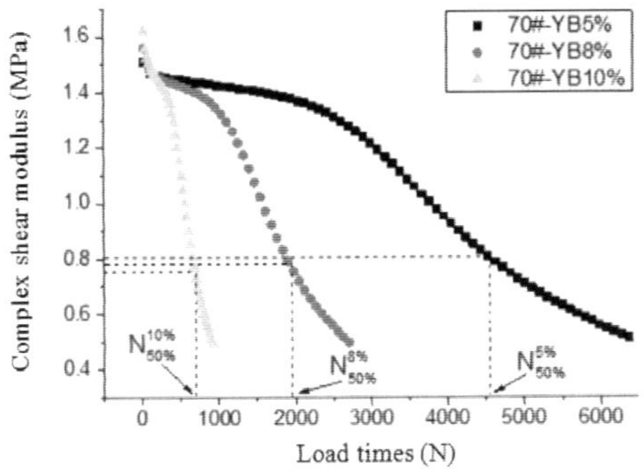

Figure 4. # 70 common asphalt curve.

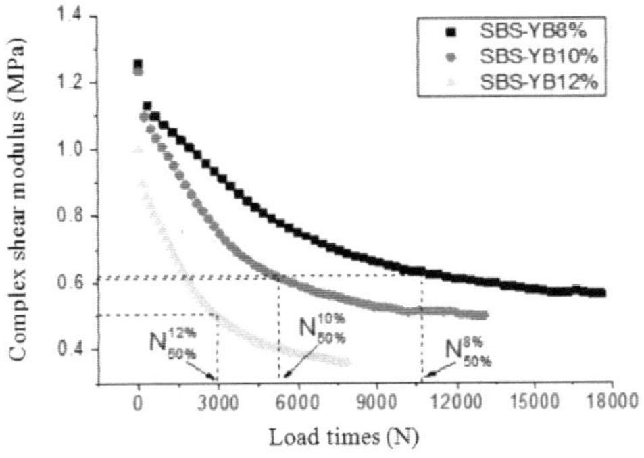

Figure 5. the curve of the SBS modified asphalt.

Can be seen from the graph:

1) under the condition of different strain levels, the plural of 70 # ordinary asphalt and SBS modified asphalt shear modulus with the increase of the number of loading and gradually decreases, but the two kinds of bitumen is obviously different from the shape of the curve.70 # ordinary asphalt is roughly s-shaped curve shape, but in the late loading, plural shear modulus decay gradually slow down and stable. So the curve of the strain mode 70 # ordinary asphalt can be subdivided into four stages, namely initial stage, stable stage, accelerating phase and stable phase. And the curve shape of the SBS modified asphalt is roughly C (arc), not like a 70 # ordinary asphalt has half a stable stage.

2) with the increase of the strain level, the initial complex shear modulus of asphalt are reduced, and the plural shear modulus attenuation speed, specimen damage can bear before the loading times and complex shear modulus decay to 50% when the corresponding loading times were significantly reduced.

3.3. Different fatigue equation of asphalt

Related studies have shown [6,7] that asphalt or fatigue life of asphalt rubber paste DSR fatigue test and loading strain obey power relations, usually can use the type 12 said.

$$N_f = K\gamma^{-b} \tag{12}$$

Type: N_f — the fatigue life of asphalt;

γ — test applied strain size;

K、b regression coefficient, K associated with the shear strength of the material, b the fatigue life of asphalt strain sensitivity.

According to the test data, can return to get two constant strain fatigue equation of asphalt is as follows:

70 # ordinary asphalt:

$$N = 1.473059 \cdot \gamma^{-2.72894} \tag{13}$$

SBS modified asphalt:

$$N = 3.361244 \cdot \gamma^{-3.19648} \tag{14}$$

It can be seen that the SBS modified asphalt fatigue performance is better than that of 70 # ordinary asphalt, but compared with the ordinary asphalt, SBS modified asphalt fatigue life more sensitive to the strain level.

4. Conclusions

This paper based on the DSR time scanning mode, investigated constant strain fatigue test for different type of asphalt. Analyzing the often strain fatigue process and its fatigue life, main conclusions are as follows:

1) Under the condition of different strain levels, the complex shear modulus of 70 # ordinary asphalt and SBS modified asphalt was increased when the number of loading was gradually decreased, but the shape of the curve $G^* - N$ of two kinds of bitumen is obviously different.

2) The curve $G^* - N$ of the 70 # ordinary asphalt can be subdivided into four stages, namely initial stage, half stable stage, accelerating stage and stable stage. However, the SBS modified asphalt does not exist a half stable stage.

3) Constant strain fatigue equation of the 70 # ordinary of asphalt and SBS modified asphalt was obtained respectively, and the fatigue performance of SBS modified asphalt is better than that of 70 #

ordinary asphalt. However, compared with the ordinary asphalt, the fatigue life of SBS modified asphalt is more sensitive to the strain level.

References

[1] Tan Y, Shan L and Li X. Fatigue Characteristic of Asphalt 2008 *J. Geotechnical Special Publication* **182** 98-107.
[2] Li X M. 2006 Study on Evaluation Methods and Performance of Asphalt Mortar Based on Rheological Properties[D]. Harbin Institute of Technology
[3] Link R E, Shenoy A, and Link R E. 2002 Fatigue Testing and Evaluation of Asphalt Binders Using the Dynamic Shear Rheometer *J. Test Eval.* **30**(4) 303-12.
[4] Yuan Y. 2005 Evaluation of the Fatigue Properties of Modified Asphalt Mortar[D]. South China University of Technology
[5] Bonnetti K, Nam K and Bahia H. Measuring and Defining Fatigue Behavior of Asphalt Binders 2002 J. *Transportation Research Record: Journal of the Transportation Research Board,* **1810** 33-43.
[6] Liu Q T. 2008 Research on the Fatigue Properties of Asphalt Mortar[D]. Wuhan University of Technology.
[7] Wu S P, Jiang C J, Lin J T, etc. 2013 Research on Fatigue Characteristic of Aged Asphalt Binder J. *Wuhan Univ. Technol.* **37** 451-55.

IOP Conf. Series: Materials Science and Engineering **170** (2017) 012004 doi:10.1088/1757-899X/170/1/012004

Low temperature sintering and microwave dielectric properties of $BaZn_2Ti_4O_{11}$ ceramics with Li_2O-B_2O_3-SiO_2 glass

S X Duan, E Z Li, Y W Chen, B Tang, Y Yuan, S R Zhang

School of Microelectronics and Solid State Electronics, University of Electronic Science and Technology of China, Chengdu 610054, People's Republic of China

Email: lienzhu@uestc.edu.cn

Abstract. The effects of Li_2O-B_2O_3-SiO_2 glass (LBS) on the sintering behavior, phase evolution, microstructure and microwave dielectric properties of the $BaZn_2Ti_4O_{11}$ ceramic were studied in detail. The results show that LBS significantly lowers the sintering temperature of the $BaZn_2Ti_4O_{11}$ ceramic from 1200 °C to 850 °C. Small amount of LBS promotes the sintering process of the $BaZn_2Ti_4O_{11}$ ceramic and obtains excellent microwave dielectric properties. However, excessive LBS addition leads to the change of phase and growth of abnormal grain, destroying the compact microstructure and deteriorating the microwave dielectric properties of the BaZn2Ti4O11 ceramic. $BaZn_2Ti_4O_{11}$ ceramic with 4~6 wt.% LBS can be well sintered at 850 °C and shows excellent microwave dielectric properties: εr = 22.95~24.92, Q×f = 27181~31301 GHz.

1. Introduction

With the rapid growth in wireless communications industry, low temperature cofired ceramics (LTCC) has been playing a significant role due to the miniaturization of multilayer devices with high electric performance using highly conductive metallic electrode such as silver[1-2]. In BaO-TiO_2 ceramic system, many compounds such as $BaTi_4O_9$ and $BaTi_5O_{11}$ present desirable microwave dielectric properties [3-4]. However, the high sintering temperature limits their applications in LTCC industry. Additives such as ZnO can effectively promote sintering and optimize the microwave dielectric properties of the BaO-TiO_2 ceramic[5-9]. For example, $BaZn_2Ti_4O_{11}$ sinters at 1200 °C with microwave dielectric properties: ε_r = 30, Q×f = 68000 GHz, and τ_f = -30 ppm/°C [7]. However, a sintering temperature of less than 950 °C is generally required in LTCC technology, considering the melting temperature of Ag (961 °C) that is usually used as the internal electrode. A common method to reduce the sintering temperature is adding the sintering aid with low melting points, such as glasses or oxides[10]. Many kinds of sintering aids, such as ZnO–B_2O_3–SiO_2 glass[1-2], Li_2O–ZnO–B_2O_3 glass[11], B_2O_3-ZnO-La_2O_3 glass[12], BaO-ZnO-B_2O_3 glass[13], Li_2O-B_2O_3-SiO_2 glass[14-15], $BaCu(B_2O_5)$[16-17], have been used to decrease the sintering temperature successfully, among which Li_2O-B_2O_3-SiO_2 glass (LBS) is one of the widely used sintering aids.[18-20] The purpose of this work is to lower the sintering temperature of $BaZn_2Ti_4O_{11}$ below the silver melting point without damaging its microwave dielectric properties. In this work, the effects of LBS on the sintering behaviors, phase structures, microstructures and the microwave dielectric properties of $BaZn_2Ti_4O_{11}$ ceramics have been hence investigated in detail.

2. Experimental

To synthesize the $BaZn_2Ti_4O_{11}$ powders, $BaCO_3$, ZnO and TiO_2 with purity higher than 99% were weighted as the mole ratio of the $BaZn_2Ti_4O_{11}$, and ball-milled with deionized water in a nylon jar using zirconia balls for 3 h. The resulting mixtures were dried and calcined at 950 °C for 3 h. The LBS was fabricated by a conventional glass fabrication process: high-purity powders of Li_2CO_3, H_3BO_3, and SiO_2 were weighed as the raw materials according to the composition of 56.92 wt.%, 37.59 wt.%, and 5.49 wt.%. The batches were dry-milled with zirconia balls for 2 h and then melted in a platinum crucible at 1400 °C for about 2 h. The glass melt was then quickly poured into a copper plate to quench, and then pulverized in a vibration mill for 4 h. The calcined $BaZn_2Ti_4O_{11}$ powders and LBS powders were weighed with the ratio of $BaZn_2Ti_4O_{11}$ + x wt.% LBS, where x = 3~8, and re-milled with alcohol in a nylon jar using zirconia balls for 3 h. The mixtures were dried and added with acrylic acid to form pellets (15 mm in diameter and 8 mm in height), and then sintered at 850 °C for 1 h at a heating rate of 5 °C/min in air. The bulk density of the sintered sample was measured by the Archimedes method. For the experiments of thermomechanical analyses (TMA) (NETZSCH, Germany), samples were fired at a heating rate of 5 °C/min in flowing air. The phase composition was analyzed by X-ray diffraction (XRD) using CuKα radiation (DX–1000 CSC, Japan). The microstructures of the materials with polished surfaces were determined using scanning electron microscopy (SEM, FEI Inspect F, the United Kingdom). The compositional and elemental analysis was carried out with an energy dispersive spectroscopy (EDS). The microwave dielectric characteristics were measured by the Hakki-Coleman dielectric resonator method in the TE011 mode using a network analyzer (HP83752A, the United States), and the τ_f value was determined from the difference between the resonant frequencies obtained at 25 °C and 85 °C using the equation:

$$\tau_f = \frac{f_{85} - f_{25}}{f_{25} \times 60} \times 10^6 \text{ (ppm / }^\circ\text{C)}$$

where f_{25} and f_{85} are the resonant frequencies at 25 °C and 85 °C, respectively.

3. Results and discussion

The effect of LBS on linear shrinkage of BZT ceramic was also determined in Figure 1. The onset of shrinkage takes place at about 1050 °C for pure $BaZn_2Ti_4O_{11}$ ceramic(a), while the onset dramatically decreases to about 800 °C in the presence of 6 wt.% LBS additive(b), indicating that the LBS effectively enhances the sintering process of $BaZn_2Ti_4O_{11}$ ceramic.

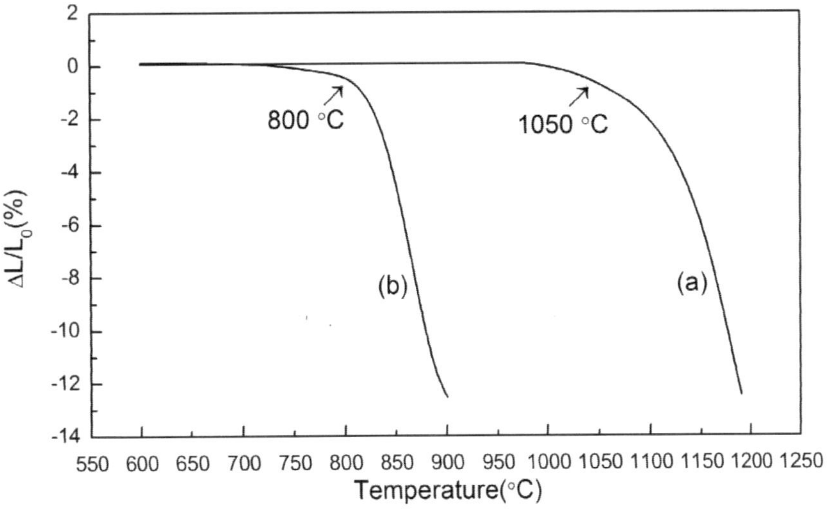

Figure 1. The linear shrinkages curves and onset temperatures of shrinking for $BaZn_2Ti_4O_{11}$ ceramic (a) and $BaZn_2Ti_4O_{11}$ ceramic with 6 wt.% LBS (b) fired at a heating rate of 5 °C /min in flowing air.

Figure 2 shows the XRD patterns of $BaZn_2Ti_4O_{11}$ ceramic and $BaZn_2Ti_4O_{11}$ ceramics with different amount of LBS. Only $BaZn_2Ti_4O_{11}$ phase was formed and no secondary phase could be observed after pure $BaZn_2Ti_4O_{11}$ powders calcined at 950 °C as shown in Figure 2 (a), which indicates that the $BaZn_2Ti_4O_{11}$ ceramic with monoclinic phase can be obtained. Figure 2 (b~d) shows the effect of different amounts of LBS on the phases development of $BaZn_2Ti_4O_{11}$ ceramics sintered at 850 °C. With 3 wt.% LBS added (b), three phases exist: $Zn_2Ti_3O_8$, $BaZn_2Ti_4O_{11}$, and $BaTi(BO_3)_2$. When LBS increases from 3 wt.%(b) to 7 wt.%(d), the peaks of $BaZn_2Ti_4O_{11}$ weaken and the peaks of $Zn_2Ti_3O_8$ and $BaTi(BO_3)_2$ become more obvious. It indicates that with the LBS added, the $BaZn_2Ti_4O_{11}$ decompose to $Zn_2Ti_3O_8$ and $BaTi(BO_3)_2$.

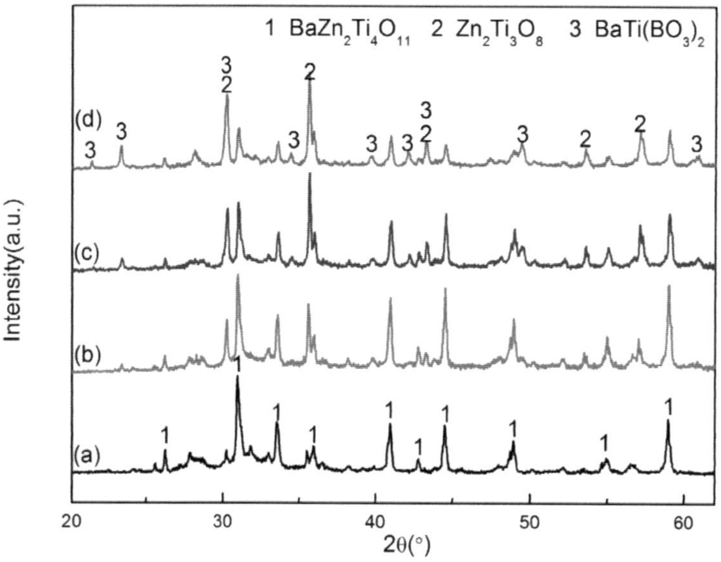

Figure 2. The XRD patterns of $BaZn_2Ti_4O_{11}$ ceramic calcined at 950 °C for 3 h (a), with 3 wt.% (b), 5 wt.% (c), and 7 wt.% (d) LBS sintered at 850 °C for 1 h.

Figure 3 shows the SEM micrographs of the polished surfaces of $BaZn_2Ti_4O_{11}$ ceramics with different amounts of LBS sintered at 850 °C. The samples sintered at 850 °C become more compact with the LBS increased from 3 wt.% to 5 wt.% (Figure 3 (a)~Figure 3 (b)). However, with the increasing of LBS, abnormal grain growth observed and become more obvious. (7 wt.% LBS Figure 3 (c)).

Figure 3. The SEM micrographs of the polished surfaces of $BaZn_2Ti_4O_{11}$ ceramics with 3 wt.% (a), 5 wt.% (b), and 7 wt.% (c) LBS sintered at 850 °C for 1 h.

To further confirm the composition of the abnormal grains, EDS was carried out to determine the composition of the normal (spot A in Figure 3 (a)), square abnormal (spot B in Figure 3 (c)) and strip abnormal (spot C in Figure 3 (c)) big grains of $BaZn_2Ti_4O_{11}$ ceramics with 3 wt.% and 7 wt.% LBS. As shown in Table 1, the molar ratio of Ba:Zn:Ti for spot A is close to $BaZn_2Ti_4O_{11}$. Spot B and Spot C have a composition poor in Ba and Zn element respectively and the molar ratio of Ba:Zn:Ti is similar to $Zn_2Ti_3O_8$ and $BaTi(BO_3)_2$ as shown in XRD. These EDS results are in good agreement with the XRD results discussed above.

Table 1. The composition ratio of normal grain (A) and abnormal grain (B and C).

Element	A		B		C	
	wt%	mol%	wt%	mol%	wt%	mol%
O	20.46	51.40	33.04	65.3	29.42	67.81
Ti	35.93	30.15	27.72	18.3	22.74	17.51
Ba	25.96	7.60	10.16	2.34	41.61	11.17
Zn	17.66	10.86	29.08	14.06	6.23	3.52

The bulk densities of $BaZn_2Ti_4O_{11}$ ceramics with different amounts of LBS sintered at 850 °C were determined, as shown in Figure 4. The densities increase first, reach maximum values, and then decrease with the increase of LBS additions. The $BaZn_2Ti_4O_{11}$ ceramic well densified with addition of 3~5 wt.% LBS as shown in Fig. 3. However, the density reach maximum values of 4.53 g/cm^3 with 3 wt.% LBS and then decrease with the increase of LBS additions, which probably attribute to the decomposition of $BaZn_2Ti_4O_{11}$ as shown in XRD. The densities of $Zn_2Ti_3O_8$ (4.52 g/cm^3)[21] and $BaTi(BO_3)_2$ (4.217 g/cm^3)[22] are much lower than that of $BaZn_2Ti_4O_{11}$ (5.07 g/cm^3) [9]. Moreover, the decomposition of $BaZn_2Ti_4O_{11}$ lead to the appearance of abnormal grain of $ZnTi_3O_8$ and $BaTi(BO_3)_2$ with the increasing of LBS, which destroy the compact microstructure as shown in Figure 3 (c). All the results demonstrate that the LBS is an effective sintering aid for $BaZn_2Ti_4O_{11}$ ceramic. The aiding mechanism can be attributed to the formation of liquid phases caused by the low melting point of LBS at about 845 °C[15] that promotes the low temperature sintering process and the densification of $BaZn_2Ti_4O_{11}$ ceramic. However, excess LBS leads to abnormal grain growth and the obvious glass phase as shown in Figure 3, which has a negative effect on the compact microstructure. Additionally, the low density of LBS glass (2.23 g/cm^3)[18], also reduces the density of $BaZn_2Ti_4O_{11}$ ceramic when excess amount added.

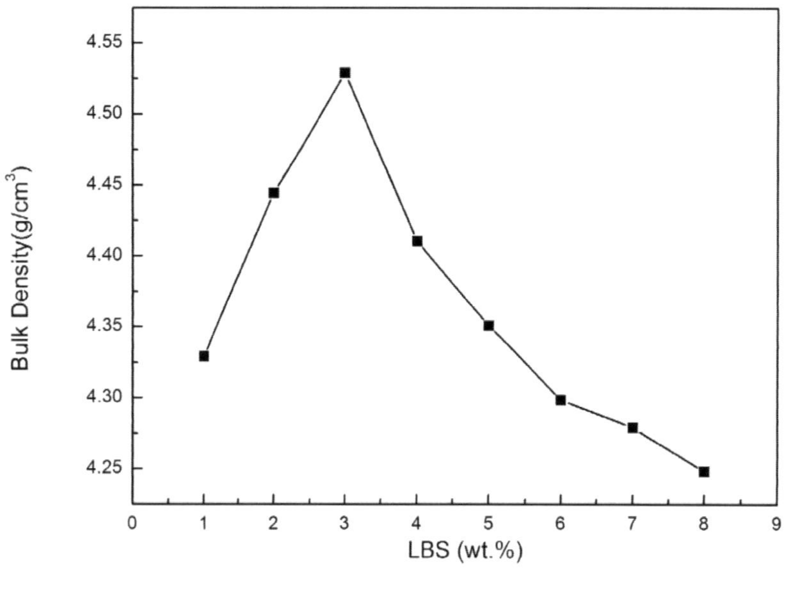

Figure 4. The bulk density of $BaZn_2Ti_4O_{11}$ ceramics with different amounts of LBS sintered at 850 °C for 1 h.

Figure 5 shows the ε_r and Q×f values of the $BaZn_2Ti_4O_{11}$ ceramics with different amounts of LBS sintered at 850 °C. The ε_r value increases firstly as the amount of LBS increasing from 1 wt.% to 3 wt.% and then decreases with more LBS added. The results are consistent with the trend of ceramic densities (Figure 4), which probably caused by the decomposition of $BaZn_2Ti_4O_{11}$ and the appearance of abnormal grains. The Q×f value increases firstly, reach maximum value with 6 wt.% LBS, and then decreases with the increasing of LBS. The increasing of Q×f value attribute to the compact microstructure with increasing of LBS as shown in Figure 3. On the other hand, the decreasing of Q×f value is likely due to the obvious appearance of abnormal grains in the ceramic, which destroy the compact microstructure and have a negative effect in Q×f value.

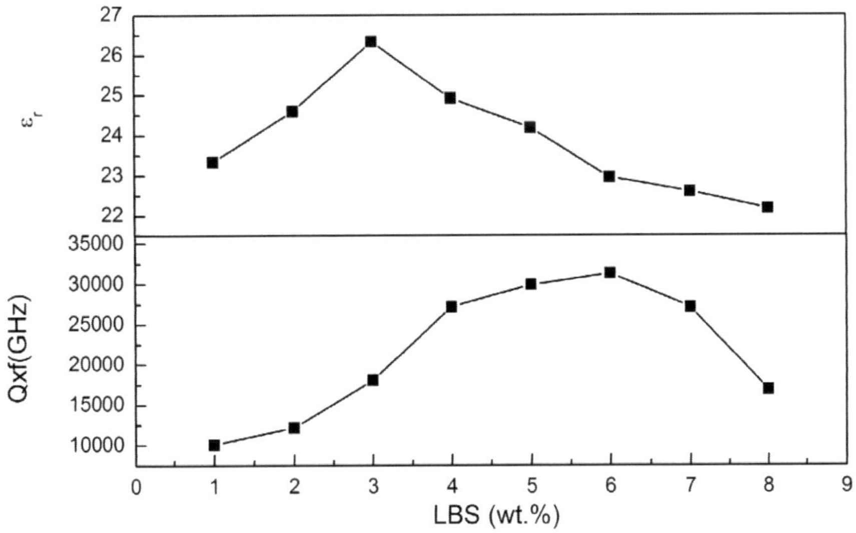

Figure 5. The ε_r and the Q×f value of $BaZn_2Ti_4O_{11}$ ceramics with different amounts of LBS sintered at 850 °C for 1 h.

The results show that the microwave dielectric properties are related to the density, microstructure, and composition of ceramics. The microwave dielectric properties of the $BaZn_2Ti_4O_{11}$ ceramic with LBS in this work are summarized in Table 2.

Table 2. The microwave dielectric properties of the $BaZn_2Ti_4O_{11}$ ceramic with LBS

Composition	Sintering Temperature (°C)	ε_r	Q×f (GHz)
$BaZn_2Ti_4O_{11}$+4 wt.% LBS	850	24.92	27181
$BaZn_2Ti_4O_{11}$+5 wt.% LBS	850	24.18	29919
$BaZn_2Ti_4O_{11}$+ 6 wt.% LBS	850	22.95	31301

4. Conclusions

In this study, the effect of LBS on the sintering behaviors and microwave dielectric properties of $BaZn_2Ti_4O_{11}$ ceramic sintered at low temperature were studied. The results indicate that the LBS significantly reduces the sintering temperature of $BaZn_2Ti_4O_{11}$ ceramic. The $BaZn_2Ti_4O_{11}$ phase decompose to $Zn_2Ti_3O_8$ and $BaTi(BO_3)_2$ phase when LBS added, which is corresponding to the abnormal large grains. The excellent microwave dielectric properties and densification microstructure can be obtained with appropriate amount of LBS added. However, excessive LBS addition causes the obvious formation of abnormal grain growth, deteriorating microwave dielectric properties. $BaZn_2Ti_4O_{11}$ ceramics with LBS can be well sintered at 850 °C and shows excellent microwave dielectric properties: ε_r = 22.95~24.92, Q×f = 27181~31301 GHz.

Acknowledgements
This work is partly supported by 'Specialized Research Fund for the Doctoral Program of Higher Educ ation of China 20110185120004', 'National Natural Science Funds of China (Grant No. 51202021)' a nd 'Scientific Research Foundation for the Returned Overseas Chinese Scholars'.

References

[1] Lu X P, Zheng Y, Dong Z W and Huang Q 2014 Low temperature sintering and microwave dielectric properties of 0.6Li2ZnTi3O8–0.4Li2TiO3 ceramics doped with ZnO–B2O3–SiO2 glass *Mater. Lett.* **131** 1-4.
[2] Li E Z, Zhang P, Duan S X, Wang J, Yuan Y and Tang B 2015 Low temperature sintering of low-loss ZnTiO3 microwave dielectric ceramics with Zn-B-Si glass *J. Alloy. Compd.*. **647** 866-872.
[3] Mhaisalker S G, Readey D W and Akbar S A 1991 Microwave Dielectric Properties of Doped BaTi4O9 *J. Am. Ceram. Soc.* **74 (8)** 1894–1898.
[4] Zhou H F, Wang H, Chen Y H, Li K C and Yao X 2008 Microwave Dielectric Properties of BaTi5O11 Ceramics Prepared by Reaction-Sintering Process with the Addition of CuO *J. Am. Ceram. Soc.* **91 (10)** 3444–3447.
[5] Gormikov Y I, Makarova Z Y, Belous A G, Gavrilova L G, Paskov V M and Chalyi V P 1984 Effect of Zinc Oxide Additions on the Phase Compositions and Dielectric Properties of Barium Tetratitanate *Sov. Prog. Chem.* **50 (12)** 11–12.
[6] Lee M J, You B D and King D S 1995 Properties of Mn-Doped BaTi4O9-ZnO-Ta2O5 Ceramics Part I Reaction Sequence, Microstructural Analysis and Microwave Dielectric Properties *J. Mater. Sci.-Mater. El.* **6 (3)** 165–172.
[7] Belous A G, Ovchar O V, Marjeta M K and Valant M 2006 The Homogeneity Range and the Microwave Dielectric Properties of the BaZn2Ti4O11 Ceramics *J. Eur. Ceram. Soc.* **26 (16)** 3733–3739.
[8] Chen L Y, Wu S H, Wang S and Wang G Q 2008 Microwave Dielectric Ceramics in the BaO–TiO2–ZnO System Doped with Nb2O5 *J. Mater. Sci.-Mater. El.* **19 (5)** 463–465.
[9] Roth R S, Rawn C J, Lindsay C G and Wong-Ng W 1993 Phase Equilibria and Crystal Chemistry of the Binary and Ternary Barium Polytitanates and Crystallography of the Barium Zinc Polytitanates *J. Solid State Chem.* **104 (1)** 99–118.
[10] Imanaka Y 2005 Multilayered Low Temperature Cofired Ceramics (LTCC) Technology *Springer U.S.* .
[11] Shih Y T and Jean J H 2013 Low-Fire Processing of Microwave BNBT-Based High-k Dielectric with Li2O–ZnO–B2O3 Glass *J. Am. Ceram. Soc.* **96 (12)** 3849–3856.
[12] Lu X P, Zheng Y, Dong Z W, Cheng P and Lin R 2014 Low temperature sintering and microwave dielectric properties of Li2ZnTi3O8 ceramics doped with ZnO–La2O3–B2O3 glass *Ceram. Int.* **40** 7087–7092.
[13] Jhou M Z and Jean J H 2006 Low-Fire Processing of Microwave BaTi4O9 Dielectric with BaO-ZnO-B2O3 Glass *J. Am. Ceram. Soc.* **89 (3)** 786–791.
[14] Choi Y J, Park J H, Ko W J, Park J H, Nahm S and Park J G 2005 Low Temperature Sintering of BaTi4O9-Based Middle-k Dielectric Composition for LTCC Applications *J. Electroceram.* **14** 157–162.
[15] Duan S X, Li E Z, Chen H T, Tang B, Yuan Y and Zhang S R 2016 Influence of Li2O-B2O3-SiO2 glass on the sintering behavior and microwave dielectric properties of BaO-0.15ZnO-4TiO2 ceramics *Ceram. Int.* **42** 7943-7949.
[16] Tang B, Guo X, Yu S Q, Fang Z X and Zhang S R 2015 The shrinking process and microwave dielectric properties of BaCu(B2O5)-added 0.85BaTi4O9–0.15BaZn2Ti4O11 ceramics *Mater. Res. Bull.* **66** 163–168.
[17] Chen Y H, Chen Y H, Zeng Y and Wang H 2012 Microwave dielectric properties of BaO–4.3TiO2–0.5ZnO ceramics with BaCu(B2O5) *Ceram. Int.* **38S** S147–S152.
[18] Park J H, Choi Y J, Park J H and Park J G 2004 Low-fire dielectric compositions with permittivity 20–60 for LTCC applications *Mater. Chem. Phys.* **88** 308–312.
[19] Sebastian M T and Jantunen H 2008 Low loss dielectric materials for LTCC applications: a review *Int. Mater. Rev.* **53 (2)** 57-90.
[20] Choi Y J, Park J H, Park J H, Nahm S and Park J G 2007 Middle- and high-permittivity dielectric

compositions for low-temperature co-fired ceramics *J. Eur. Ceram. Soc.* **27** 2017–2024.
[21] Bartram S F and Slepetys R A 1961 Compound Formation and Crystal Structure in the System ZnO-TiO2 *J. Am. Ceram. Soc.* **44** 493-499.
[22] Zhang S Y,Wu X, Chen X L,He M, Cao Y G, Song Y T and Ni D Q 2003 Phase relations in the BaO–B2O3–TiO2 system and the crystal structure of BaTi(BO3)2 *Mater. Res. Bull.* **38** 783–788.

Effect of Point and Linear Defects on Band-gap Properties in Triangular-Honeycomb Structure Photonic Crystals

D H Ge[1, a], J H Zhang[1, b], L Q Zhang[1, 2, c], L Lu[1, d] and X K Huang[1, e]

[1]Nano/micro science and technology center, School of Mechanical Engineering, Jiangsu University, Zhenjiang 212013, PR. China

[2]Laboratory of Span-Scale Design and Manufacturing for MEMS/NEMS/OEDS, School of Mechanical Engineering, Jiangsu University, Zhenjiang, 212013, PR China.

E-mail:[a]email: gedaohan@mail.ujs.edu.cn, [b]email: zhang_jinhua163@163.com, [c]email: zhanglq4158@ujs.edu.cn, [d]email: lulemailbox123@163.com, [e]email: huangxiukang123@163.com

Abstract. Hybrid photonic crystals, namely by combining two lattices to create hybrid lattice geometries, which produce larger photonic band gaps, can effectively improve the performance of photonic devices. In this paper, the photonic band gap of the hybrid-lattice photonic crystal was calculated by considering the point and linear defects in the triangular-honeycomb structure based on the plane wave expansion method. The influence of the sizes of air holes on the photonic band gap was simulated by changing the radius ratio of two kinds of air holes in triangular and honeycomb lattices. Three complete band gaps occurred in the hybrid photonic crystal with the triangular-honeycomb structure with the increase of the radius of the small holes in honeycomb lattices, while the width of the complete band gap decreases gradually. Two kinds of defects (point and linear defects) were introduced and the effect on the optical properties in triangular-honeycomb structure photonic crystals was studied. The results show that the width of the photonic band gap reduced after introducing the linear defect, whereas the point defect kept the width of the band gap and the frequency range of the photonic crystal unchanged. Moreover, the limit of the passing rate of the light wave was reduced, and most of the light wave could be transmitted directly through the linear defect.

1.Introduction

Photonic crystal is a dielectric structure [1, 2], which is composed of different dielectric materials in a certain period of space, with a photonic band and photonic band gap. Its appearance makes it possible for people to manipulate and control the photonic dream. It is great significant to explore the photonic crystal structure with a wider photonic band gap (PBG). The complex periodic structure is formed by the superposition of multiple periodic structures with different cell sizes in the same or different lattice. The symmetry of the structure is very important for controlling complete forbidden band gap (CBG) [3]. When the new structure is added in a single cycle [4], several methods have been suggested to reduce symmetry of the photonic crystal structure and the hybrid photonic crystals, which are combining two kinds of lattices together, could produce larger PBG. Besides, different point [5, 6] and linear defects [7, 8] introduced in the photonic crystal also can destroy the symmetry of original crystal structure and change the PBG of the photonic crystal.

In this paper, the PBG of the hybrid-lattice photonic crystal with triangular-honeycomb structure is calculated by using the plane wave expansion method. The effect of point defects or linear defects on the CBG of hybrid photonic crystal is investigated, which presents a numerical simulation of photonic bands for 2D hybrid triangular and honeycomb lattices.

2.Photonic Crystal with Triangular Honeycomb Structure

2.1Physical Model. The small air holes (R_2 is the radius) are embed in the triangular lattice photonic crystal (ε=12.096, a=1μm, R_1=0.48 μm), which change the symmetry of the perfect triangular lattice photonic crystal and form hybrid triangular-honeycomb lattice photonic crystal structures, as shown in figure 1. The influence of the change of the small air hole radius on the photonic band gap is calculated by the plane wave expansion method. As the radius R_2 of the small air hole scanning variable, we employ the complete photonic band gap. The scanning range of the radius R_2 is 0.00~0.09 μm, and the size of scanning step is 0.00118 μm. For the PBG of photonic crystal with triangular-honeycomb structure, the width of TE mode gradually reduces with the increase of the radius of the small holes, and the normalized frequency moves upper. TM mode width is also gradually increasing and the normalized frequency also moves upper. The CBG appears when the radius of the air hole in the range of 0-0.071 μm, and three CBGs appear in the 0-0.021μm range. The first CBG is the widest. When the radius of the small hole is in the range of 0-0.066 μm, the width of the first CBG gradually reduces in the 0-0.25 μm range. Therefore, those paramaters (ε=12.096, a=1 μm, R_1=0.48a, R_2=0.01a) are chosen as the research conditions.

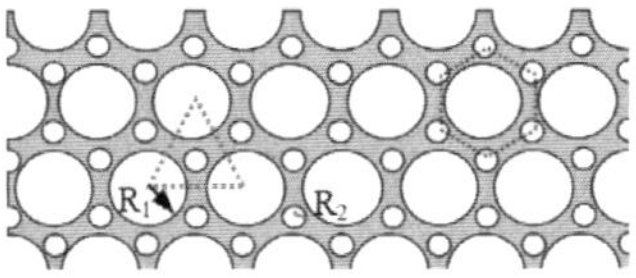

Figure 1. Triangular-honeycomb structure photonic crystal

2.2Defect models. The band structures of photonic crystals with point defects are calculated by plane wave expansion method [9], and the band structure is compared with the complete PBG of the photonic crystal with triangular-honeycomb structure. Then, we summarize the effect of point defects on the PBG of photonic crystal. As shown in figure 2(a) (ε=12.096, R_1=0.48a, R_2=0.01a, a=1 μm). The plane wave expansion method is only suitable to calculate the band gap of perfect photonic crystal. Therefore, it is necessary to adopt the principle of super unit cell structure [10] to calculate the photonic crystal with defect when the method of plane wave expansion is used. The structure of the super cell must contain the defect structure and part of the complete photonic crystal structure. Besides, linear defect [11] is formed by removing the periodic complete photonic crystals on a row or a column of air holes. Figure 2(b) shows the linear defect which removed the middle of a row of air holes. 1×7 super cell structure was used to calculate the PBG of the linear defect photonic crystal (ε=12.096, a=1 μm, R_1=0.48a, R_2= 0.01a)

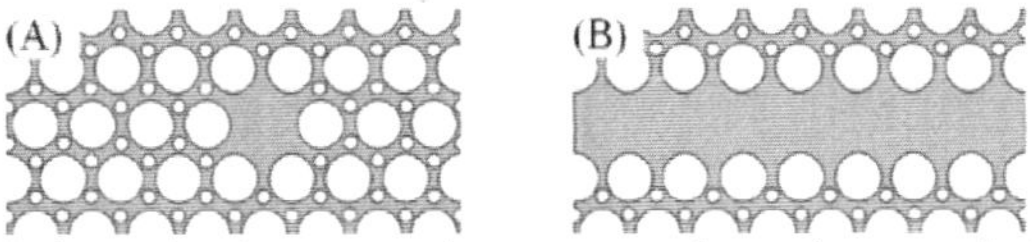

Figure 2. Defect models of triangular-honeycomb structure photonic crystal; (a) Point defect model; (b) Linear defect model.

3.Results and discussions

Figure 3 shows the band gap diagram of photonic crystal without any defects. Figure 4 shows the band gap diagram of photonic crystal with one point defect. The comparison of TM and TE model shows that the frequency range of the first PBG is 0.375-0.535 between the first band and the second band of TE mode before defect and the width of the band gap is 0.16. The band gap frequency range of TE mode is 0.375-0.41, 0.41-0.44, 0.44-0.47, 0.47-0.52, 0.52-0.535 after defect. Between the second energy band and the third energy band of the TM mode before defect, a band gap is formed, whose frequency range of band gap is 0.445-0.54 and the width of band gap is 0.09. While band gap frequency range of TE mode is 0.445-0.470, 0.470-0.490, 0.490-0.54 after defect. The forbidden band gap and width of TE and TM before and after the introduction of point defects are compared. The range and width of band gap of TE/TM mode have little change. But after the defect, there are five conduction bands occurring in the band gap of TE mode and two conduction bands occurring in the band gap of TM mode. Many conduction bands are relatively flat, which can be approximated as a single frequency. Those provide the conditions for controlling the transmission of optical waves of different wavelengths.

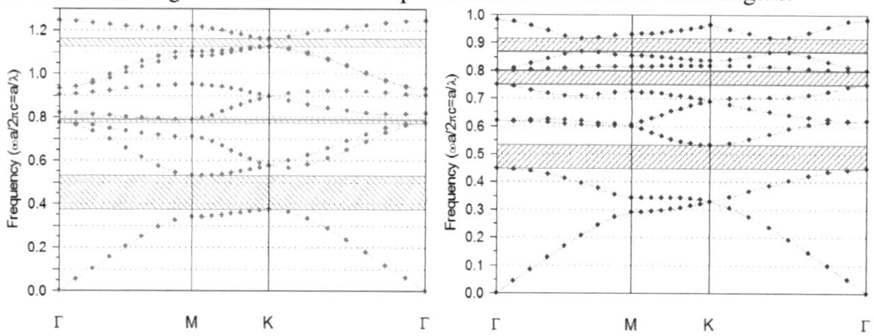

(a) TE mode (b) TM mode

Figure 3. TE/TM forbidden band gap of triangular-honeycomb structure

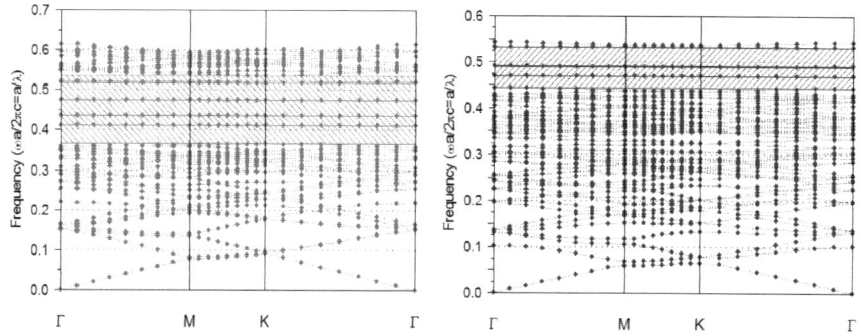

(a) TE mode (b) TM mode

Figure 4. TE/TM band gap of triangular-honeycomb structure with point defect.

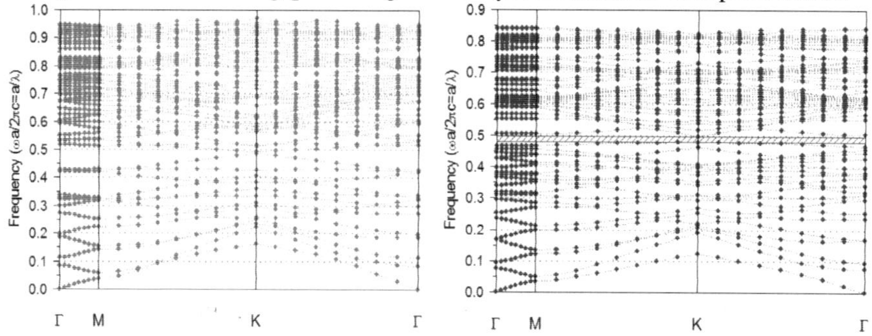

(a) TE mode (b) TM mode

Figure 5. TE/TM band gap of triangular-honeycomb structure with linear defect.

Compring with figure 5 and figure 3, it indicates that there is no photonic band gap in the TE mode of the linear defect photonic crystal or in the second and third band gaps of the TM mode, which becomes defective mode. The width of the first photonic band gap of the TM mode is only 0.023 μm, and the band gap of the photonic crystal is narrow compared with the band gap of the complete photonic crystal. The local mode of the line defect is different from the local mode produced by the point defect. The local mode of the point defect is very flat, while the local mode of the line defect is not flat. In the photonic crystal waveguide (linear defect photonic crystal), the local mode is the transmission mode that is transmitted along the waveguide, and the limitation of the line defect to the light wave passing rate is reduced.

4.Summary

The plane wave expansion method is used to calculate the PBG of hybrid photonic crystals. A small air hole is embedded in the triangular-lattice photonic crystal to form the triangular-honeycomb structure. The introduction of point defects produce some conduction bands in the triangle -honeycomb photonic crystal, and can allow the light to pass through a certain frequency of light wave. However, the local mode of the linear defect is not as smooth as that of the point defect, and the light wave can be transmitted directly through the linear defect, with the limitation of the passing rate of the optical wave reduced.

5.Acknowledgments

This work was supported by the National Natural Science Foundation of China (11404146), China Postdoctoral Science Foundation (2014M561575), the Natural Science Foundation of Jiangsu Province of China (BK20140556 & BK20130537), the Research Fund for the Doctoral Program of Higher Education of China (20133227120022), the support of Jiangsu Postdoctoral Science Foundation (No. 1301047C) and the Starting Foundation of Jiangsu University Advanced Talent (13JDG021 & 13JDG020).

References

[1] Yablonovitch E 1987 Inhibited spontaneous emission in solid-state physics and electronics *Phys. Rev. Let.* **58** 2059-2062.

[2] John S 1987 Strong localization of photons in certain disordered dielectric super lattices *Phys. Rev. Let.* **58** 2486-2489.

[3] Cassagne D, Jouanin C and Bertho D 1996 Hexagonal photonic-band-gap structures *Phys. Rev. B.* **53** 7134-7142.

[4] Melo E G and Alayo M I 2015 Photonic band gaps of wurtzite GaN and AlN photonic crystals at short wavelengths *Photonics and Nanostructures - Fundamentals and Applications.* **14** 35-45.

[5] Feng S and Wang Y 2013 Unidirectional wavelength filtering characteristics of the two-dimensional triangular-lattice photonic crystal structures with elliptical defects *Optical Materials.* **12** 2166-2170.

[6] Stem R 2004 Periodically arranged point defects in two-dimensional photonic crystals *Phys. Rev. B* **70** 3352-3359.

[7] Noori M and Soroosh M 2015 A comprehensive comparison of photonic band gap and self-collimation based 2D square array waveguides *Optik-Inter. J. Light and Electron Optics.* **126** 4775 -4781.

[8] Jha A K, Sandeep U, Samad S A and Srinivas T 2011 Analysis of line defects in 2D Photonic Band Gap structures of dielectric rods in air for single waveguide modes *IEEE India Conference IEEE.* 1-6.

[9] Anderson C M and Giapis K P 1996 Larger Two-Dimensional Photonic Band Gaps *Phys. Rev. Let.* **77** 2949-2952.

[10] Benisty H 1996 Modal analysis of optical guides with two-dimensional photonic band-gap boundaries *J. Appl. Phys.* **79** 7483-7492.

[11] Prather D W, Murakowski J, Shi S, Venkataraman S and Sharkawy A 2002 High-efficiency coupling structure for a single-line-defect photonic-crystal waveguide *Optics Letters* **27** 1601-1603.

Analysis on flow control forming of magnesium alloy wheel

M J Zhao[1], Z L Wu[1] and Z R Chen[2], X B Huang[1]

[1]The School of Mechanical Engineering, Nanjing University of Science & Technology, Nanjing, China

[2]Nanjing Hub Electromechanical Technology Co., Ltd, Nanjing, China

E-mail:wuruinan-1994@mail.njust.edu.cn

Abstract. Deformation force and forming quality are the important research focus in process of Magnesium (Mg) alloy wheel. In this paper, a process model of flow control forming (FCF) for the wheel was built, and its finite element model was given to simulate the FCF process at different extrusion speed (5 mm/s, 10 mm/s, 15 mm/s and 20 mm/s). The simulated results show that the FCF method for the wheel can reduce deformation force to less than 22000 kN, while there are cracks on rim of the wheel. The results are verified by the experiment and the cracks predicted by simulation are in good agreement with the experimental results.

1. Introduction

Mg alloy is one of the most attractive structural light metal, especially for automobile applications, due to its appealing properties including low density excellent specific strength and specific stiffness, high damping capacity, high thermal conductivity, reducing fuel consumption and decreasing pollution [1,2]. Because of the above merits, there is an increasing trend in automotive in automotive industries [3-5]. Currently, Mg alloy can be divided into cast Mg alloy and wrought one in terms of difference in processing[6]. Wrought Mg alloy had greater advantages over cast ones in structural components[7,8]. However, it usually exhibits poor plastic formability at room temperature because of their hexagonal close-packed (HCP) structures and the limited number of slip systems, which restricts its large-scale applications [9], particularly in complicated parts, such as Mg alloy wheel.

The current studies on plastic forming of Mg alloy wheel were mainly as follows. [10] presented a kind of plastic forming technology: roughness forging, blocker forging, finishing forging, T6 treatment and spinning process, of which the billets was the cast Mg alloy with average crystalline particle size of not greater than 100 μm. Meanwhile, the forging was carried out at a temperature within the range of 300—420℃. [11] provided a process of pre-forging, finishing forging and flaring, and [12] provided a more effective technique with only including roughness forging and finishing forging, but the above methods focused basically on forming efficiency of Mg alloy wheel. For reducing the forming load, [13,14] showed a new extrusion process of Mg alloy automobile wheels which can reduce the maximum forming load, but the method was only achieved through processing billet. In addition, [15,16] adopted the technology of combined extrusion and expanded forming, which saved working procedure and investment in equipments. [17] applied roughness extrusion, finishing extrusion and expanded forming, and further analyze the mechanical performance of the Mg alloy wheel. However, it is still one of difficulties to reduce the deformation force. Meanwhile, with the thin and bending wall of the rim being easy to crack, guaranteeing forming quality of the rim is also one of the difficulties in plastic forming processes of Mg alloy wheel.

To solve these problems, the above forming methods of Mg alloy wheel shared one common feature that the rim was formed by extrusion and flaring, while a new alternative method: the Flow Control Forming (FCF) will be introduced in the present paper, and then was analyzed and investigated by the finite element method and experiment. The good result is acquired in reducing forming force, besides, the simulated results is well agree with the experimental results.

2. FCF process of Mg alloy wheel

In this study, the FCF method of Mg alloy wheel consisted of two main steps: forming of the rim and the forming of spokes. According to the stress characteristics of forming process for the wheel, the FCF method can be subdivided into relatively simple process: cylindrical compression (CC), a process from Fig.1(a) to Fig.1(b), backward extrusion (BE), a process from Fig.1(b) to Fig.1(c), extrusion-shearing (ES), a process from Fig.1(c) to Fig.1(d) and closed-die forging (CDF), a process from Fig.1(d) to the end, as shown in Fig.1. In actual forming processes, Mg alloy cast was prepared and heated to a given temperature; next, billet was extruded to the Fig.1(b), and then form the rim, as shown in Fig.1; the last spokes was forged successfully through closed die forging. The die assembly of the Mg alloy wheel is shown schematically in Fig.2. And the arrow represented the direction of punch movement and female die movement respectively.

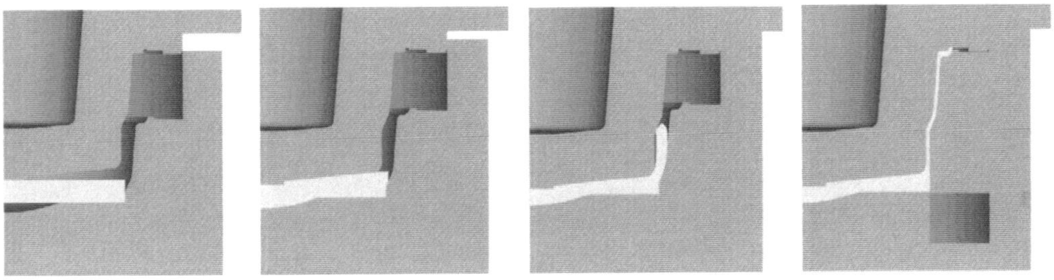

(a) (b) (c) (d)

Fig. 1 Schematic diagram of FCF process: ((a) CC; (b) BC; (c) ES; (d) CDF)

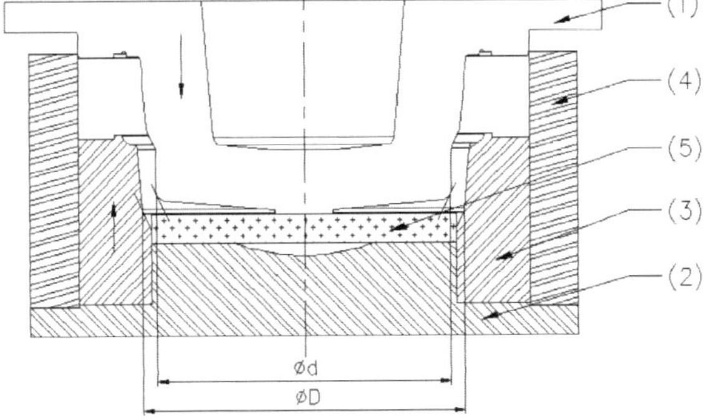

(1) Punch; (2) Bottom die; (3) Female die; (4) Die sleeve; (5) Billet

Fig. 2 Schematic diagram of FCF die

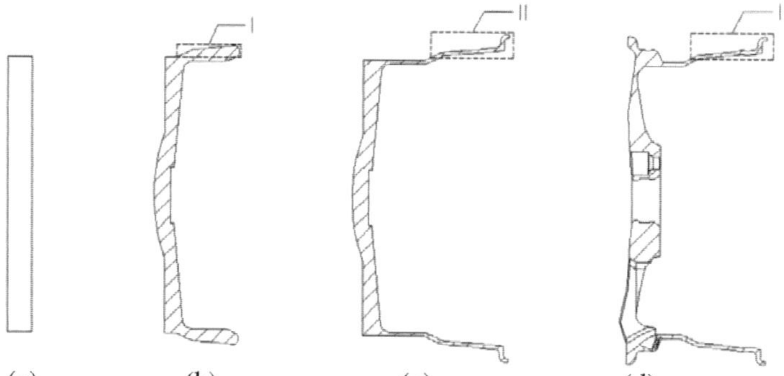

Fig. 3 Schematic diagram of billet changing: (a) Billet; (b) Pre-forming; (c) Forming the rim; (d) Forming spokes

It was the key point of FCF method to control the metal flow direction, the forming sequence and the forming force [18]. During the forming process of Mg alloy wheel, the rim material flow of the wheel was firstly controlled and forced to flow into a certain position, as shown in Fig. 3(b), and then the billet was further extruded to final rim shape, as shown in Fig. 3(c). Besides, because of the weight loss in isothermal forming, it should guarantee that the volume of the portion "I" was slightly more than portion "II", and the shape of portion "II" was nearly to shape of portion "III", which shorted the flow path of Mg alloy during the forming process, reduced the flow resistance, and was beneficial to forming of the wheel.

3. FEM simulation on forming process of the wheel

The materials of the punch, bottom die, female die, die sleeve were made of H13 steel and the billet was Mg alloy AZ80, of which physical characteristics referred to [19]. And the constitutive equation given in [20] was adopted to describe the hot compression of magnesium alloy AZ80, as follows:

$$\dot{\varepsilon} = 2.9\times10^9[\sinh(0.0174\sigma)]^{6.905}\exp[\frac{-(154.6\times10^3)}{8.314T}]$$

Due to the rapid forming process, the model of FEM simulation was assumed to be isothermal process and the thermal boundary condition was set at 415℃. Since the normal pressure is commonly greater than the yield stress of billet during high pressure processes[21], constant friction model was adopted in this paper, and the fiction factor was set to 0.21, according to [22]. And the simulation parameters was arranged in table 1.

Table 1 Simulation parameters

Simulation No.	Initial temperature (℃)	Extrusion speed (mm/s)	Friction factor	Outer diameter (mm)	Height of the wheel (mm)
Simulation 1	415	5.0	0.21	440.6	200.4
Simulation 2	415	10.0	0.21	440.6	200.4
Simulation 3	415	15.0	0.21	440.6	200.4
Simulation 4	415	20.0	0.21	440.6	200.4

Punch, bottom die, female die, die sleeve were rigid parts, and the billet was plastic part in the simulation model. During the simulated processes, the first step was the process from Fig.3(a) to Fig.3(c), which the punch move down, and the second step was the process from Fig.3(c) to Fig.3(d), which the female die move up. The simulations was conducted by DEFORM-3D software, and the model of billet was meshed by tetrahedron element with the maximum size being equal to 4 mm. The sum of element was 78544.

4. Results and analysis

It is quite significant work to predict the formability of Mg alloy wheel. The load-stroke curves for the forming processes of Mg alloy wheel were shown in Fig.4, 5. And the curves at different extrusion speed showed the same change trend, that is, the forming force on punch appeared three wave crests with stroke, as shown in Fig.4. The first crest appeared in backward extrusion, the second crest in extrusion-shearing, and the third in closed-die forging. Besides, the maximum force all appeared during closed-die forging and the minimum force all during extrusion-shearing. The forming force on female die increased slightly, smoothly and increased rapidly in the later stage, as shown in Fig.5.

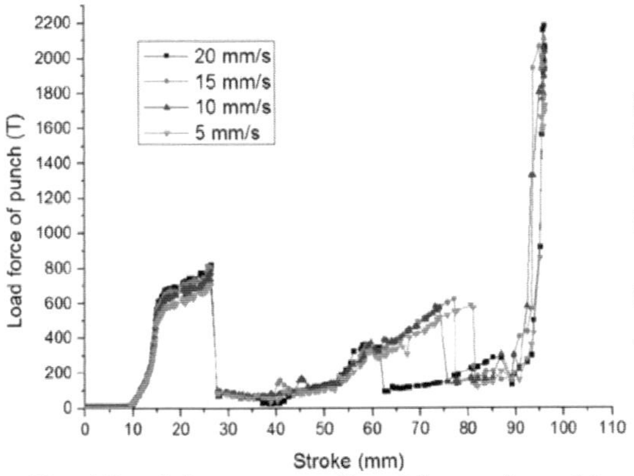

Fig. 4 Load-force curves (Load force of punch)

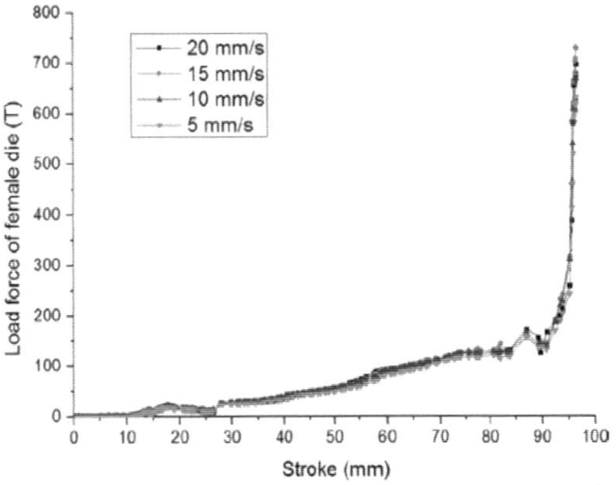

Fig. 5 Load-force curves (Load force of female die)

Forming force on punch increased slightly with the increase of the extrusion speed, which accorded with the law of stress-true strain in [23]: the stress is higher with higher strain rate at elevated temperature. while the higher the extrusion speed is, the higher strain rate is. And the forming force on female die was in line with the above results.

Because of thin wall of the rim and its complicated shape, it often happened to crack in its forming process. The simulation found that the rim of Mg alloy wheel could appear to crack on the near end course of forming process. It can be illustrated by Fig.6 and Fig.7. On the one hand, the effective stress in the rim was more than 56.3 MPa at elevated temperature, while according to GB/T 5153-2003, the yield stress was less than 35 MPa. Therefore, the rim was in the state of plastic, meanwhile, there were maximum stresses value in the point "A", "B" and "C", nearly to 113 MPa, as shown in Fig.6. On the other hand, the maximum principle stress in the rim was more than 75 MPa at elevated temperature, while according to GB/T 5153-2003, the tensile strength was less than 65 MPa, as shown in Fig.7. Therefore, the maximum principle stress was more than the tensile strength at points "A", "B" and "C". A forming experiment of Mg alloy wheel was conducted to validate the above analysis, and the experimental condition was set in line with the simulation condition, the experimental equipment was shown in Fig. 8. The experimental results showed that there existed three cracks on the rim of Mg alloy wheel, as shown in Fig.9,10, and these agreed well with the predicted point by the finite element simulation.

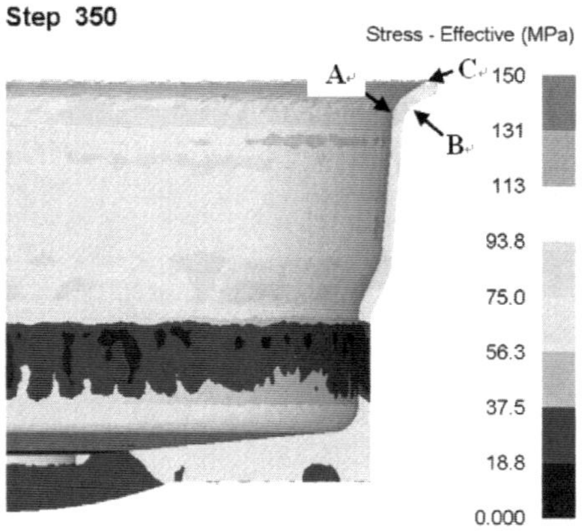

Fig. 6 Effective Stress distribution

Fig. 7 Max principal Stress distribution

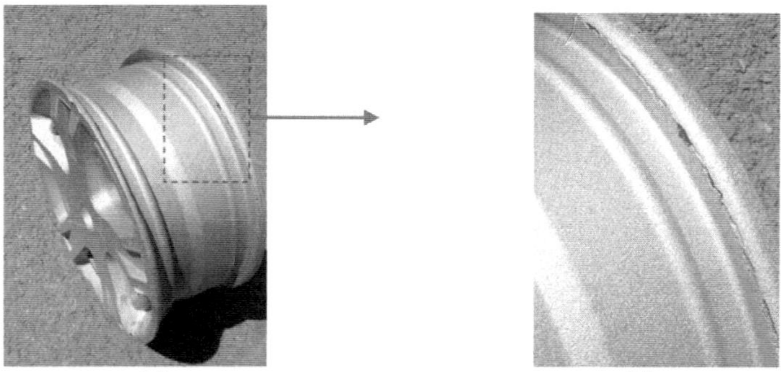

(a) (b)

Fig. 9 Photo of plastic forming product ((a) Mg alloy wheel; (b) Crack point "B" on the rim)

Fig.8 The forming equipment of Mg alloy wheel

Fig. 10 Photo of plastic forming product
(Crack
points "A" and "C" on the rim)

5. Conclusions

In the present paper, the FCF method for Mg alloy wheel was introduced, besides, the process was theoretically subdivided into cylindrical compression, backward extrusion, extrusion-shearing and closed-die forging in terms of its stress characteristics. Based on the above analysis, a finite element simulation was carried to study the deformation force and formability of Mg alloy wheel. The simulated results show that the maximum deformation force is only less than 2200T in the whole processes, and slightly changes with extrusion speed. Meanwhile, the cracks of the wheel were predicted, of which accuracy have been validated by the actual experiment for Mg alloy wheel. On the whole, the FCF method for Mg alloy wheel has effect in saving deformation force, but need still improve the formability of the rim. In addition, the influences of extrusion-shearing on the wheel are in need of further study.

Foundation item: Projects(2012ZX04010101) supported by National Science and Technology Major Project.

References

[1] Li S S, Jin Y, Tang J L et al 2005 *Special Casting & Nonferrous Alloys* p. 60.
[2] Musfirah A H, Jaharah A G 2012 *Journal of Applied Sciences Research* **8** p. 4865.
[3] LUO A A 2013 *Fundamentals of Magnesium Alloy Metallurgy* p. 266.
[4] Kulekci M K 2008 *The International Journal of Advanced Manufacturing Technology* **39** p.851.
[5] Sameer Kumar D, Tara Sasanka C, Ravindra K et al 2015 *American Journal of Materials Science and Technology* **4** p. 12.
[6] Polmear I J, *Mater. Sci. Technol.* **10** p. 114.
[7] Doege E, Droder K 2001 *Journal of Materials Processing Technology* **115** p. 14.
[8] Behrens B A, Schmidt I 2007 *Journal of Materials Processing Technology* **187–188** p. 761.
[9] Wu W, Qiao H, An K et al 2014 *International Journal of Plasticity* **62** p. 105.
[10] Fujita M, Yamamoto Y, Sakate N and Hirabara S 1999 *US5902424 [P]* 1999−11−05.
[11] Sun H X, Wang Z S, Guo P et al 2013 *202861315U[P]*. 2013-4-10.
[12] Liu S D 2010 *201516493U[P]*. 2010-6-30.
[13] Zhang Z M, Wang Q, Zhang X et al 2006 *1864915[P]*. 2006-11-12.
[14] WANG Q, ZHANG Z M, ZHANG X et al 2010 *Trans. Nonferrous Met. Soc. China* **20** p. 599.
[15] Yin Y Y, Yang Y S, Guo J Q 2010 *Hot Working Technology* **39** p. 77.
[16] Yang Y S, Guo J Q and Chen F X 2011 *FORGING & STAMPING TECHNOLOGY* **36** p. 20.
[17] Li J, Sun H X, Guo P et al 2013 *Journal of Yanshan University* **37** p. 494.
[18] Wang X Y, Wu Y S, Xia J C et al 2005 Trans. *Nonferrous Met. Soc. China* **15** p. 452.
[19] Fereshteh-Saniee F, Fallah-Nejad K, Beheshtiha A S and Badnava H 2013 *Materials and Design* **50** p 702.
[20] Zhou H T, Li Q B, Zhao Z K, Liu Z C, Wen S F et al 2010 *Materials Science and Engineering A* **527** p. 2022.
[21] Fereshteh-Saniee F, Pillinger I, Hartley P, *Journal of Materials Processing Technology* Wen p. 151.
[22] Gontarz A, Dziubinska A and Okon L 2011 *Archives of Metallurgy & Materials* **56** p.379.
[23] Wang Q, Zhang Z M, Cao Y Q 2008 *JOURNAL OF PLASTICITY ENGINEERING* **15** p. 28.

Adsorption of hexavalent chromium onto sisal pulp/polypyrrole composites

Y Y Tan[1,2], C Wei[1], and Y Y Gong[1], L L Du[1]

[1]College of Materials Science and Engineering, Guilin University of Technology, Guilin, China
[2]Shenzhen Youpuhui Pharmaceutical Co.,Ltd.,Shenzhen,China

E-mail: 1986024@glut.edu.cn

Abstract. Sisal pulp/polypyrrole composites(*SP/PPy*) utilized for the removal of hexavalent chromium [Cr(VI)] from wastewater, were prepared via *in-situ* chemical oxidation polymerization approach. The structure and morphology of the *SP/PPy* were analyzed by polarizing optical microscopy (POM), field-emission scanning electron microscopy (SEM)), Energy-dispersive X-ray spectroscopy (EDS) and X-ray photoelectron spectroscopy (XPS), the results indicated *SP* could be efficient dispersion of PPy. The hexavalent chromium adsorption results indicate adsorption capacity of the *SP/PPy* were dependent on the initial pH, with an optimum pH of 2.0. The sorption kinetic data fitted well to the pseudo-second order model and isotherm data fitted well to the Langmuir isotherm model. The maximum adsorption capacity determined from the Langmuir isotherm is 336.70 mg/g at 25° C.

1.Introduction

Chromic (Cr) compounds are widely used in chemical industry, such as electroplating, metallurgy, electronics, industrial dyeing, leather tanning, etc. Consequently, large amounts of industrial wastewater containing Cr(VI) are produced. The direct discharge of waste-water has serious adverse effects on aquatic life and human health. Chromium generally exists in trivalent and hexavalent oxidation states. Cr(VI) compounds are more toxic due to their carcinogenicity. The World Health Organization (WHO) has proposed regulations on the limits of Cr(VI) in groundwater and drinking water, which are 0.1 mg/L and 0.05 mg/L, respectively. Hence, waste-water containing Cr(VI) must be strict treated before being discharged. Commonly used methods for the removal of metal ion from water include chemical precipitation, electrochemical reduction method[1,2], ion exchange[3,4], membrane separation[5,6], complexation and adsorption[7]. Among these methods, adsorption is a promising alternative because it is simple, efficient and economic.

Polypyrrole (PPy) containing nitrogen have shown high binding affinity to metal ion. Moreover, PPy are easily coating on different substrates through in-situ polymerization of pyrrole, and their coatings can be exchanged with ambient ions under solution states, this properties render PPy promising candidates as metal ion absorption materials. Many sorbing materials based PPy have been reported for Cr(VI) removal. For examples, montmorillonite /PPy[8], sepiolite fiber/PPy[9], halloysite/PPy nanotube clay[10] and polyaniline/PPy [11]. The maximum of adsorption capacity (q_m) of bamboo-like PPy nanotubes[12], PPy-graphene oxide composites[13] and polyacrylonitrile/PPy core-shell nanofiber membrane[14] for Cr(VI) adsorption were 482.60 mg/g, 625 mg/g and 61.80 mg/g, respectively. However, for practical utility there is still a requirement for new materials showing a highly removal capacity.

The development of PPy based toxic metal ions absorption materials are of both scientific and technological interest. However, an obstacle is the notorious aggregation of PPy. This problems must be properly tackled, because of aggregation highly effect on absorptivity. In order to solve aggregation problem and enhance the adsorption capacity, various chemical, physical, and engineering approaches have been taken to frustrate PPy aggregation.

In this paper, we utilized renewable and biodegradable sisal pulp, prepared *SP/PPy* via a simple in-situ chemical oxidation polymerization. PPy well distributed on *SP* surface, and the *SP/PPy* with large surface area and enhance adsorption capacity due to the microstructures and good dispersion of *SP*.

2. Experimental details

2.1.Materials

Sisal fiber was obtained from Guangxi Sisal Group Co. Ltd. pyrrole was vacuum distillated before used. Other commercially available reagents were used without further purification. Sisal pulp(*SP*) was prepared following the sulfate digestion process[15]. A stock Cr(VI) solution was prepared by dissolving a known amount of K2Cr2O7 with deionized water. The pH of the solution was regulated by HCl and NaOH solution.

2.2.Preparation of SP/PPy.

SP/PPy were prepared by an in-situ chemical oxidation polymerization approach, as follows: 1000 mg of dry *SP*, 926 mg of $FeCl_3 \bullet 6H_2O$ and 100 mL of 0.5 M HCl were placed into a 500 mL three neck round bottom flask and stirred 30 min; 105 uL pyrrole was dropsies at room temperature under and continually stirred for 5 h. After the reaction, the solid was collected, and rinsed with deionized water and ethanol alternately until the filtrate was clear, dried at 55° C for 24 h. The molar ratio of pyrrole to $FeCl_3 \bullet 6H_2O$ was 1:2.3. Different pyrrole concentrations were denoted as *SP/PPy-x* (x was pyrrole the concentration, g/L). The preparation process is illustrated in **Figure 1.**

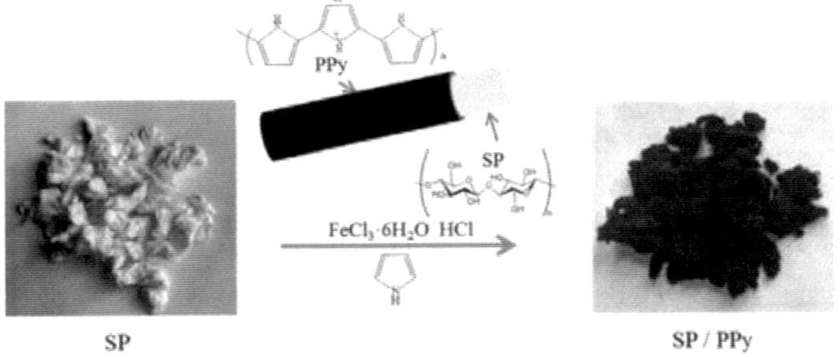

Figure 1. Preparation of *SP/PPy*.

2.3.Adsorption studies

Effects of pH. The effects of the initial pH on Cr(VI) adsorption were investigated in the range of 2~10. 20 mg *SP/PPy* were added to 20 mL of solution with 100 mg/L, after 24 h, the adsorbent was separated from the solution by filtration, and the filtrate was analyzed by UV-vis spectroscopy.

Kinetic study and adsorption isotherms. An amount of 100 mL of different Cr(VI) solutions (initial concentrations from 100~400 mg/L) and 100 mg *SP/PPy* were added until reaching the adsorption equilibrium at room temperature (25° C). Time was taken for the samples to be analyzed. The concentrations of the Cr(VI) samples were analyzed by spectrophotometry using 1,5-diphenylcarbazide as the complexing agent at the wavelength of 540 nm (GB 7467-87). At the end of the experiments, the adsorbent was separated from the solution by filtration. The removal and adsorption capacity for Cr(VI) were obtained as follows:

$$\% \text{ of removal} = \frac{c_o - c_t}{c_o} 100\% \quad (1)$$

$$q_t = \frac{(c_o - c_t)}{m} v \quad (2)$$

$$q_e = \frac{(c_o - c_e)}{m} v \quad (3)$$

Where C_0, C_t and C_e were the initial concentration, the concentration at t and equilibrium concentrations ($mg \bullet L^{-1}$) of Cr(VI), respectively; q_t and q_e were the adsorption capacity at t and equilibrium ($mg \bullet g^{-1}$); m is the mass of adsorbent (mg); and V is the initial volume (mL) of the Cr(VI) solution.

The surfaces of *SP/PPy* were observed using a polarized optical microscope (POM; ECLPSE E200, Japan). The morphologies of composites were characterized by field-emission scanning electron microscopy (FE-SEM; JSM6380-LV, Japan) and the elemental components were characterized using distribution energy dispersive X-ray analysis (EDS; JSM-5600LV, Japan). The concentrations of Cr(VI) ions in the aqueous solutions were measured by UV-vis spectroscopy (TU-1901; Beijing). X-ray photoelectron spectroscopy (XPS; ESCALAB 250XI, USA) equipped with an Al Ká radiation source was employed to determine the element of *SP/PPy*.

3. Results and discussion

3.1.Morphology of SP/PPy

Figure 2 shows SEM micrographs of *SP, PPy, SP/PPy-7.0* and *Cr@SP/PPy-7.0* (*SP/PPy-7.0* after Cr(VI) adsorption). **Figure 2** (a) shows the pure SP had an uneven surface with wrinkles and were covered with flocculent fiber. **Figure 2** (b) shows the PPy had a cauliflower-like structure[16] with a particle size of ~1 μm. **Figure 2** (c) show a PPy layer was adhered to the surfaces of the *SP/PPy-7.0* with diameters of ~100 nm, indicating that the flocculent fiber around the SP acted as a template while the pyrrole was polymerized [9,17]. This perhaps that PPy carried positive charges on the N atoms when it propagated in the reaction medium. Due to the large amount of -OH groups and the negative charges on the SP, positively charged PPy was oriented on the fibrous surface of the SP, and the *SP/PPy* was formed during oxidation polymerization. After Cr(VI) absorption, the micrographs of the *Cr@SP/PPy-7.0* composites hardly changed, as shown in **Figure 2** (d).

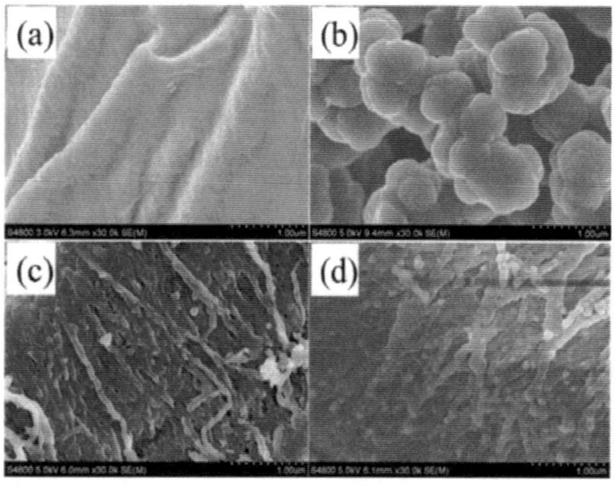

Figure 2. SEM images of (a)*SP*; (b)PPy; (c)*SP/PPy-7.0*; (d)*Cr@SP/PPy-7.0*.

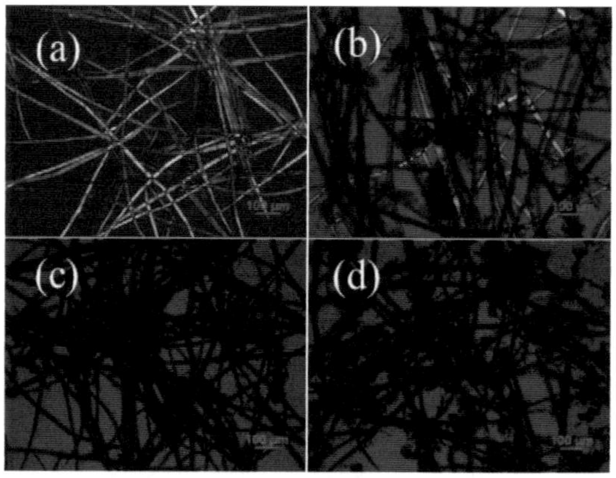

Figure 3. POM images of (a) *SP*; (b) *SP/PPy-1.0*; (c) *SP/PPy-5.0*; (d) *SP/PPy-7.0*.

Figure 3 shows POM images of the *SP* and *SP/PPy* composites. The *SP* showed a typical birefringence colorful pattern under the polarizing microscope (**Figure 3** (a)) because of their crystalline. Increase of pyrrole content, the birefringence colorful pattern of *SP/PPy* fade away (**Figure 3** (b) ~(d)), especially, *SP/PPy-5.0* and *SP/PPy-7.0*, the displayed views are dark and the birefringence have been completely disappeared[18]. This results indicates that PPy have been effective coating on sisal pulp surface. The average diameter and lengths of the *SP* and *SP/PPy* were 15 μm and 200 to 300 μm.

3.2.Batch adsorption studies

The concentration and pH of the aqueous solution greatly affected the Cr(VI). When pH<1, Cr(VI) mainly existed as H_2CrO_4. When 2<pH<6, Cr(VI) was converted into $HCrO4^-$ (main) and $Cr_2O_7^{2-}$. When 6<pH<8, Cr(VI) existed as $Cr_2O_7^{2-}$ and $CrO4^{2-}$. When pH>8, Cr(VI) existed as CrO_4^{2-} [19].

The pH of the solution was an important factor for metal ion sorption, because it not only affected the diffusion of dichromate ions but also the surface charge of the adsorbent. Moreover, it may have participated in the speciation of the adsorbent during reaction [20]. **Figure 4** (a) shows the Cr(VI) removal efficiency with different adsorbents at various pH values. The figure shows that Cr(VI) removal with the *SP/PPy* and PPy were pH-dependent and optimum removal for both occurred at 2. The percentage of Cr^{6+} removal with the *SP/PPy* and the PPy were 99.2% and 55.6%, respectively. These values may have been due to the absorbents being protonated and the strong attraction between $HCrO_4^-$ at a low pH level, resulting in the Cr(VI) being easily removed. As the pH increased, more OH^- competed with the $HCrO_4^-$, resulting in a decrease of Cr(VI) removal [9,13,14]. Moreover, the adsorption capacity of Cr(VI) with the *SP* was very weak and pH-independent. The Cr(VI) adsorption was shown to be more successful with the *SP/PPy* than the PPy. This was probably due to the increase in the number of active sites because the SP of the *SP/PPy* acted as a support matrix material that effectively prevented the PPy from agglomerating during polymerization.

The *SP/PPy* prepared with different pyrrole concentrations showed different adsorption performances. **Figure 4** (b) shows the Cr(VI) removal efficiency of the *SP/PPy* prepared with various pyrrole concentrations.

It was observed that the removal efficiency increased from 33.7% to 97.9% as the pyrrole concentration increased from 1.0 to 7.0 g/L after 24 h absorption. The increase in performance as the pyrrole concentration increased was probably due to an increased number of active sites provided by the *SP/PPy*. However, when the monomer concentration was more than 7.0 g/L, the removal efficiency decreased sharply to 64.5% [21]. It could have been the *SP/PPy* prone to agglomeration as the pyrrole concentration increased resulted in a reduction of the density of effective adsorption sites for Cr(VI) removal. In summary, the *SP/PPy-7.0* showed the best adsorption capacity for Cr(VI) removal.

The dosage of sorbent is an important parameter for metal ion removal, so the amount of *SP/PPy-7.0* was investigated to establish the appropriate value for a given initial concentration of heavy metal ion solution. The results illustrated in **Figure 4** (c) show how the different *SP/PPy-7.0* amounts affected Cr(VI) removal in 20 mL of 200 mg/L solution. The removal efficiency increased from 32.1% to 98.0% as the sorbent dosage increased from 5.00 mg to 20.00 mg. This was because the increased amount of adsorbent provided more adsorption sites for Cr(VI) removal. But if the amount of *SP/PPy-7.0* was increased to more than 20 mg/20 ml, the Cr(VI) removal began to slow. This suggested that the optimal amount of adsorbent was 20 mg/20 ml for the treatment of a solution with an initial Cr(VI) concentration of 200 mg/L.

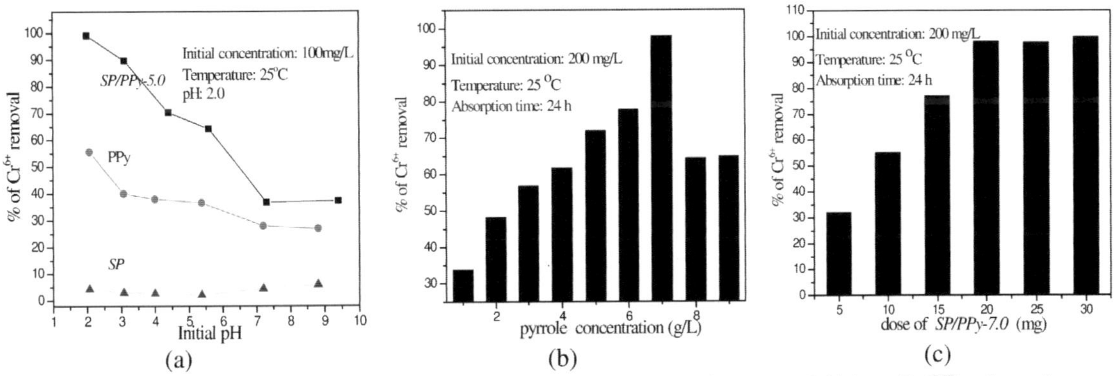

Figure 4. Effects of pH(a), pyrrole concenreation(b), and dose of *SP/PPy-7.0*(c) on Cr(VI) adsorption.

3.3.Adsorption kinetics

To study the adsorption process of Cr(VI), the adsorption kinetics of the *SP/PPy-7.0* with different initial concentrations of Cr(VI) were studied (show in **Figure 5** (a)). It was observed that when the initial Cr(VI) concentration increased from 100 to 400 mg/L, the adsorption capacity of the *SP/PPy* also increased. When the initial concentration of Cr(VI) was 100 mg/L, the absorb equilibrium time was 12 h. It was extended to 32 h with the initial concentration at 200 mg/L. A higher initial concentration of Cr(VI) caused longer adsorption equilibrium time. The equilibrium adsorption capacity was 100.37, 200.33, 291.01, and 335.00 mg•g^{-1} when the initial concentrations were 100, 200, 300, and 400 mg/L, respectively.

The kinetic data were further interpreted using the pseudo-second order kinetic models [22], which can be expressed as follows:

$$\frac{t}{q_t} = \frac{1}{k_2 q_e^2} + \frac{t}{q_e} \quad (4)$$

where K_2 was the rate constant (g mg^{-1}•min^{-1}) of the pseudo-second-order kinetic model.

The pseudo-second order rate parameters for chromium sorption with four initial concentrations were calculated from the linear plots of t/q_t versus t (**Figure 5** (b)). The values of the correlation coefficients (R^2) were >0.99, and calculated equilibrium capacity values were consistent with the experimental values, which indicated that the kinetics of sorption of Cr(VI) ions with the *SP/PPy* followed the pseudo-second order mechanism[16,23]. In other words, this adsorption process was chemisorption involving valency forces through the sharing or exchange of electrons between the *SP/PPy* and Cr(VI) ions as covalent forces and ion exchanges[22,24,25].

3.4.Adsorption isotherm

To determine the capacity parameters of Cr(VI) adsorption on the *SP/PPy-7.0*, adsorption isotherm was obtained at 25 °C, 35 °C, 45 °C. **Figure 5** (c) illustrates that the adsorption capacity of the *SP/PPy-7.0* increased with the increasing temperature. And the adsorption equilibrium data was fitted with a linearized Langmuir model (5), which is shown in **Figure 5** (d). The expression of the Langmuir model is as follows:

$$\frac{c_e}{q_e} = \frac{c_e}{q_m} + \frac{1}{q_m b} \quad (5)$$

where q_m is the maximum adsorption capacity (mg/g) and b is the adsorption equilibrium constant (L/mg). Then, q_m and b can be calculated, as shown in **Figure 5** (d) and listed in **Table 1**.

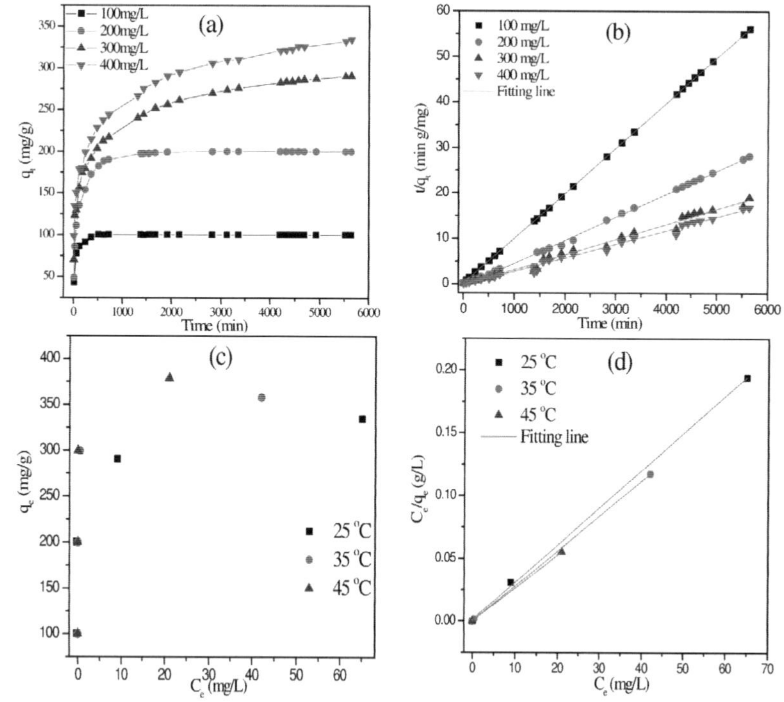

Figure 5. (a) Effects of adsorption time on Cr(VI) adsorption capacity with different concentrations, (b)Pseudo-second-order kinetic model ,(c) Plot of adsorption capacity vs initial Cr(VI) concentration and (d) Langmuir adsorption isotherm for Cr(VI) onto *SP/PPy-7.0*.

Table 1. Parameters of the Langmuir isotherm model

T(k)	b(mg/L)	q_m(mg/g)	R^2
298	1.4854	326.7974	0.9988
308	5.7724	358.4229	0.9998
318	109.8423	377.3585	1.0000

The Langmuir adsorption isotherm model was assumed as monolayer adsorption onto the surface of the adsorbent and there was no interaction between the adjacent adsorption sites. The values of the correlation coefficients (R^2) were > 0.999, indicating that the equilibrium data of the Cr(VI) removal on the *SP/PPy-7.0* was as described by the Langmuir isotherm. The maximum adsorption capacities calculated from the Langmuir isotherm of the *SP/PPy-7.0* were 336.80, 358.42 and 377.36 mg/g at 25 °C, 35 °C and 45 °C, respectively. Compared to the other adsorbents listed in **Table 2**, SP/PPy composites have relatively high adsorption capacity.

Table 2. Adsorption performances of various adsorbents and *SP/PPy-7.0*.

Adsorbent	q_m(mg/g)	pH	T/K	Ref.
Glycine doped PPy	217.39	2.0	298	[26]
PPy-sepiolite nanofibers	302.00	2.0	298	[9]
Bamboo-like PPy nanotubes	482.60	3.0	-	[12]
PPy-PANI nanofiber	227.00	2.0	298	[11]
PPy/Fe_3O_4 nanocomposites	169.49	2.0	298	[27]
PPy/MMT nanocomposites	166.70	2.0	303	[8]
PAN/PPy nanofiber mat	61.80	2.0	298	[14]
SP/PPy composites	334.45	2.0	298	this

These results indicated *SP/PPy-7.0* have important and potential application in environmental protection filed due to their excellent adsorption capacity, simple procedure, low cost, and easy to separate from the solution after the adsorption process.

3.5.EDS and XPS of SP/PPy before and after Cr(VI) adsorption

Energy dispersive X-rays (EDS) were used to analysis the elemental constituents of *the SP/PPy* before and after the adsorption of Cr(VI) . As shown in Fig.6., there were obvious intense Cr peaks after the adsorption, while there weren't observed before the adsorption. This results clearly indicated that Cr species had been efficient adsorbed by the *SP/PPy* [8,13,16].

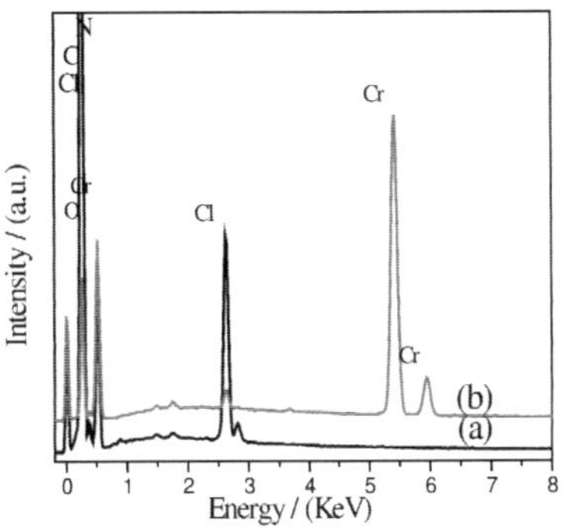

Figure 6. EDS spectra of *SP/PPy-7.0*: (a) Before and (b) After Cr(VI) adsorption.

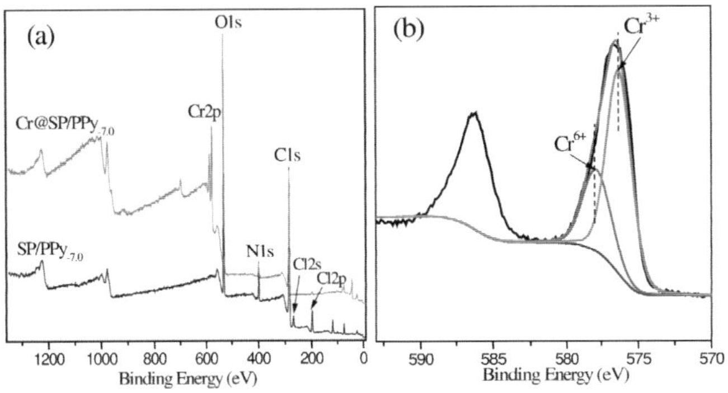

Figure 7. (a) XPS survey spectra of *SP/PPy-7.0*; (b) XPS Cr2p region of *SP/PPy-7.0* after Cr(VI) adsorption.

In order to study the adsorption mechanism, XPS was used to investigate the surface element composition of the *SP/PPy* before and after the adsorption. **Figure 7 (a)** shows the XPS survey spectra of the *SP/PPy-7.0*. It was observed that a new Cr element emerged on the *SP/PPy-7.0* after the adsorption of Cr(VI) that was not observed before. Simultaneously, an increase in oxygen content was noted, while the signal of the Cl was significantly reduced after the Cr(VI) adsorption. The Cr(VI) ions most likely replaced the Cl ions on the *SP/PPy* during the Cr(VI) adsorption by anion exchange[28,29]. **Figure 7 (b)** shows the XPS Cr2p region of the *SP/PPy-7.0* after the Cr(VI) adsorption. The significant bands appearing at the binding energies of 577.0 ~ 578.0 and 586.0 ~ 588.0 eV appeared after the Cr(VI) adsorption, which corresponded to Cr 2p1/2 and Cr 2p3/2, respectively[30]. The Cr 2p3/2 can be split into two peak components with BEs of 576.48 eV and 578.09 eV, corresponding to Cr(III) and Cr(VI), respectively[30-32]. In summary, the Cr(VI) on the surface of the *SP/PPy* attributed to the anion-exchange of the PPy by replacing the doped Cl^- with either $HCrO_4^-$ or $Cr_2O_7^{2-}$. And the presence of Cr(III) was due to the reduction property of the PPy on the Cr(VI)[12,30].

4. Conclusions

SP/PPy were prepared *via* simple *in-situ* chemical oxidation polymerization approach. The adsorption of Cr(VI) on the *SP/PPy* depended on the initial pH, with an optimum pH of 2.0. Under the same conditions, the adsorption efficiency of SP, pure PPy and *SP/PPy* are 4%, 55.6% and 99.2%, respectively. This result indicates PPy could be more efficient absorption ability through SP dispersion. The interaction between the Cr(VI) and the *SP/PPy-7.0* followed the pseudo-second order mechanism. The adsorption equilibrium data fitted well with the Langmuir isotherm model, maximum adsorption capacity up to 336.70 mg/g at 25 °C was achieved with the *SP/PPy-7.0*. To the best of our knowledge, *SP/PPy* are one of best absorption materials based PPy for removal Cr(VI) ions.

Acknowledgment

This study was carried out by the foundation of 21264005, 2013GXNSFDA019008, 2014GXNSFAA118321 in China.

References

[1] Vasudevan S, Sozhan G, Mohan S, Balaji R, Malathy P, Pushpavanam S, 2007 *Ind. Eng. Chem. Res.* **46** 2898-2901.
[2] Rodríguez F J, Gutiérrez S, Ibanez J G, Bravo J L, Batina N, 2000 *Environ. Sci. Technol.* **34** 2018-2023.
[3] Bhaumik M, Maity A, Srinivasu V V, Onyango M S, 2011 *J. Hazard. Mater.* **190** 381-390.
[4] Naushad M, Alothman Z A, 2015 *Des. & Water Treat.* **53** 2158-2166.

[5] Scindia Y M, Pandey A K, Reddy A V R, 2005 *J. Membr. Sci.* **249** 143-152.
[6] Walkowiak W, Bartsch R A, Kozlowski C, Gega J, Charewicz W A, Amiri-Eliasi B, 2000 *J. Radioanal. Nucl. Chem.* **246** 643-650.
[7] Awual M R, Eldesoky G E, Yaita T, Naushad M, Shiwaku H, AlOthman Z A, Suzuki S, 2015 *Chem. Eng. J.* **279** 639-647.
[8] Setshedi K Z, Bhaumik M, Onyango M S, Maity A, 2014 *J. Ind. Eng. Chem.* **20** 2208-2216.
[9] Chen J, Hong X, Xie Q, Li D, Zhang Q, 2014 *J. Chem. Eng. Data.* **59** 2275-2282.
[10] Ballav N, Choi H J, Mishra S B, Maity A, 2014 *Appl. Clay Sci.* **102** 60-70.
[11] Bhaumik M, Maity A, Srinivasu V V, Onyango M S, 2012 *Chem. Eng. J.* **181–182** 323-333.
[12] Li S, Lu X, Li X, Xue Y, Zhang C, Lei J, Wang C, 2012 *J. Colloid Interface Sci.* **378** 30-35.
[13] Setshedi K Z, Bhaumik M, Onyango M S, Maity A, 2015 *Chem. Eng. J.* **262** 921-931.
[14] Wang J, Pan K, He Q, Cao B, 2013 *J. Hazard. Mater.* **244–245** 121-129.
[15] Lv J, Zeng D, Wei C, 2015 *Adv. Polym. Tech.* **34** n/a-n/a.
[16] Mthombeni N H, Onyango M S, Aoyi O, 2015 *J. Taiwan Inst. Chem. Eng.* **50** 242-251.
[17] Richardson M J, Johnston J H, Borrmann T, 2006 *Curr. Appl Phys.* **6** 462-465.
[18] Dall'Acqua L, Tonin C, Varesano A, Canetti M, Porzio W, Catellani M, 2006 *Synth. Met.* **156** 379-386.
[19] Scancar J, Milacic R, 2014 *J. Anal. At. Spectrom.* **29** 427-443.
[20] Demiral H, Demiral İ, Tümsek F, Karabacakoğlu B, 2008 *Chem. Eng. J.* **144** 188-196.
[21] Lei Y, Qian X, Shen J, An X, 2012 *Ind. Eng. Chem. Res.* **51** 10408-10415.
[22] Ho Y-S, 2006 *J. Hazard. Mater.* **136** 681-689.
[23] Al-Othman Z A, Ali R, Naushad M, 2012 *Chem. Eng. J.* **184** 238-247.
[24] Ho Y S, McKay G, 1999 *Process Biochem.* **34** 451-465.
[25] Setshedi K Z, Bhaumik M, Songwane S, Onyango M S, Maity A, 2013 *Chem. Eng. J.* **222** 186-197.
[26] Ballav N, Maity A, Mishra S B, 2012 *Chem. Eng. J.* **198–199** 536-546.
[27] Abda M, Gavra Z, Oren Y, 1981 *J. Appl. Electrochem.* **21** 734.
[28] Tabačiarová J, Mičušík M, Fedorko P, Omastová M, 2015 *Polym. Degrad. Stab.* **120** 392-401.
[29] Kang E T, Neoh K G, Ong Y K, Tan K L, Tan B T G, 1990 *Synth. Met.* **39** 69-80.
[30] Park D, Yun Y-S, Park J M, 2004 *Environ. Sci. Technol.* **38** 4860-4864.
[31] Park D, Yun Y-S, Park J M, 2008 *J. Colloid Interface Sci.* **317** 54-61.
[32] Sun X-F, Ma Y, Liu X-W, Wang S-G, Gao B-Y, Li X-M, 2010 *Water Res.* **44** 2517-2524.

IOP Conf. Series: Materials Science and Engineering **170** (2017) 012008 doi:10.1088/1757-899X/170/1/012008

Properties of Cement-based Composite Materials under Different Storing Environment Temperature

T L Weng[1,a], S H Weng [2,b], S W Cho [2,c]

[1] Physics Division, Tatung University, Taipei City 104, Taiwan.

[2]Department of Architecture, China University of Science and Technology, Taipei City 115, Taiwan

E-mail:[a]wengabc@ttu.edu.tw, [b] s.h.weng0816@gmail.com, [c]swcho@ms19.hinet.net

Abstract. This study reports on the properties of cement-based composite materials (mortars) under different storing environment temperature, as determined using the accelerated chloride migration test (ACMT). Mortars with a water/cement ratio of 0.45 and five fine aggregate volume fractions (0%, 15%, 30%, 50% and 60%) under various environment temperatures (25, 40, 60 and 80°C) were evaluated according to the passage of chloride ions through the specimens using ACMT. Calculate chloride migration coefficients on the steady-state. Cement-based composite materials with 60 % fine aggregate presented a migration coefficient higher than that of other specimens, whereas mortar with 30 % fine aggregate was lower, due to the effects of dilution and tortuosity.

1. Introduction

Interfacial Transition Zone (ITZ) has a higher concentration of calcium hydroxide crystals and porosity far exceeding that of the matrix paste [1,2,3], Reduces resistance to the diffusion of chloride ions.

Temperature is one of the fundamental factors affecting the pore structure of cement-based composite materials (mortars), which makes it an important factor in ion transport. The size of pores and extent of micro-cracking has been shown to increase in specimens stored at temperatures of 40 °C, 60 °C, or 80 °C over a period of 56 days, even after undergoing curing in water for 28 days. These effects have been shown to increase the ion transport properties of mortar, due to the effects of dilution, tortuosity, and percolation between the ITZ and the surrounding paste [4].

The Steady-state chloride migration coefficient, rate (K) was calculated using linear regression based on experiment data, in order to determine the extent of chloride ion penetration, as follows:

$$C = Kt + c \text{, for steady state} \tag{1}$$

The Nernst–Planck equation was used to calculate the steady-state migration coefficient(M_s), as follows [5]:

$$M_s = \frac{RT}{\left(|z|C_0FE\right)}\frac{VK}{A} \tag{2}$$

where V is the volume of the solution in the anodic cell (m^3), and A is the surface of the specimen exposed to chloride ions (m^2). The chloride migration rate (K).

This study used the ACMT to investigate the ion transport properties of cement-based composite materials (mortars) stored at 40 °C, 60 °C, and 80 °C for 56 days [2-4]. Our aim was to determine the steady-state migration coefficients in unheated (air, 25 °C) and heated mortar specimens.

2. Experiments

2.1. Composite materials (Mortars)and specimen preparation

In this study, mortar specimens with a water/cement ratio of 0.45, Were a cement-based composite materials comprising fine aggregate embedded within a matrix of cementitious paste [6]. In characterizing the influence of aggregate on the chloride migration coefficient of mortar, the volume fraction of fine aggregate in the mix proportions (V_f, the volume of aggregate/ mortar) was varied between 0 % and 60 %, as shown in Table 1.

Table 1. Mix design including volume fraction of aggregate

Unit: Kg/m^3				Fine Aggregate	
Mix	Water	Cement	SP*	Sand	V_f
M0	574.6	1276.9	0	0	0
M15	486.6	1081.3	0	390.3	0.15
M30	398.8	886.2	0	779.6	0.3
M50	278.2	618.2	6.18	1299.8	0.5
M60	217.7	483.8	9.68	1559.8	0.6

* The dosage of superplasticizer (SP) was adjusted to obtain a slump closed to 150 mm.

Cast cylindrical mortar specimens (ø100 × 200 mm) were cured in water (23 ± 2 degree) for 28 days and then stored at 40 °C, 60 °C, or 80 °C for 56 days in an oven. Samples were cut from the middle section of the cylindrical specimens and the thickness of the specimen was 30 mm due to reduction the Joule effect [2,3,6], whereupon the lateral surface was coated with epoxy before being placed in vacuum desiccators at a pressure of less than 1 mm Hg (133 MPa) for 3 h. The specimens were then immersed in water 18 h prior to ACMT, in accordance with ASTM C1202 specifications.

2.2. Experimental procedure

The ACMT was used to measure the cumulative concentration of chloride ions that passed through the specimens, as shown in Figure 1. We employed a modified version of the RCPT method in accordance with ASTM C1202 [7,8].

The specimens were placed between two acrylic cells with #20 brass mesh screen electrodes (10 cm diameter) placed against the sides of the specimens. In ACMT, electrical fields are manipulated to pass through specimens. In this case, one cell was filled with 4500 ml of 0.30N NaOH solution and the other cell was filled with 4500 ml of 3.0 % NaCl solution. As shown in Figure 1, the two cells were connected to a 24-V DC power source with the NaOH electrode acting as the anode and the NaCl electrode acting as the cathode [7,8]. The chloride concentration in the anode cell was measured periodically using an ion chromatograph.

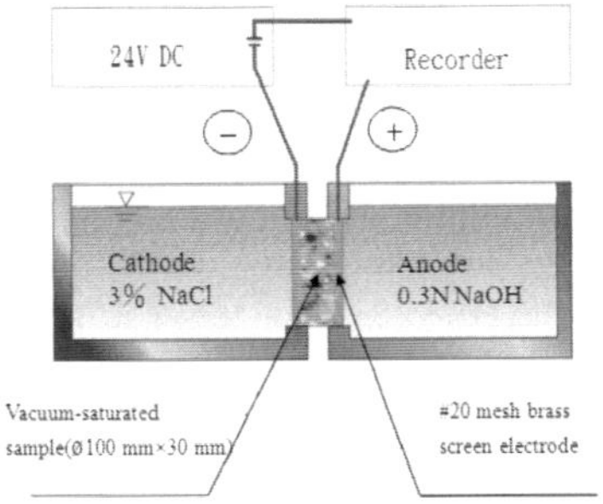

Figure 1. Schematic diagram of ACMT

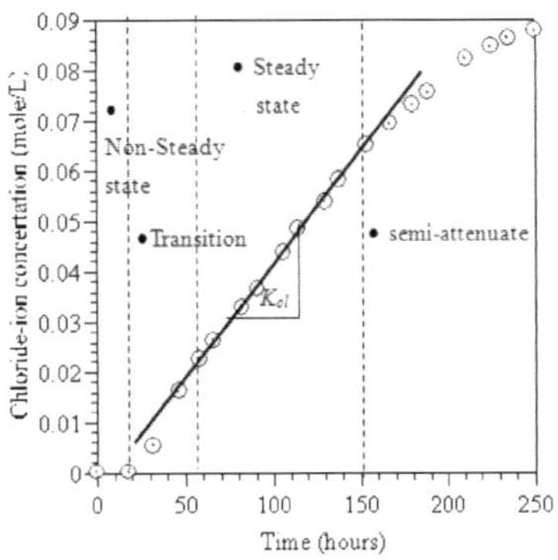

Figure 2. Typical chloride concentration levels a function of time [2,6]

3. Results and Discussion

ACMT provides a method by which to characterize the permeability of materials. The chloride concentration was determined by sampling the solution in the anodic cell of ACMT. As shown in Figure 2, this process includes three stages: non-steady-state, transition period, and steady-state [2,6].

3.1. chloride migration coefficient

The steady-state migration coefficients (Ms) for all specimens can be calculated using Eq. (2), as shown in Table 2. Cement-based Composites (Mortars) with 60 % fine aggregate presented a migration coefficient higher than that of other specimens, due to the formation of connections between ITZs. whereas mortar with 30 % fine aggregate was lower, due to the effects of dilution and tortuosity. The chloride ion transport behavior of cement-based composite materials (mortar) is affected by the temperature at which it is stored. Table 2 shows that the steady-state migration coefficient of mortar increases with temperature due to the effects of micro-cracking caused by an increase in temperature associated with percolation the formation of ITZs [9]. In cement-based composite materials (mortar) with a fine aggregate volume fraction of 0.3, the steady-state migration coefficient in Ms45_40 was 1.86 × that of Ms45_25. The steady-state migration coefficient of Ms45_60 was 4.11 × that of Ms45_25 and the steady-state migration coefficient of Ms45_80 was 4.37 × that of Ms45_25. In mortar with a fine aggregate volume fraction of 0.6, the steady-state migration coefficient of Ms45_40 was 1.57 × that of Ms45_25, the steady-state migration coefficient of Ms45_60 was 2.75 × that of Ms45_25, and the steady-state migration coefficient of Ms45_80 was 2.79 × that of Ms45_25.

Table 2. Steady-state migration coefficient ($M_S \times 10^{-12} m^2/sec$) under 25°C, 40°C 60°C and 80°C

Migration coefficient of Chloride Migration in steady-state, $M_S(\times 10^{-12} m^2/sec)$					Rs		
M45	25℃ (a)	40℃ (b)	60℃ (c)	80℃ (d)	b/a	c/a	d/a
Vf=0%	11.6	16.8	38	48	1.45	3.28	4.14
Vf=15%	11.2	15.6	33.3	39.7	1.39	2.97	3.54
Vf=30%	7.3	13.6	30	31.9	1.86	4.11	4.37
Vf=50%	10.3	27	40	45	2.62	3.88	4.36
Vf=60%	20.4	32	56	57	1.57	2.75	2.79

3.2. Influence of aggregate volume fraction on chloride migration coefficient

3.2.1 Dilution and tortuosity. Even if the bond between the aggregate and cement paste were perfect, the migration coefficient of fine aggregate would be far below that of cement paste, due to the effects of dilution. The blockage of flow paths by aggregate reduces the area of the mortar that provides permeability (as measured in cross-section) [10]. The dilution effect depends on the volume fraction of aggregate, wherein the relative impermeability of the aggregate results in a linear parallel chloride ion flow [2].

Taking into account the effects of dilution, the chloride migration coefficient of mortar Ms can be expressed as follows [10]:

$$Ms = M_0(1-V_f) \quad (3)$$

where M_0 is the migration coefficient of cement paste, and V_f is the volume fraction of fine aggregate. The tortuosity effect can be attributed to the impermeability of fine aggregates, which forces migrating chloride ions to navigate around aggregate particles, thereby increasing the length of the flow paths and reducing the flow rate. The effects of tortuosity and dilution can be combined to express the chloride migration coefficient of mortar using the Bruggeman equation, as follows [10]:

$$Ms = M_0(1-V_f)^{3/2} \quad (4)$$

Figure 3 present specimens at 25 °C , 40 °C , 60 °C , and 80 °C , showing that when the volume fraction of the fine aggregate is below 0%,15%,30%,50%, and 60% any increase in the volume fraction of the aggregate reduces the migration coefficient(Steady-state) of the mortar. Figure 4 present a comparison of the theoretical chloride migration coefficient of specimen with a w/c ratio of 0.45 and M_s values calculated using Eq.(3) and (4). The theoretical chloride migration coefficient M_s/M_0 is defined as the ratio of the chloride migration coefficients of mortar and cement paste.

A comparison of experiment data with chloride migration coefficient for various volume fraction of fine aggregate in Figure 4 illustrates how the chloride migration coefficient at 25℃ is influenced dilution and tortuosity when the volume fraction of the fine aggregate is below 0.3. At V_f=0.3, taking into account only the effects of dilution reduces the theoretical chloride migration coefficient from 1.0 to 0.7, whereas taking into account the effects of tortuosity reduces the theoretical chloride migration coefficient from 1.0 to 0.65. Increasing the volume fraction of fine aggregate to beyond 0.3 led to an increase in the length of the flow paths. As indicated by the experiment data in Figure 4, the chloride migration coefficient is influenced by the effects of dilution as well as tortuosity.

As shown in Figure 4, at temperatures of 40℃, 60℃, and 80℃, increasing the volume fraction of aggregate leads to a decrease in the migration coefficient of mortar, in cases where the volume fraction of the aggregate is below 30%.

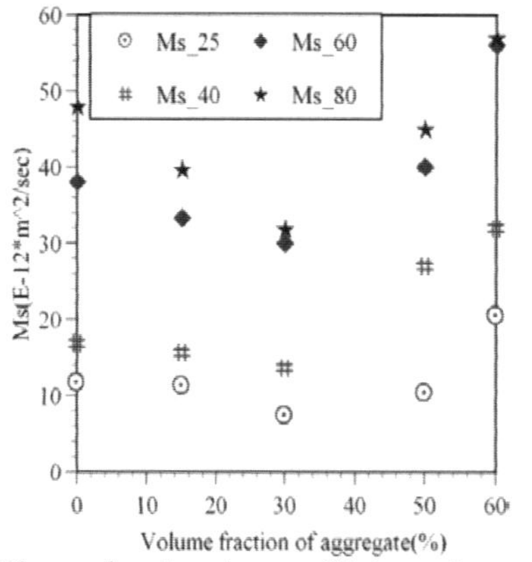

Figure 3. migration coefficient of mortar

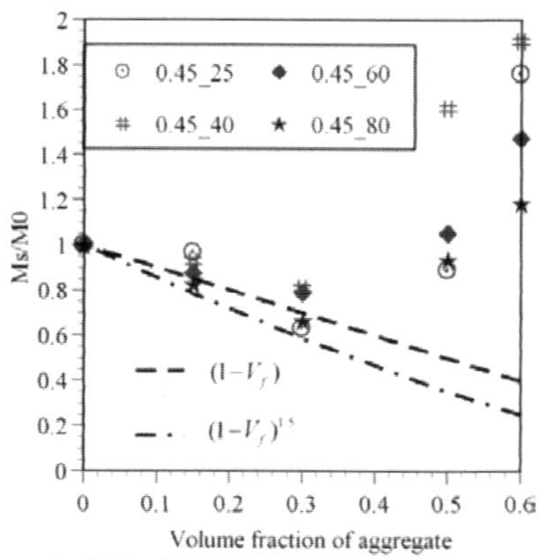

Figure 4. Chloride migration coefficient (M_s/M_0)

*3.2.2 ITZ and percolation effects.*In this study, mortar is regarded as a three-phase composite materials (cement paste, fine aggregate, and ITZ). The ingress of chloride ions leads to the formation of an ITZ between the fine aggregate and the matrix, thereby increasing the chloride migration coefficient of the mortar. The connected-pore paths between ITZs would allow harmful ions such as chlorine ion to penetrate the concrete quickly [9].

As shown in Figure 4, mortar with a volume fraction of the fine aggregate exceeding 0.4~0.5 can is susceptible to percolation of ITZs at 25℃. The apparent width of the ITZs in which the higher porosity is interconnected observed as proposed by the previous study [11].

As shown in Figure 4, samples with a volume fraction of 0.3~0.4 at 25℃ presented a chloride migration coefficient of 0.5~0.7. This means that the ITZ did not exert a significant influence, due to the fact that percolation does not occur at such low volume fractions of fine aggregate. As shown in Figure 4, increasing the volume fraction of fine aggregate to beyond 0.3 led to a sharp increase in the chloride migration coefficient, particularly at 0.6. In samples with a fine aggregate volume fraction of 0.3 stored at temperatures exceeding 40℃, the appearance of micro-cracking is an indication of percolation resulting from the overlap of porous regions and increased tortuosity in ITZs [12]. This is consistent with previous findings based on prediction models [13,14]. In addition, a higher storage temperature also increased the amount of shrinkage and subsequent microcracking. It also indicated that the aggregate size and volume fraction on shrinkage induced micro-cracking as well as consequently greatly increased tortuosity and permeability in ITZ regions correspond with the results of previous findings [15]. Generally, constituent phases and their geometric arrangement are the primary factors determining ion migration in concrete. Dilution and tortuosity tend to decrease the permeability of mortar, whereas the formation of connections between ITZs tend to increase permeability.

4. Conclusions

This study demonstrated how the transport properties of composite materials (mortar) are influenced by temperature (25, 40, 60 and 80℃). Based on the results obtained in these experiments, the following conclusions can be drawn:

1. Specimens with a 30 % volume fraction of fine aggregate presented the lowest steady-state migration coefficient among the mixes in this study. whereas mortar with 30 % fine aggregate was lower, due to the effects of dilution and tortuosity.
2. Cement-based composite materials (Mortars) with 60 % fine aggregate presented a migration coefficient higher than that of other specimens, due to the formation of connections between ITZs.
3. at temperatures of 40℃, 60℃, and 80℃, increasing the volume fraction of aggregate leads to a decrease in the migration coefficient of mortar, in cases where the volume fraction of the aggregate is below 30%.
4. constituent phases and their geometric arrangement are the primary factors determining ion migration in concrete. Dilution and tortuosity tend to decrease the permeability of mortar, whereas the formation of connections between ITZs tend to increase permeability.

Reference

[1] Diamond, A S. and Mindess, S 1985 The microstructure of the steel fiber-cement interface, *Journal of Materials Science,* **Vol. 20,** pp. 3610-3620.
[2] Yang, C C 2003 The relationship between charge passed and the chloride concentrations in anode and cathode cells using the accelerated chloride migration test, *Materials and Structures*, **Vol. 36, No. 10**, pp.678-684.
[3] Yang, C C. and Huang, R 1996 Double inclusion model for approximate elastic moduli of

concrete material, *Cement and Concrete Research*, **Vol. 26, No. 1**, pp. 83-91.

[4] Weng, S H., Yang, C C, Cho, S W and Yang, K C 2012 The study of chloride ion transport behavior of mortar under different storing environment temperatures, *Journal of Marine Science and Technology,* **Vol. 20, No. 3**, pp. 290-294.

[5] Andrade, C 1993 Calculation of chloride diffusion coefficients in concrete from ionic migration measurements, *Cement and Concrete Research*, **Vol. 23, No. 3**, pp. 724-742.

[6] Yang, C C and Liang, C H 2009 The influence of medium–high temperature on the transport properties of concrete by using accelerated chloride migration test, *Materials Chemistry and Physics*, **Vol. 114, No. 2–3,** pp. 670-675.

[7] Yang, C C, Chiang, S C and Wang, L C 2007 Estimation of the chloride diffusion from migration test using electrical current *Constructionand Building Material* pp. 1560-1567.

[8] Tong, L and Gjørv, O E 2001 Chloride diffusivity based on migration testing, *Cement and Concrete Research*, pp.973-982.

[9] Mohd Zain, M F and Yusof, K M 1999 The influence of medium temperature environments on the water permeability of high performance mortar, *Cement and Concrete Research*, **Vol. 29, No. 5,** pp. 785–788.

[10] Yang, C C and Cho, S W 2003 Influence of aggregate content on the migration coefficient of concrete materials using electrochemical method, *Materials Chemistry and Physics*, **Vol. 80, No. 3,** pp. 752-757.

[11] Karen, L S and Kamran, M N 1996 The percolation of pore space in the cement paste/aggregate interfacial zone of concrete, *Cement and Concrete Research*, **Vol. 26, No. 1**, pp. 35–40.

[12] Delagrave, A, Bigas, J P, Ollivier, J P, Marchand, J and Pigeon, M 1997 Influence of the interfacial zone on the chloride diffusivity of mortars, *Advanced Cement Based Materials,* **Vol. 5, No. 3**, pp. 86-92.

[13] Caré, S, and Hervé, E 2004 Application of a n-phase model to the diffusion coefficient of chloride in mortar. *Transport in Porous Media*, **Vol. 56, No. 2**, pp. 119-135.

[14] Oh, B H and Jang, S Y 2004 Prediction of diffusivity of concrete based on simple analytic equations, *Cement and Concrete Research*, **Vol. 34, No. 3**, pp. 463-480.

[15] Grassl, P, Wong, H S and Buenfeld, N R 2010 Influence of aggregate size and volume fraction on shrinkage induced micro-cracking of concrete and mortar, *Cement and Concrete Research*, **Vol. 40, No. 1**, pp. 85-93.

IOP Conf. Series: Materials Science and Engineering **170** (2017) 012009 doi:10.1088/1757-899X/170/1/012009

Ultrathin Passivation of P-type silicon Surface by Atomic Layer Deposited Gallium Oxide Thin Films

J Wen[1], L Q Guo[1,2], J Tao[1]

[1] Micro/Nano Science & Technology Center, Jiangsu University, Zhenjiang, 212013, China.

[2] Jiangsu Collaborative Innovation Center of Photovolatic Science and Engineering and the Breeding Construction Point of State Key Laboratory of Photovolatic Engineering Science, Changzhou University, Changzhou, 213164, China.

E-mail:guoliqiang@ujs.edu.cn

Abstract: Surface passivation performances of Ga_2O_3 thin films deposited on p-type Czochralski Si wafers by atomic layer deposition (ALD) were investigated as a function of post-deposition annealing conditions. Minority carrier lifetimes were characterized by Semilab WT-2000PVN lifetime tester. The average effective minority carrier lifetime can reduce to 218.09 μs after annealed treatment at the temperatures of 650 ℃ for 3 min. It found that the surface recombination velocities of Ga_2O_3 thin film (3.4 nm and 7.6 nm) decrease to ~30 cm/s and are saturated with the annealing temperature of 600 ℃~700 ℃. Such results indicate that Ga_2O_3 thin film show the functions of surface passivation for photovoltaic applications.

1.Introduction

Environmental and energy problems have become a bottleneck for the development of the entire society and sustainable economy.[1] Seeking clean and renewable energy has been a most important problem which is also common concern and must face. Recently, solar energy is known as an inexhaustible clean energy and photovoltaic technology has attracted great attention. While Silicon solar cell as the main photovoltaic production is confronting with high cost and low conversion efficiency. To decrease the Silicon solar cell cost and increase its conversion efficiency, the thickness of silicon wafer must be reduced and surface recombination velocity is also kept the low value. Passivating treatment for Silicon solar cell can decrease surface activity and surface recombination velocity of semiconductor materials.[2] The conventional passivating treatment methods are low-temperature surface passivation and hydrogen post-annealing.[3,4] After low-temperature surface passivation, effective surface recombination velocities of 55 cm/s and 24 cm/s were obtained on 2.1 Ω·cm p-type and 3.3 Ω·cm n-type c-Si wafers, respectively. However, when the thickness of Si wafer is greater than 180 μm, electrical properties of Silicon solar cells will decrease. Therefore, low-temperature surface passivation is not good for conversion efficiency of Silicon solar cells. The surface recombination velocity obtained in this hydrogen-radical annealing was much lower than that in forming gas annealing or hydrogen plasma annealing, but a longer high temperature process is needed.[5] An important high-κ dielectric material can be used as passivation layer and such materials can play an important role in anti-reflectance.[6-10] In this contribution, we focus on good passivation performances for the ALD deposited Ga2O3 thin films. A average effective minority carrier lifetime of 218.09 μs is addressed after annealed treatment at the temperatures of 650 ℃ for 3 min. Physical origins for the passivation performances are discussed. The results indicate that Ga_2O_3 thin films have the function of surface passivation for potential photovoltaic applications.

2. Experiment

Boron-doped p-type Czochralski single crystalline silicon (c-Si) wafers (shiny-etched, (100)-oriented, 400mm thick, 30 μm, 15 cm in diameter) purchased from QL Electronics Corporation were used as the substrates. The wafers were cleaned using RCA method and followed by a diluted HF dip to remove the native oxide layer before the deposition. Ga_2O_3 thin films were deposited by using $Ga(CH_3)_3$ and water as precursors in a thermal atomic layer deposition (ALD) reactor with a 100 sccm back ground flow of N_2. A cycle of the reaction consisted of a 0.3 s injection of $Ga(CH_3)_3$ vapor followed by 7 s N_2 purge and a 0.1 s injection of H_2O vapor followed by 7 s N_2 purge. The deposition rate is estimated to be about 0.13 nm/cycle. To study passivation performances, post-deposition annealing was performed in a quartz furnace at different temperature for different time under atmosphere ambient. Spectroscopic ellipsometry measurement was employed to obtain the thickness of Ga_2O_3 thin films on shiny-etched Si. Minority carrier lifetimes were characterized by Semilab WT-2000PVN lifetime tester.

3. Result and discussion

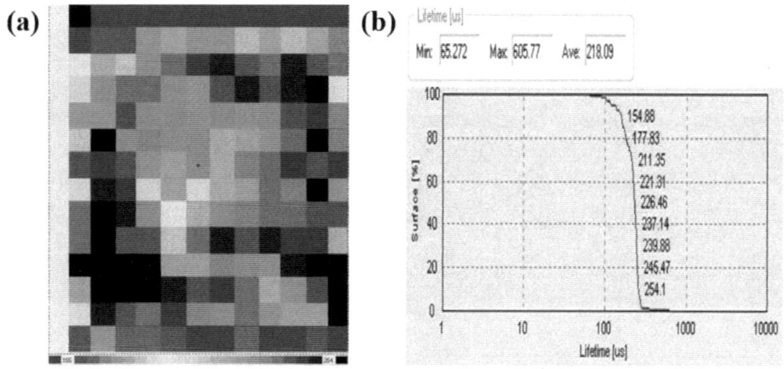

Fig.1 the effective minority carrier lifetime of Ga_2O_3 thin film annealed at the temperatures of 650 ℃ for 3 min

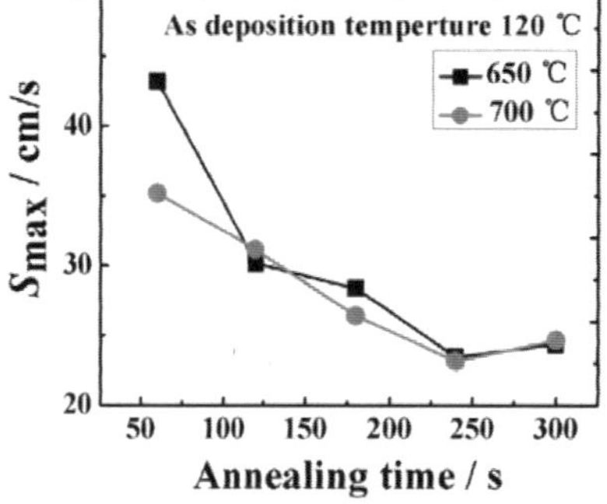

Fig.2 the surface recombination velocity of samples prepared at the temperature of 120 ℃ as a function with the annealing time

ALD process is based on sequential, self-limiting surface chemical " half-reaction". The main driving force for the efficient reactions is the formation of a very strong Ga-O bond. Therefore, Ga_2O_3 film thickness could be controlled accurately by controlling the number of reaction cycles. It should be noted here that there are some residual Ga-OH bonds during the reaction. O-H bonds would be easily broken, resulting in the interstitial H atoms within the Ga_2O_3 matrix and H atoms may play an important role in the passivation performance. In order to measure the effective minority carrier lifetime (τ_{eff}), clean c-Si wafers were coated by Ga_2O_3 thin films symmetrically as lifetime sample and were annealed at different temperatures in atmosphere ambient. These samples were characterized by Semilab WT-2000PVN lifetime tester, as shown in Fig.1. Ga_2O_3 thin film was annealed at the temperatures of 650 ℃ for 3 min. The range of the effective minority carrier lifetime is from 65.27 μs to 605.77 μs and the average effective minority carrier lifetime is 218.09 μs. The surface recombination velocity (S_{eff}) could be calculated by the formula (1):

$$\frac{1}{\tau_{eff}} = \frac{1}{\tau_{bulk}} + \left[\frac{W}{2S} + \frac{1}{D_n}\left(\frac{W}{\pi}\right)^2\right]^{-1} \quad (1)$$

Where τ_{eff} and τ_{bulk} are the effective minority carrier lifetime and the bulk minority carrier lifetime, respectively. W is the thickness of c-Si wafers, and D_n is the diffusion coefficient.

Fig.2 shows the surface recombination velocity of samples prepared at the temperature of 120 ℃ as a function with the annealing time. The surface recombination velocity of Ga_2O_3 thin film after annealed at the temperatures of 650 ℃ with the annealed time for 50 s is about 43 cm/s and is bigger than that (~36 cm/s) of Ga_2O_3 thin film after annealed at the temperatures of 700 ℃ with the annealed time for 50 s. With the increase of the annealed time from 50 s to 300 s, both surface recombination velocities of Ga_2O_3 thin film after annealed at the temperatures of 650 ℃ and 700 ℃ decrease to 25 cm/s. It indicates that annealing process pay a role for decreasing surface recombination velocities of Ga_2O_3 thin film. In addition, after the annealing time of 300 s, surface recombination velocities would be saturated. Fig.3 shows the surface recombination velocity of samples prepared at the temperature of 120 ℃, 200℃, 280 ℃ as a function with the annealing time, respectively. The annealing temperature is 500 ℃. From Fig.3, it can be seen that with the increase of the annealing time, the surface recombination velocities of samples prepared at the temperature of 120 ℃, 200℃, 280 ℃ decrease. After the annealing time of 300 s, surface recombination velocities would be also saturated.

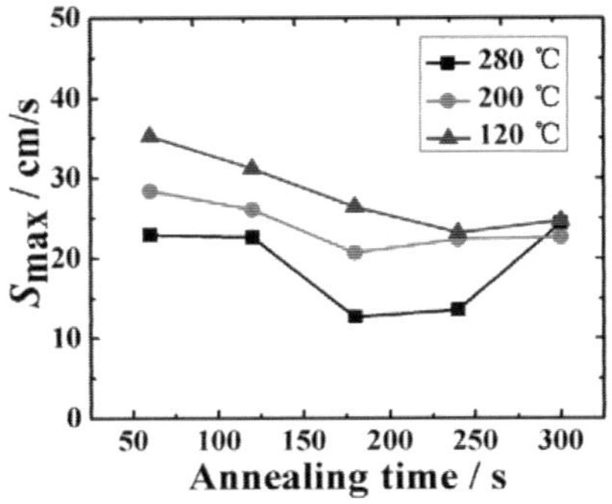

Fig.3 the surface recombination velocity as a function with the annealing time

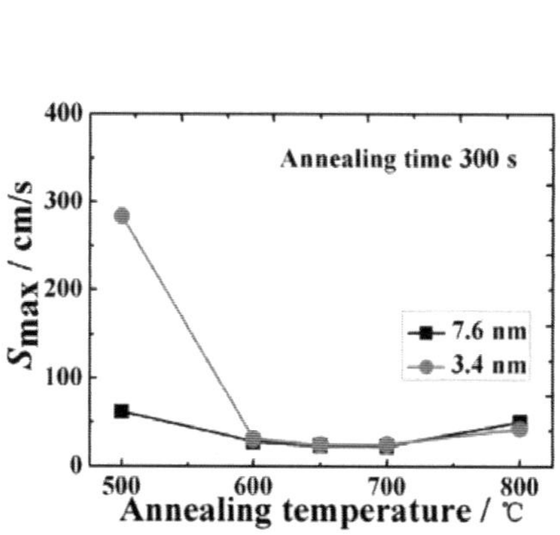

Fig.4 the surface recombination velocity of samples thickness ranged between 3.4 nm and 7.6 nm as a function with the annealing temperature

To help lower costs of the passivation of Ga_2O_3 thin film, the thickness of 7.6 nm and 3.4 nm were deposited. Fig.4 shows the surface recombination velocity of samples as a function with the annealing temperature. After annealing time of 300 s at the temperature of 500 ℃, surface recombination velocities are 61.2 cm/s and 283.6 cm/s at the temperature of 500 ℃, respectively. With the increase of annealing temperature, surface recombination velocities decrease to ~30 cm/s and are saturated. When annealing temperature is 800 ℃, surface recombination velocity increases a little. Such passivation degradations are due to the deteriorated interface properties. However, the passivation effect is obtained with ultrathin Ga_2O_3 film. It may be that interstitial H within Ga_2O_3 thin film resulted in the negative fixed charges, while the Si dangling bonds passivated partially by H atoms resulted in the decreased the interfacial state density, which leaded to the high passivation performances Therefore, Ga_2O_3 film thickness could be decreased for Si solar cell effectively.

4. Summary

In conclusion, Ga_2O_3 thin films were deposited by atomic layer deposition on n-type c-Si wafers. Surface passivation performances were studied. The average effective minority carrier lifetime can reduce to 218.09 μs. Wide annealing time and wide annealing temperature are addressed for obtaining good passivation performances with surface recombination velocity about 25 cm/s. The surface recombination velocity of ~30 cm/s was obtained for Ga_2O_3 coated textured Si with Ga_2O_3 thickness ranged between 3.4 nm and 7.6 nm. Such passivation performances were related to the released H atoms from Ga–OH bonds and the formation of Ga vacancies and O interstitials in Ga_2O_3 films. Such results indicate that Ga_2O_3 have the functions of surface passivation for potential photovoltaic applications.

Acknowledgements

This work is supported by Postdoctoral Research Funding Plan of Jiangsu Province (1402013B), Research Fund of Jiangsu Province Cultivation Base for State Key Laboratory of Photovoltaic Science and Technology (SKLPSTKF201503) and Starting Foundation of Jiangsu University Advanced Talent (14JDG049).

References

[1] Chapin D M, Fulle C S and Pearson G L 1954 *J. Appl. Phys.* **25** 676-7.
[2] Liang Z C, Chen D, Feng C K, Cai J T and Shen H 2011 *Phys. Procedia* **18** 51-5.
[3] Gatz S, Bothe K, Müller J, Dullweber T and Brendel R 2011 *Energy Procedia* **8** 318-23.
[4] Lin F, Hoex B, Koh Y H, Lin J J and Aberle A G 2012 *Energy Procedia* **15** 84-90.

[5] Kaminski P M, Abbas A, Bass K and Claudio G 2011 *Energy Procedia* **10** 71-5.
[6] Ikeda M, Nagayoshi H, Onozawa Y, Saitoh T and Kamisako K 1997 *Solar energy materials and solar cells* **48** 109-15.
[7] Thi T C, Koyama K, Ohdaira K and Matsumura H 2012 *Sol Energ Mat & Sol C* **100** 169-73.
[8] Repo P, Talvitie H, Li S, Skarp J and Savin H 2011 *Phys. Procedia* **8** 681-7.
[9] Xiao H Q, Zhou C L, Cao X N, Wang W J, Zhao L, Li H L and Diao H W *Chin. 2009 Phys. Lett.* **8** 088102.
[10] Hoex B, Gielis J J H, van de Sanden M C M and Kessels W M M *J. 2008 Appl. Phys.* **104** 113703.

Synthesis and properties of a novel bio-based polymer from modified soybean oil

Y T Li[1, 2], L T Yang[1], H Zhang[1], Z J Tang[1]

[1] School of Chemistry and Environment, South China Normal University, Guangzhou P.R. China
[2] MBS Environmental Technology Corporation

E-mail: yanglt@scnu.edu.cn

Abstract. Maleated acrylated epoxidized soybean oil (MAESO) was prepared by acrylated epoxidized soybean oil (AESO) and maleic anhydride. AESO were obtained by the reaction of epoxidized soybean oil (ESO) with acrylic acid as the ring-opening reagent. The polymer was prepared by MAESO react with styrene. The structures of the products were studied by Fourier transformation infrared spectrometer (FT-IR), and were consistent with the theoretical structures. Swelling experiment indicated that the crosslinking degree increased with increasing epoxy value of ESO. Thermal properties was tested by thermo-gravimetric analysis (TG) and differential scanning calorimetry analysis (DSC), indicating that glass transition temperature (T_g) of the polymer increased with increasing epoxy value of ESO, and thermal stability of polymer have a good correlation with the crosslinking degree. Mechanical properties analysis presented that tensile strength and impact strength affected by epoxy value of ESO. With the increase of epoxy value, the tensile strength increase, while the impact strength decrease. The property of the polymer ranged from elastomer to plastic character depended on the functionality of the ESO.

1. Introduction

Problems of environmental pollution caused by polymer materials have recently attracted increasing attention. Research into environmentally friendly raw materials substitute for preparing polymeric materials has become crying needs in the development of materials science. Many raw materials available in nature can be used to prepare polymer. Such as bio-based polymers were gaining interest, they were applited to many life and industry fields. Triglycerides from plant oil can be utilized [1-4]. Polymer materials obtained from the plant are becoming extensively and important used to take place of petroleum based polymer, because of their environmental advantages and economic. These have been widely used in various engineering and commodity applications [5-7].

Plant oil is one of the most interesting environmental resources for the synthesis of monomers or polymers, because of the aliphatic chain and functional groups contribute as a primary element to the backbone of polymer [8]. Plant oil has received much attention as a renewable resource for preparing chemicals and polymers [9]. Soybean oil is one of the plant oil and widely used for bio-based polymers production. Double bonds in soybean oil molecule are difficult to take radical-initiated polymerization reaction to form the polymer. Thus the soybean oil must be functionalized; the usual method is to transform the soybean oil into epoxidized soybean oil [10]. In order to be use as a monomer for more polymerization, epoxidized soybean oil has to be further functionalized [11]. Many chemical methods have been applied to modify soybean oil, such as hydroxylation reaction [12], acrylated reaction [13], epoxidation reaction [14], maleic anhydride esterification reaction [15], and so on.

The properties of polymer were controlled by the crosslinking density, functional group and flexibility of the macromolecule chain. Hydroxyl groups, ester groups and double bonds in the plant oil molecule make it available for modifications and reactions [16]. For AESO, hydroxyl groups on the triglyceride chain make it enable to be further modified, such as take mono-esterification reaction with maleic anhydride to form the maleated acrylated epoxidized soybean oil (MAESO). Afterwards, MAESO was free-radically copolymerized with vinyl monomers or polyfunctional monomers to form the polymers. The introduction of maleic anhydride increasing the degree of crosslinking of the final materials and introducing functional groups into the network of the polymers, providing more favorable performance for the materials.

For many publications, AESO was usually used in paints or inks. As its functionality not high enough, AESO was limited in the use of thermosetting resin to some extent. Nevertheless, AESO was modified by maleic anhydride greatly improved the degree of unsaturation of the functional monomer. Meanwhile, the carboxyl groups were introduced in the final products. Many attentions have been focused on the research of polyols based on soybean oil, and seldom studied the soybean oil based epoxy acrylate or maleate. Most of the reported materials based on soybean oil were made by ESO with high epoxy value. Few discussions about the relationship between epoxy value of ESO and properties of the polymer.

In this work, ESO with different epoxy values has been synthetized. AESO with different saturation was obtained by the reaction of ESO and acrylic acid. Then the synthetic AESO took mono-esterification reaction with maleic anhydride to form the MAESO. Finally, MAESO took free-radically copolymerized with vinyl monomers to produce the polymer.

2. Experimental

2.1. Materials

Soybean oil was obtained from Nanhai Oil Co. Ltd (Guangzhou, China). Glacial acetic acid, acrylic acid, sulfuric acid, hydrogen peroxide (30%), hydroquinone, ethyl ether, sodium carbonate, sodium bicarbonate, sodium chloride, maleic anhydride, methylbenzene, styrene , benzoyl peroxide (BPO), all the reagents were AR grade and purchased from Damao Chemical Reagent Co. (Tianjin, China). Triphenyl phosphine was GR grade and purchased from Aladdin Chemistry Co. Ltd (Shanghai, China).

2.2. Methods

Epoxy value was determined by the ASTM D1652 testing standard. Hydroxyl value was determined by the ASTM D1957 testing standard. Degrees of crosslinking of the polymer was detemined by swelling equilibrium method. FTIR spectra of the samples were recorded using a Prestige-21 spectrometer (Shimadzu, Japan). DSC analysis was recorded on a 200PC instrument (Netzsch, Germany). Sample (8-10 mg) was used each measurement with nitrogen was use as purge gas. The sample was heated from -50 to 200℃ at a heating rate of 20℃/min. Thermo-gravimetric analysis was recorded on a STA 409PC instrument (Netzsch, Germany). The sample was heated from 30 to 800℃ at a heating rate of 10℃/min with air as the purge gas. Tensile strength of the sample was determined on a WSM-100 tensile testing machine (Changchun, China). The rate of extension was 10 mm/min. Impact strength of the sample was determined on a JJ-20 impact testing machine (Changchun, China).

2.3. Preparation of modified soybean oil

ESO with different epoxy value have been prepared from the epoxidation reaction of SBO. Peroxyacetic acid was used as oxidizing agent and sulphuric acid was used as catalyst. The reaction took place at 65℃ and different reaction time. The extent of epoxidation was 0.25 mol/100g, 0.33 mol/100g, 0.35 mol/100g, 0.38 mol/100g, respectively.

AESO was obtained from the reaction of ESO and acrylic acid. The molar ratio of epoxy group to carboxyl was 1 to 1.2. Triphenyl phosphine was used as catalyst at concentration of 1.5% in weight. Hydroquinone was used as inhibitor at concentrations 0.1% in weight. The reaction took place in a round bottomed flask and kept in 110 ℃ for 8 hours. After finishing reaction, the mixture was cooled to room temperature. And then it was dissolved in ethyl ether. Saturated $NaHCO_3$ sodium and NaCl sodium were used to wash the mixture. After removing the water and solvent, the samples were obtained. The hydroxyl values of the AESO were 119 mgKOH/g, 151 mgKOH/g, 164 mgKOH/g, 178 mgKOH/g, respectively. The samples were used for preparing MAESO.

MAESO was obtained by reaction of AESO and maleic anhydride. The molar ratio of maleic anhydride to hydroxy group is 2:1, ensuring that the hydroxy group completely reacted. AESO and methylbenzene were placed into a 500 mL round bottom flask and kept in 110℃for 1 hour to ensure that there was no water in the mixture. Then the maleic anhydride was added into the mixture. The reaction kept in 110℃ for 5 hours. After the reaction, the mixture was cooled down. Deionized water was used to wash the mixture. After removing the water and solvent, the samples were obtained.

2.4. Synthesis of polymer

The polymer was prepared by mixing the MASEO with styrene. BPO was used as initiator at a concentration of 3% in weight. The mass fractions of styrene and MAESO were 40% and 60%, respectively. BPO was stirred to dissolve in styrene, and blended with MAESO. The mixture was put into vacuum oven to eliminate air bubbles, and poured into a mold. Then it was heated to 60 ℃ for 1 hour, and 100℃ for 3 hours. The degrees of crosslinking of the polymers were characterized by the swelling experiments.

3. Results and discussion

3.1. Characterization of MAESO

Synthesis routes of MAESO and polymer were provided in Figure 1. A series of ESO and AESO were synthesized and shown in Table1, indicating that the hydroxyl value increases with the epoxy value. The degrees of crosslinking of the polymers were characterized by the swelling experiments; the results indicating that gel fraction of the polymers increase with the epoxy value. The polymer with an epoxy value of 0.25 mol/100g had low gel fraction, due to the low average functionality of the MAESO, presenting the elastomer character. While the polymer with higher gel fraction displayed rigid plastic properties.

Table 1. properties of polymer derived from MAESO

Sample	a	b	c	d
Epoxy value of ESO/(mol/100g)	0.25	0.33	0.35	0.38
Hydroxyl value of AESO /(mgKOH/g)	119	151	164	178
Gel fraction of polymer /%	75.2	89.9	93.4	97.0
Character of the polymer	elastomer	plastic	plastic	plastic
Glass transition temperature /℃	10.5	47.2	52.9	54.6

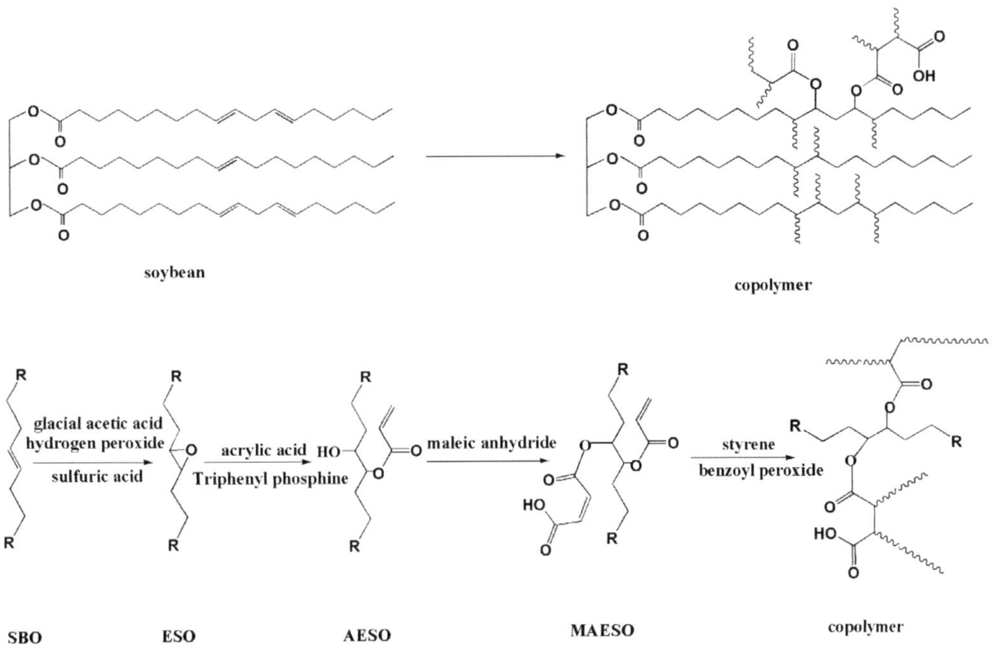

Figure 1. Synthesis routes of MAESO and polymer

3.2. FTIR spectra analysis

FTIR spectra were showed in Figure 2. FTIR spectrum of SBO showed a peak at 3010 cm^{-1} due to the carbon-hydrogen stretching vibration in the carbon-carbon double bond. In the ESO spectrum, there was no peak in the same wave number, and the peak at 823 cm^{-1} due to the epoxy group, illustrating that the epoxidation reaction has took place. In the spectrum of AESO, the peak at 1653 cm^{-1} due to stretching vibrations of carbon-carbon double bond. The stretching vibrations of hydroxyl appeared at 3467 cm^{-1}. The peak of epoxy group at 823 cm^{-1} not appeared, indicating that the reaction of ESO with acrylic acid has taken place extensively. The peak at 1743 cm^{-1} showed the stretching vibration carbonyl. In the spectrum of MAESO, The peak of hydroxy group at 3467 cm^{-1} widened, and presents dispersion characteristics, indicating that the AESO has taken mono-esterification reaction with maleic anhydride, and generated carboxyl group. In the spectrum of the polymer, the peaks at 3064 and 3034 cm^{-1} due to the carbon-hydrogen stretching vibrations of the benzene ring. The peaks at 1584, 1494, 1453 and 1402 cm^{-1} are attributed to the bending vibrations of carbon-carbon double bond of benzene ring. The stretching vibration peak of the carbon-carbon doubles at 1653 cm^{-1} almost disappeared, indicating that the reaction of MAESO and the styrene has occurred.

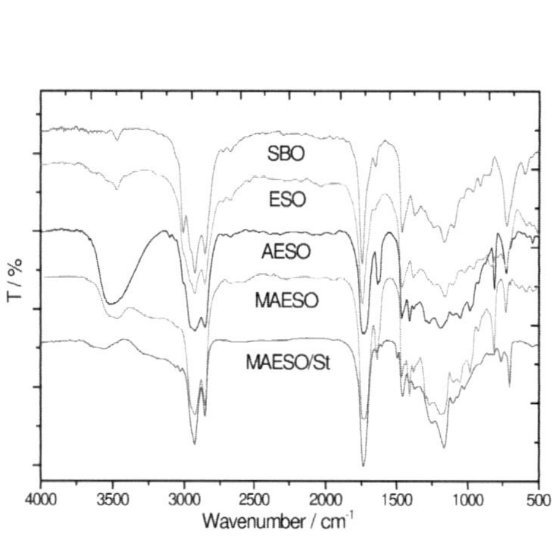

Figure 2. Infrared spectra of expoxidized soybean oil and its products

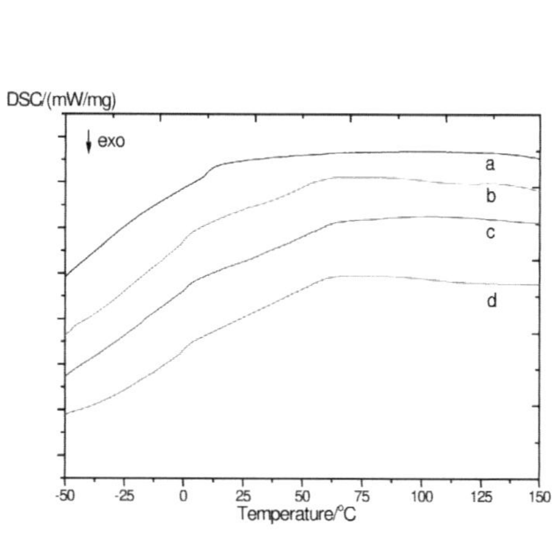

Figure 3. DSC curves of polymer derived from MAESO

3.3. Thermal properties analysis

Properties of polymer affected by the cross linking degree, which correlated with the degree of functionality of the monomer molecules, such as the hydroxyl group and the epoxy group. DSC thermograms were given in Figure 3. Glass transition temperatures of the polymers (T_g) were showed in Table 1. The result indicating that T_g of the polymers increased with the increasing hydroxyl value and epoxy value. Due to the crosslinking degree of the polymer increased with the degree of functionality, as the number of active carbon-carbon double bond increased. Also, carbon-carbon double bond in the triglyceride chain effect flexibility of the macromolecule chain. Glass transition temperature of the polymer with an epoxy value of 0.25 mol/100g was 10.5℃, which low and under the room temperature, because of low crosslinking degree and gel fraction. The polymer was conjectured an imperfect networks. While the epoxy value of the ESO was 0.38 mol/100g, the resulting material was glassy; the glass transition temperature was 54.6℃. This result was proved by the s mechanical properties as provided in Table 2.

TG and DTG thermograms were provided in Figure 4. The 5% mass loss temperatures of all the polymers ranged from 183.1℃ to 231.9℃. All the polymers have been found to start degradation below 250℃, due to the thermolabile groups of carboxyl and allyl ester. Thermal stability material showed a correlation with the epoxy value. The decomposition temperature decreases with epoxy value at lower epoxy value, and increases at higher epoxy value. Due to the crosslinking degree and stability of functional group. While at low epoxy values, the resulting materials were mainly affected by the increasing of carboxyl group. Whereas the active carbon-carbon double bond introduced from maleic anhydride and acrylic acid increases obviously at higher epoxy value, thus the crosslinking degree of the polymer was extremely high and controlling the thermal stability. The variation of the decomposition temperatures was obviously at low temperature, and verging to unconspicuous with the increases of the temperature, because of most of decompositions synchronized at high temperature.

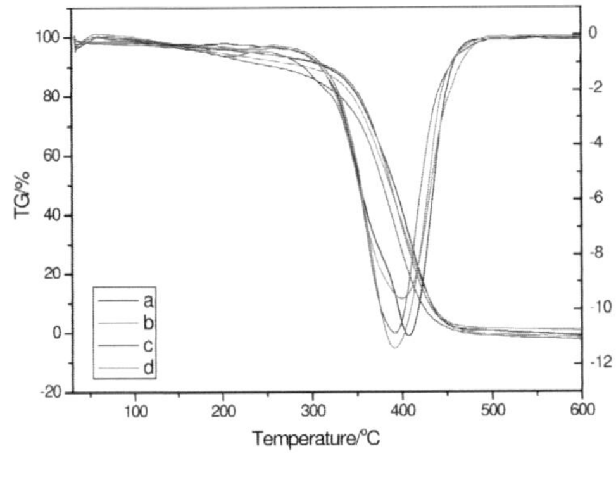

Figure 4. TG and DTG curves in air of polymers derived fromMAESO

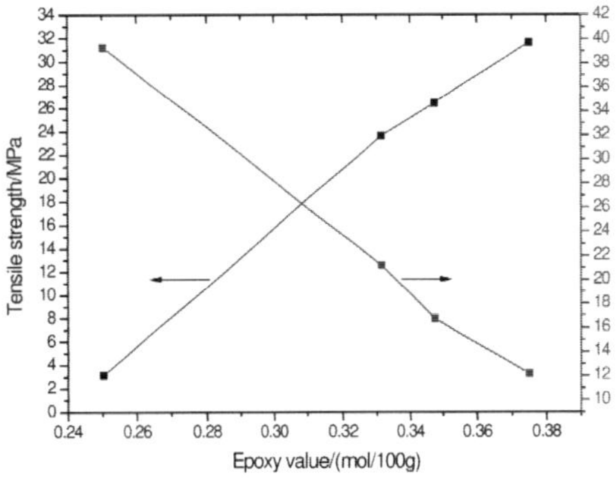

Figure 5. Relationship betweem epoxy value and mechanical properties

3.4. Mechanical properties analysis

Mechanical properties of the polymers presented in Figure 5 and Table 2. The result indicated that tensile strength increases with the increasing epoxy value and the crosslinking degree of the polymers, yet the elongation at break decreases. The active carbon-carbon double bond introduced from maleic anhydride and acrylic acid increases with the increasing epoxy value, thus the crosslinking degree increase and the flexibility of the polymer chain decrease. It was confirmed by the DSC analysis. The polymer obtained from the ESO with an epoxy value of 0.25 mol/100g has very low tensile strength of 3.17 MPa, which was characteristic of elastomer, because of the low crosslinking degree and imperfect networks of the polymer. The fact was related to the low glass transition temperature of the polymer, under the testing temperature of 25℃. While epoxy value of ESO was 0.33 mol/100g, the final material presents a moderate tensile strength of 23.62 MPa, and the tensile strength was higher (31.65MPa) for the polymer from the ESO with an epoxy value of 0.38 mol/100g, revealing rigid plastic properties to the materials. The result was supported by the elongation at break. The elongation at break was shown in Table 2, which decreases with the increasing epoxy value of the ESO.

Table 2. Mechanical properties of polymer derived from MAESO

Sample	Epoxy value/ (mol/100g)	Tensile strength/ MPa	Elongation at break/ %	Impact strength/ KJ/m^2
a	0.25	3.17±0.21	27.86±3.22	39.32±1.46
b	0.33	23.62±2.49	19.04±2.54	21.19±2.93
c	0.35	26.47±1.11	10.84±1.67	16.75±3.01
d	0.38	31.65±2.30	5.74±0.89	12.16±1.39

Impact strength of the polymers was given in Figure 5 and Table 2. The result showed that impact strength decreases with the increasing epoxy value, due to decreasing of flexibility of the polymer chains. The variation trend of the impact strength was similar to trend of the elongation at break, whereas countering to the tensile strength. The flexibility of the polymer chain was low at high epoxy value of the ESO, whereas the crosslinking degree of the network was high. In this case, the effect of the flexibility of the polymer chain was not obvious, because of the strong cross-linked network mainly controlled the strengths, thus the material also have a certain degree of impact strength. Soybean oil molecular chain also makes a contribution to the flexibility of the final material.

4. Summary

A series of bio-based polymers were 1. Relationship of the structure and property on the polymer has been studied. Mechanical and thermal properties depended on the functionality of ESO. Wide distribution of the epoxy values of ESO was reflected in different crosslinking degree of polymer. Crosslinking degree of the polymer increased with the increasing epoxy value. The polymer with an epoxy value of 0.25 mol/100g had low gel fraction, due to the low average functionality of the MAESO, presenting the elastomer character. While the others with higher gel fraction displayed rigid plastic property. T_g of the polymers increased with the increasing hydroxyl value and epoxy value. Thermal stability displayed a well correlation with the epoxy value. Glassy polymer displayed excellent mechanical strength reached up to 31.65 MPa and glass transition temperature of 54.6℃. The bio-based polymer showed a range of attractive properties and a wide range of uses. Also, these polymers were prepared from renewable environmentally friendly resource.References

References

[1] Wang C S, Yang L T, Ni B L and Shi G 2009 Polyurethane networks from different soy-based polyols by the ring opening of epoxidized soybean oil with methanol, glycol, and 1,2-propanediol, *J. Appl. Polym. Sci.* **114** 125–131.

[2] Dai H H, Yang LT, Lin B, Wang C S and Shi G 2009 Synthesis and characterization of the different soy-based polyols by ring opening of epoxidized soybean oil with methanol, 1,2-ethanediol and 1,2-propanediol, *J. Am. Oil. Chem. Soc.* **86** 261-267.

[3] Hong C K and Wool R P 2005 Development of a bio-based composite material from soybean oil and keratin fibers, *J. Appl. Polym. Sci.* **95** 1524-1538.

[4] Bakhshi 2013 Synthesis and characterization of antibacterial polyurethane coatings from quaternary ammonium salts functionalized soybean oil based polyols, *Mater. Sci. Eng.*, **C 33** 153-164.

[5] Nagarjuna R P, Smita M and Nayak S K 2015 Mechanical and thermal properties of sisal fiber reinforced acrylated epoxidized castor oil toughened diglycidyl ether of bisphenol A epoxy nanocomposites, *J. Reinf. Plast. Comp.* **34** 1476-1490.

[6] Reddy P N, Mohanty S and Nayak S K 2014 Synthesis and modifications of epoxy resins and their composites: a review, *Polym. Plast. Technol. Eng.* **16** 1723-1758.

[7] Sapuan F M and AL-Oqla S M 2014 Natural fiber reinforced polymer composites in industrial applications: feasibility of date palm fibers for sustainable automotive industry, *J. Cle. Prod.* **66** 347-354.

[8] Audrey L, Patrick-Kurt D and Marc von C 2016 Renewability is not Enough: Recent Advances in the Sustainable Synthesis of Biomass-Derived Monomers and Polymers, *Chem. Eur. J.* **22** 11510-11521.

[9] Muhammad Y A and Muhammad I 2015 Synthesis and properties of cross-linked polymers from epoxidized rubber seed oil and triethylenetetramine, *J. Appl. Polym. Sci.* **10** 42591-42603

[10] Tan S G, Ahmad Z and Chow W S 2013 Relationships of cure kinetics and processing for epoxidized soybean oil bio-thermoset. *Ind. Crops. Prod.* **43** 378-385.

[11] Wool R P and Sun X S 2005 Bio-Based Polymers and Composites, Elsevier Academic Press, San Diego, California.

[12] Eren T and Kusefoglu S H 2004 Hydroxymethylation and polymerization of plant oil triglycerides, *J. Appl. Polym. Sci.* **91** 4037–4046.

[13] Behera D and Banthia A K 2008 Synthesis, characterization, and kinetics study of thermal decomposition of epoxidized soybean oil acrylate, *J. Appl. Polym. Sci.* **109** 2583–2590.

[14] Yang L T, Dai H H, Yi A H and Lin B 2008 Structure and properties of partially epoxidized soybean oil, *J. Therm. Anan. Cal.* **93** 875–879.

[15] La Scala J J and Wool R P 2005 Rheology of chemically modified triglycerides, *J. Appl. Polym. Sci.* **95** 774–783.

[16] Wang H J, Rong M Z and Zhang M Q 2008 Biodegradable Foam Plastics Based on Castor Oil, *Biomacromolecules* **9** 615–623.

IOP Conf. Series: Materials Science and Engineering **170** (2017) 012011 doi:10.1088/1757-899X/170/1/012011

Microwave Dielectric Properties of $Na_{0.5+x}Nd_{0.5}TiO_3$ Ceramics (x=-0.05~0.1)

B Tang[1], Z X Fang[1], L Zhou[1], S R Zhang[1]

1 State Key Laboratory of Electronic Thin Films and Integrated Devices, University of electronic Science and Technology of China, Chengdu 610054, China

E-mail: ZXFANG2015@163.com

Abstract: The influence of nonstoichiometry of Na on micronmicrowave dielectric properties of $Na_{0.5+x}Nd_{0.5}TiO_3$ (NxNT) ceramics was systematically investigated in this research. The main phase of samples was indexed as cubic perovskite structure, and $Nd_{0.667}TiO_3$ was detected as secondary phase($-0.05 \leq x \leq 0$), which was caused by the evaporation of Na ions. As x was increased from 0 to 0.05, the $Nd_{0.667}TiO_3$ phase was effectively eliminated and the densification process was improved. Therefore, the ε_r and $Q\times f$ initially increased, later they decreased, which was accordance to the change of density. The τ_f value didn't show a dramatic change, and a relatively low value of 248.2 ppm/°C was obtained at x=0.05. Excellent microwave dielectric properties of ε_r=110, $Q\times f$=8115 GHz and τ_f=248.2 ppm/°C) were

1. Introduction

With the wireless communication technology evolving from 3G to 4G, microwave dielectric ceramics have always been studied and utilized as key components of filter, oscillator and antenna components[1]. The search of ceramics with high dielectric constant (ε_r), high quality factor ($Q\times f$) and tunable temperature coefficient of resonant frequency (τ_f) and the research on improving the comprehensive properties have attracted considerable attention. To date, lots of microwave ceramic systems have been extensively researched, such as $CaO-Li_2O-Ln_2O_3-TiO_2$ and $BaO-Ln_2O_3-TiO_2$ (Ln=rare earth) [2, 3]. But their dielectric permittivities were mainly distributed from 80 to 100, which are not competitive to satisfy the requirements of miniaturization sufficiently. As a result, many efforts were made to develop microwave ceramics with higher dielectric constant.

Takahashi et al. first investigated the microwave dielectric properties of $Na_{0.5}Nd_{0.5}TiO_3$ ceramics with perovskite structure, as well as its special super-lattice structure[4]. Zhou et al. [5] reported that the Sn doped $Na_{0.5}Nd_{0.5}TiO_3$ ceramics possessed a good combination properties of ε_r >100, $Q\times f$ >7000 GHz and τ_f=260 ppm/°C. However, in the Na containing compounds, the Na contents can seriously evaporate at high temperature. And the phase composition and microwave dielectric properties of NST ceramics are highly dependent on the Na contents[6]. As is known to all, the microwave properties would be greatly affected by the variation of microstructure and phase composition. The present study of NST ceramics are mainly focus on how to tune their τ_f by the compositional method and few studies are devoted to deal with the nonstoichiometry of Na ion at A-sites. Thus, it is quite necessary to optimize the concentrations of Na_2CO_3 to accommodate evaporation of Na ion. In the present work, the densification, microstructure and microwave dielectric properties were systematically investigated with the variation of Na content in A-site cation.

2. Experimental Procedure

In this work, the samples were conventionally synthesized by using solid state methods.
The raw materials were Na_2CO_3, Nd_2O_3 and TiO_2 with at least 99.9% purity. After drying, the starting materials were weighed corresponding to nonstoichiometric ratio of $Na_{0.5+x}Nd_{0.5}TiO_3$ (N*x*ST), where *x*= -0.05, 0, 0.05 and 0.1. They were milled in a nylon jar with zirconia balls in ethanol for 10 h, and the mixtures were dried and calcined at 1000 °C for 3 h. Later, the calcined powders were mixed with a 6 wt% of a 10% solution of PVA as a binder and granulated. The resulting powder was axially pressed into cylindrical disks with a thickness of 6.5 mm and a diameter of 14 mm under a pressure of 200 kg/cm^2. The pellets of N*x*NT were sintered at 1300 °C for 2 h in air.

The apparent densities of the sintered samples were measured using the Archimedes' method. The powder phase composition was examined by X-ray diffraction (XRD) using CuKα radiation (Philips x'pert Pro MPD, Netherlands). Scanning electron microscopy (SEM) (FEI Inspect F, United Kingdom) was employed to study the thermally etched surface morphology of the specimens. The dielectric characteristics at microwave frequencies were measured by the Hakki–Coleman dielectric resonator method in the TE011 mode using a network analyzer (Agilent Technologies E5071C, USA) and temperature chamber (DELTA 9023, Delta Design, USA). The τ_f measured at 2~3 GHz were obtained by the equation: $\tau_f = (f_{t2} - f_{t1}) \times 10^6/(f_{t1} \times 60)$ (1), where f_{t2} and f_{t1} are the resonant frequencies at the measuring temperature $t1$ (25 °C) and $t2$ (85 °C).

3. Results and discussion

The XRD patterns of powdery N*x*NT ceramics with different *x* values sintered at 1300°C for 2h are presented in Fig. 1. In the whole compositional range, the main phase of samples was indexed as cubic perovskite structure (JCPDS# 39-0880). However, the $Nd_{0.667}TiO_3$ (JCPDS# 49-0244) as secondary phase was detected in the range of *x*=-0.05~0.0 for N*x*NT ceramics. The appearance of $Nd_{0.667}TiO_3$ was attributable to the evaporation of Na contents when the samples were sintered in an extremely high temperature of 1300°C. The intensity of secondary phase became weaker as the x value increased. When the Na concentrations changed from 0.05 to 0.1, only the cubic perovskite structured phase could be observe, and the secondary phase disappeared. Meanwhile, the weak superlattice peaks were also observed. The double index of the above reflections are in an even–even–odd plus odd–odd–odd manner, and these weak reflections can be assigned to space group Pnma [7]. As a result, a pure NNT phase could be obtained when the Na contents was superfluous($x > 0$).

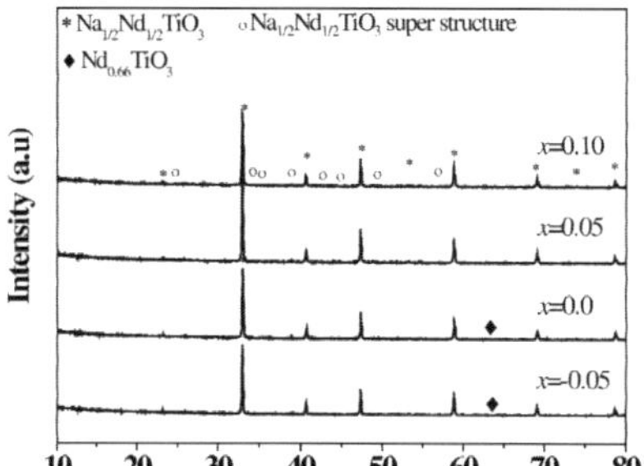

Figure 1. The XRD patterns of powdery N*x*NT ceramics with different *x* values sintered at 1300°C for 2h.

Fig. 2 displays the SEM image of N*x*NT ceramics with different *x* values sintered at 1300°C for 2h with (a) *x*=-0.05, (b) *x*=0.0, (c) *x*=0.05 and (d) *x*=0.1. As shown in Fig. 2(a), the ceramic presented a porous and inhomogeneous microstructure, had a grain size of 2 to 3 μm and two types of grains can be observed. When *x* was increased to 0.05, the grains grew to an average size of 5μm, and the pores around the grain boundaries were effectively eliminated. Obviously, the specimens showed a dense and compact microstructure with additional Na_2CO_3 dopants added (x=0.05). However, when excessive Na contents were doped ($x > 0.05$), the porous and inhomogeneous microstructure appeared again.

Figure 2. The SEM image of NxNT ceramics with different x values sintered at 1300°C for 2h with (a) x=-0.05, (b) x=0.0, (c) x=0.05 and (d) x=0.1

Fig. 3 illustrates the apparent density and microwave dielectric properties NxNT ceramics with different x values sintered at 1300°C for 2h: (a) apparent density, (b) dielectric constant, (c) quality factor and (d) temperature coefficient of resonant frequency. As shown in Fig.3 (a), the apparent density steadily increased with rising Na contents from -0.05 to 0.05, later it sharply decreased. Generally, the density was influenced by pores and secondary phase [8, 9]. Since the theoretical density of $Nd_{0.667}TiO_3$ (5.8 g/cm^{-3}) was larger than that of NNT (5.28 g/cm^{-3}), the decrease of secondary phase was not attributable to the increase of density. Thus, the improvement of densification should mainly be caused by the elimination of pores. However, with ($x > 0.05$), the appearance of pores led to the decrease of density. For the dielectric constant and quality factor, both of the dielectric constant and quality factor showed a similar varying trend to that of apparent density. This is because density, dielectric constant and quality factor are affected by the same reason, such as densification process. These changes were considered to correlate with the compactness of the samples; a higher ε_r and $Q\times f$ value were usually related to a denser microstructure, which was also confirmed by the SEM images. Fig. 3(d) depicts the variation of the τ_f as a function of Na contents. The τ_f value didn't show a dramatic change, and a relatively low value of 248.2 ppm/°C was obtained at x=0.05.

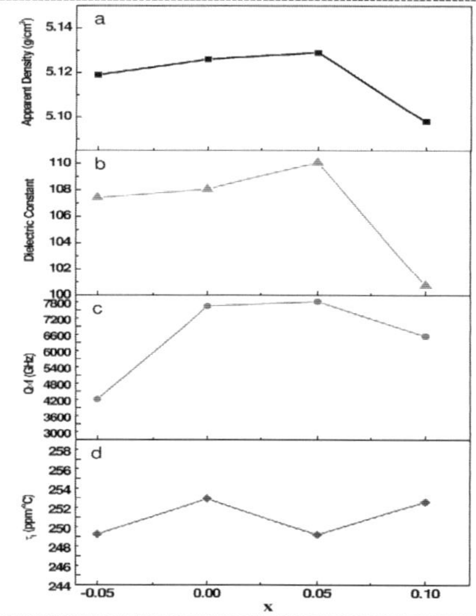

Figure 4. The apparent density and microwave dielectric properties NxNT ceramics with different x values sintered at 1300°C for 2h: (a) apparent density, (b) dielectric constant, (c) quality factor and (d) temperature coefficient of resonant frequency

4. Conclusion

The influence of nonstoichiometry of Na on micronmicrowave dielectric properties of NxNT ceramics were strongly determined by the Na contents. The main phase of samples was indexed as cubic perovskite structure, and $Nd_{0.667}TiO_3$ was detected as secondary phase($-0.05 \leq x \leq 0$), which was caused by the evaporation of Na ions. As x was increased from 0 to 0.05, the $Nd_{0.667}TiO_3$ phase was effectively eliminated and the densification process was improved. Therefore, the ε_r and $Q\times f$ initially increased, later they decreased, which was accordance to the change of density. The τ_f value didn't show a dramatic change, and a relatively low value of 248.2 ppm/°C was obtained at x=0.05. Excellent microwave dielectric properties of ε_r=110, $Q\times f$=8115 GHz and τ_f=248.2 ppm/°C) were achieved when x = 0.05 sintered in air at 1300 °C for 2h.

Acknowledgments
This work is supported by National Natural Science Funds of China (Grant No. 51402039)

References

[1] Fang Z, Tang B, Si F and Zhang S 2016 Temperature Stable and High-Q Microwave Dielectric Cera mics in the $Li_2Mg_{3-x}Ca_xTiO_6$ system (x=0.00~0.18), Ceram. Int. http://dx.doi.org/10.1016/j.cera mint.08.055

[2] Kenichi E, Yoko B, Hisakazu T, Kenichi S and Shoichi N 1993 Microwave Dielectric Properties of CaO-Li_2O-Ln_2O_3-TiO_2 Ceramics, *Jpn. J. Appl. Phys.* **32 (9S)** 4319.

[3] Long M, Tang B, Zhang S and Yu S 2012 Influence of Ba–Zn–B additives on the sintering behavior and dielectric properties of $BaNd_2Ti4O_{12}$ ceramics, *Mater. Lett.* **68** 486-489.

[4] Takahashi H, Baba Y, Ezaki K, Okamoto Y, Shibata K, Kuroki K and Nakano S 1991 Dielectric Characteristics of $(A_{1/2}{}^{1+}\cdot A_{1/2}{}^{3+})$ TiO_3 Ceramics at Microwave Frequencies, *Jpn. J. Appl. Phys.* **30 (9S)** 2339.

[5] Zhou L, Tang B and Zhang S 2015 Influence of Sn-substitution on microstructure and microwave dielectric properties of $Na_{1/2}Nd_{1/2}TiO_3$ ceramics, *J. Mater. Sci-Mater. El.* **26 (1)** 424-428.

[6] Fang Z, Tang B, Si F, Gong Y and Zhang S 2016 Effects of Zr-Substitution on Microwave Dielectric Properties of $Na_{0.5}Nd_{0.2}Sm_{0.3}Ti_{1-x}Zr_xO_3$ Ceramics (x= 0.00 ~ 0.30), Journal of Elec Materi.1-8. doi: 10.1007/s11664-016-4736-6.

[7] Glazer A 1972 The classification of tilted octahedra in perovskites, *Acta. Crystallogr. Sec. B.* **28 (11)**

3384-3392.
[8] Zhou D, Pang L X and Qi Z M 2014 Crystal Structure and Microwave Dielectric Behaviors of Ultra-Low-Temperature Fired $x(Ag_{0.5}Bi_{0.5})MoO_4–(1–x)BiVO_4$ ($0.0 \leq x \leq 1.0$) Solid Solution with Scheelite Structure, *Inorg. Chem.* **53 (17)** 9222-9227.
[9] Chen G h, Yuan C l, Zhou C R and Yang Y 2013 Low-firing high permittivity $Ca_{0.6}Sm_{0.8/3}TiO_3$–$(Li_{0.5}Nd_{0.5})TiO_3$ ceramics with $BaCu(B_2O_5)$ addition, *Ceram. Int.* **39 (8)** 9763-9766.

Thermal Performance Simulation of MWNT/NR composites Based on Levenberg-Marquard Algorithm

Z Z Yu [1,a], J S Liu [1,b]

[1]College of Electromechanical Engineering,Qingdao University of Science &Technology,China 266061

E-mail:[a]yuzhenzhu@qust.edu.cn,[b]931669075@qq.com

Abstract. In this paper, Levenberg-Marquard algorithm was used to simulate thermal performance of aligned carbon nanotubes-filled rubber composite, and the effect of temperature, filling amount, MWNTs orientation and other factors on thermal performance were studied. The research results showed: MWNTs orientation can greatly improve the thermal conductivity of composite materials, the thermal performance improvement of overall orientation was higher than the local orientation. Volume fraction can affect thermal performance, thermal conductivity increased with the increase of volume fraction. Temperature had no significant effect on the thermal conductivity. The simulation results correlated well with experimental results, which showed that the simulation algorithm is effective and feasible.

1. Introduction

Natural rubber (NR) is a natural polymer with polyisoprene as a main component, with excellent resilience, insulation, impermeability and plasticity characteristics, widely used in transportation, construction, electronics, aerospace, oil chemical industry, agriculture, machinery, military, water conservancy and information industry, people's lives and other industries. But rubber is a poor conductor of heat, then a lot of heat generated under dynamic conditions cannot be derived in time, which will lead to high-temperature that damages rubber products performance and service life, so how to reinforce rubber thermal conductivity has became a hot topic[1-3]. Carbon nanotubes's (MWNTs) unique structure makes it have ultra-high strength and toughness, excellent electrical conductivity and thermal conductivity, which make MWNTs become the focus of scientific research in recent years[4]. He Yan[5] and the group members took it as a new enhanced thermally conductive filler to improve thermal properties of rubber, and then prepared a natural rubber composites; results showed that the thermal conductivity of rubber filled with MWNTs has been significantly improved compared with pure NR.

Based on the analysis on thermal properties of rubber filled with MWNTs, this paper has simulated the thermal conductivity of composite materials with a new type of back propagation algorithm and researched the influence of factors such as the ambient temperature, volume filling fraction, MWNTs-Orientationon heat conduction, and wish to provide a reference for the study of the thermal properties of rubber composites filled with MWNTs.

2. The analysis on thermal conductivity of composite materials

In this paper, MWNTs is filled in NR to improve the thermal conductivity of composite materials. This paper has studied the influence of the MWNTs' local orientation (local orientation means that

IOP Conf. Series: Materials Science and Engineering **170** (2017) 012012 doi:10.1088/1757-899X/170/1/012012

one dimension filler is arranged in orientation in micro area element, and scatters irregularly in global scope.), whole orientation and volume filling fraction on the thermal conductivity of NR composites in comparison with pure NR under the same conditions.

This group adopted[6] the method of solution blending to fill the rubber with MWNTs, local aligned carbon nanotube (L-MWNTs), Fe3O4-MWNTs magnetic nano composite particles respectively; and prepared MWNT/NR composites, L-MWNT/NR composites, Fe3O4-MWNT/NR composites, aligned Fe3O4-MWNT/NR composites by room temperature vulcanizing; finally obtained the thermal conductivity of composites with different volume filling fraction[7] by experiments.

Fig.1 shows the thermal conductivity of 4 kinds of composites filled with 4% volume fraction. The thermal conductivity of MWNT/NR composites, L-MWNT/NR composites, Fe_3O_4-MWNT/NR composites, aligned Fe_3O_4-MWNT/NR composites are all higher than that of natural rubber; and the conductivity of MWNT/NR composites is the lowest, thermal conductivity of the worst, the conductivity of aligned Fe_3O_4-MWNT/NR composites is the highest.Fig.2 shows the thermal conductivity of MWNT/NR composites, L-MWNT/NR composites at 100°C, which all increase with the increase of the filling fraction.

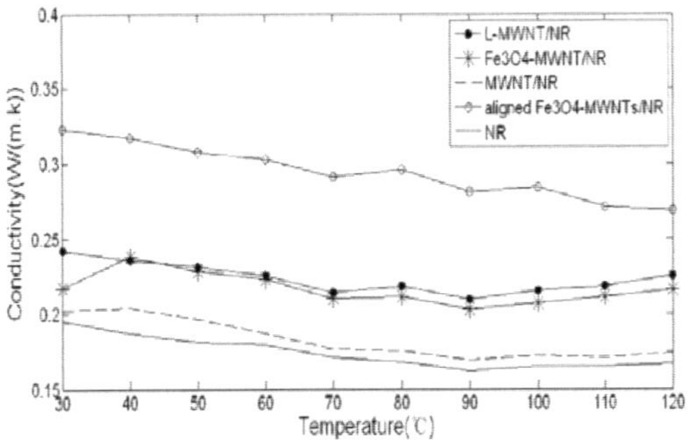

Fig. 1 Thermal conductivity of MWNT filled NR compo-sites with 4% volume fraction

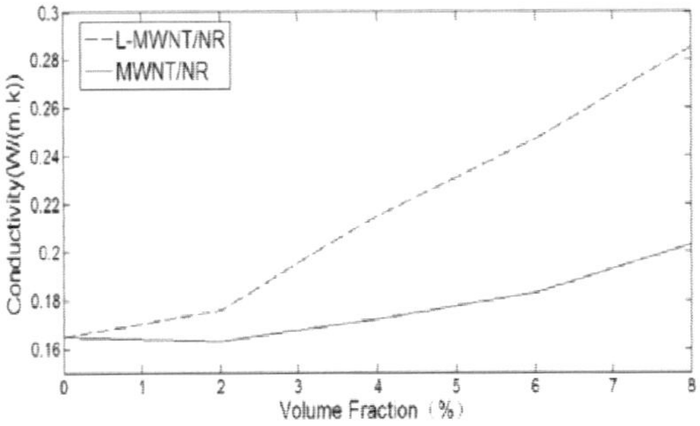

Fig.2 Thermal conductivity of L-MWNT/NR composite at 100°C

Experimental results show that 1) MWNTs can effectively improve the thermal properties of the composites, thermal conductivity of composites filled natural rubber with MWNT is greater than pure NR (see Fig.1, 2). Fe_3O_4 particles have effectively improved thermal conductivity of the composites (see in Fig.1), the thermal conductivity curve of Fe_3O_4-MWNT/NR composites was significantly higher than MWNT/NR composites. 3) MWNTs orientation have effectively improved thermal conductivity of the composites (see Fig.1), the thermal conductivity of L-MWNT /NR composites is higher than that of non-directional composites. 4) The temperature had no significant effect (see Fig.1) on the thermal properties of the composites. 5) The increase of filler fraction will improve the thermal conductivity of composites (see Fig.2), the relationship between the thermal conductivity and the filling volume fraction is non-linear.

3. Models and Methods

3.1 Heat conduction model of MWNT/NR composites

Fig.3 shows the thermal model of MWNT/NR composites, the model is a three-layer forward network with three inputs and single output (a input layer, a hidden layer and a output layer), the inputs are categories of composites (see Table 1), volume filling fraction, ambient temperature, the output is thermal conductivity, the number of hidden layer nodes is 10.

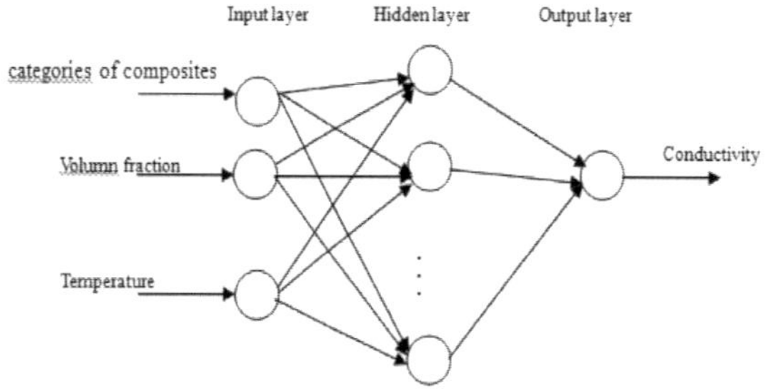

Fig.3 The thermal conductivity Model

Table 1 Sample number

No.	composites
1#	Fe_3O_4-MWNT/NR
2#	MWNT/NR
3#	Aligned Fe_3O_4-MWNTs/NR
4#	L-MWNT/NR

The input nodes are chosen linear in this paper. The hidden layer of the node is [8]:

$$s_j = \sum_{i=1}^{n} w_{ji}^{(1)} x_i - \theta_j^{(1)}; \qquad j = 1,2, \qquad (1)$$

In the formula (1), xi is the *i*th input of the model, s_j is the input of the *j*th hidden node, and $w_{ji}^{(1)}$is the weights connecting the *i*th input node and the *j*th hidden node, and the $\theta_j^{(1)}$ is the threshold of the *j*th hidden node. The output of the *j*th hidden node (O_j) is:

$$o_j = f^{(1)}(s_j); \qquad j = 1,2, \qquad (2)$$

In the formula (2), $f^{(1)}(\cdot)$ is the activation function of the hidden nodes, $f(x) = \frac{1}{1+e^{-x}}$ .The output of the model is:

$$y = \sum_{j=1}^{10} w_{kj}^{(2)} o_j - \theta^{(2)}; \qquad (3)$$

In the formula (3), $\theta^{(2)}$is the threshold of the output node, and $w_{kj}^{(2)}$ is the weights connecting the *j*th hidden node and the output node.

3.2 Levenberg-Marquard Algorithm

Levenberg-Marquard algorithm is a nonlinear optimization method, which combines the Gauss Newton method with gradient descent method. It is not sensitive to over parametric problem, and can effectively deal with nuisance parameter, so that the probability of the cost function falling into local minimum is reduced greatly. The above feature makes the LM algorithm widely used in the fields of model reconstruction, computer vision and so on. This paper used Levenberg-Marquardt algorithm to determine the parameters of thermal conductivity model. The weights of neural network are calculated as follows [9]:

$$w(k+1) = w(k) - [J_k^T J_k]^{-1} J_k^T E_k \tag{4}$$

$$E_k = \begin{bmatrix} e_1 \\ e_2 \\ \cdot \\ \cdot \\ \cdot \\ e_q \end{bmatrix} \tag{5}$$

In the formula (5), $e_i=d_i-y_i$, d_i is the expected output of the model, q is the number of experimental data sets, k is the discrete time index, N is the number of weights. J is Jacobi matrix, defined as follows [9]:

$$J = \begin{bmatrix} \frac{\partial e_1}{\partial w_1} & \frac{\partial e_1}{\partial w_2} & \cdots & \frac{\partial e_1}{\partial w_N} \\ \frac{\partial e_2}{\partial w_1} & \frac{\partial e_2}{\partial w_2} & \cdots & \frac{\partial e_2}{\partial w_N} \\ \vdots & \vdots & \vdots & \vdots \\ \frac{\partial e_q}{\partial w_1} & \frac{\partial e_q}{\partial w_2} & \cdots & \frac{\partial e_q}{\partial w_N} \end{bmatrix} \tag{6}$$

3.3 Results and discussion

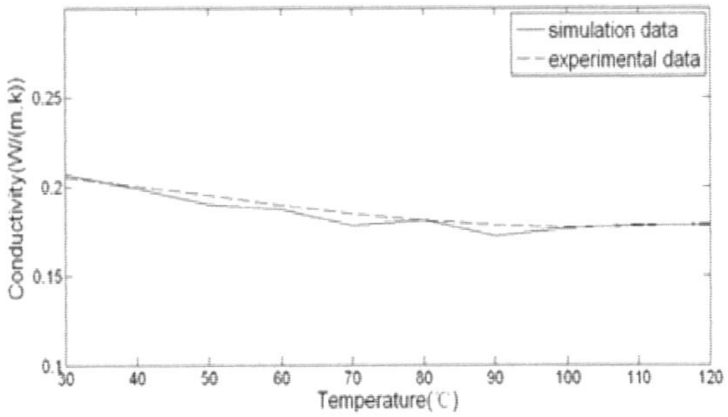

Fig.4 Thermal conductivity of L-MWNT/NR composite with 2% volume fraction

This paper used Levenberg-Marquard algorithm to construct the thermal conductivity model of MWNT/NR composites, simulated the heat transfer performance of composites. Shown in Fig.4, Fig.5, the simulation results are in agreement with the experimental results, the error is small and the maximum relative error is 3.7%. The simulation results show that simulating the thermal conductivity of composites with Levenberg-Marquard algorithm is effective and feasible.

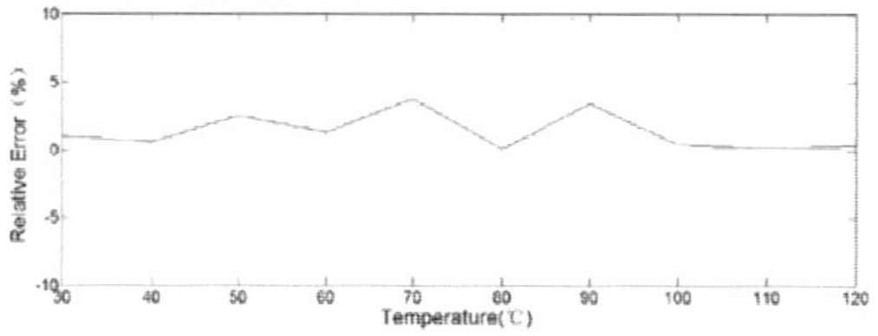

Fig.5 Relative error

4. Conclusion

Based on the analysis of the heat transfer performance of MWNT/NR composites, this paper used a new type of back-propagation algorithm to simulate the thermal performance of the composites, and investigated the influence of factors such as the ambient temperature, volume filling fraction, MWNTs-Orientation on heat conduction. The research showed:

1) MWNTs can effectively improve the thermal properties of the composites. 2) Fe_3O_4 particles have effectively improved thermal properties of the composites. 3) MWNTs orientation can effectively improve thermal properties of the composites. 4) The temperature had no significant effect on the thermal properties of the composites. 5) The increase of filler fraction will improve the thermal properties of composites, the thermal conductivity increased with the increase of volume filling fraction, the relationship between the thermal conductivity and the filling volume fraction is non-linear.6) The Levenberg-Marquard algorithm was used to simulate the thermal performance of composites, furthermore was used to research and analyse the influence of factors such as the ambient temperature, volume filling fraction, MWNTs-Orientation on heat conduction; the simulation results are accordant with the experimental results, that demonstrates the effectiveness of simulation algorithm and provides a new research approach for the study of thermal properties of rubber composites filled with MWNTs.

Acknowledgments

This work was supported by the National Natural Science Fundation of China under grant NO.51276091 and Collaborative Innovation Center of Green Tyres &Rubber NO.2014G7R0018.

References

[1] Moniruzzaman M and Winey K I 2006 Polymer nanocomposites containing carbon nanotubes,*Ma cromolecules*,**39, 16** 5194-5205.

[2] Shahil K M and Balandin A A 2012 Graphene-Multilayer Graphene Nanocomposites as Highly Efficient Thermal Interface Materials, *Nano Lett*, **12, 2** 861-867.

[3] Sabyasachi G, Ajit K Roya, David P Anderson, etal 2008 Improved thermal conductivity for chemically functionnalized exfoliated graphite /epoxy composites, *Carbon*, **46, 5**.806-817.

[4] Tas D is, Tagmatarch N is, Bianco A, etal 2006 Chemistry of carbonnanotubes,*Chem. Rev.*, **106**,31105-1136.

[5] He Y, Feng J J and Ma L X 2013 Effect of Align-Carbon Nanotubes on the Properties of Nature Rubber Composite Materials CJMR, **27, 5** 495-500.

[6] Anand K A, Sunil J T, Alex R and Joseph R 2010 Natural Rubber-Carbon Nanotube Composites through Latex Compounding, International Journal of Polymeric Materials, *Int. J. Polym. Mater.*, **59**, 1 33-34.

[7] Feng J J 2013 Experimental Study on The Thermal Conductivity of NR Composite Filled with Orientational MWNTS, Qingdao University of Science & Technology, Qingdao.

[8] Ge Z X and Shun Z Q 2007 Principles of Neural Network and MATLAB R2007 Realization, Electronics Industry, Beijing.

[9] Fredric M, Ham and Ivica K 2007 Principles of Neurocomputing for Science and Engineering,China Machine Press, Beijing.

Texture evolution of 01570 aluminum alloy sheet

X M CHEN[1,a], Y Y FAN[1,b], Q L PAN[2,c], C Y LIU[3,d]
[1]School of Electronic and Information &Mechanical Engineering, Zhaoqing University, Zhaoqing, 526061, China; [2]School of Materials Science and Engineering, Central South University, Changsha, 410083, China;[3]School of Mechanical Engineering,Guangdong applied science and technology University, Zhaoqing, 526061, China

E-mail:[a]cxm2050@163.com, [b]465040954@qq.com, [c]pql@mail.csu.edu.cn, [d]chaoying@zqu.edu.cn

Abstract: The texture configuration and evolution mechanism of 01570 aluminum alloy, under different stabilizing annealing systems, were studied by means of observations of optical microscopy and analysis of X-Ray. The results show that the major textures of 01570 alloy sheet, at cold-rolling condition, are of Brass texture{011}<211>, Copper texture {112}<111> and S texture{123}<634>, especially, the strength of Brass texture is strongest. Below the recrystallization temperature to make stabilizing annealing, the texture components change little compared with the cold-rolling sheet. When stabilizing annealing temperature reaches the recrystallization temperature, the deformation texture will disappear with the temperature rising gradually, and the recrystallization texture configuration will gradually increase; when the temperature reaches 590℃, the major textures are of the recrystallization texture of rotating cube texture and cube texture in the alloy.

1 Introdution

01570 alloy belongs to the series of containing scandium aluminum alloy, because it has high strength, good plasticity, excellent welding performance and strong corrosion resistance, so it has a broad application prospect in spaceflight, aviation and other fields[1]. 01570 aluminum alloy was first developed by the former Soviet union, and then there were quite a few countries to carry out research successively, our country began to research this alloy after entering the 21st century[2]. Now at home and abroad lots of works on microstructures and properties about this alloy have been done[3-5], but research on texture is rarely reported. The alloy to be used is mainly at the condition of stabilizing annealing after cold rolling, due to the deformation after hot and cold processing, so the texture will exist in the alloy. And the existence of the texture causes the performance difference on different directions, so the texture is used as an important reference index on engineering.

In the aluminum alloy after rolling it can form deformation texture, the deformation textures are mainly Brass texture {110}<112>, S texture {123}<634>, Copper texture {112}<111>, and they also may form Gauss texture {110} <001>and fiber texture of the surface{001} <110> [6-9]. After recrystallization annealing it mainly forms recrystallization cubic texture {001}<100> and R texture {124}<211> in aluminum alloy plate[10-11]. These textures under the different processing will exist in different configuration forms, and the evolution of each other will take place. With the development and the use of 01570 aluminum alloy, the studies on the existence form and the evolution rule under different process conditions will provide support for the design and production of the alloy. In this article the existence form and the evolution mechanism of the texture in 01570 aluminum alloy under different stabilizing annealing system will be studied.

2 Experimental

01570 experimental alloys were prepared by water chilling copper mould ingot metallurgy processing which was protected by active flux with pure Al, Mg and master alloys (Al−2.23%Sc, Al−4.48%Zr, Al−8.5%Mn), and Their nominal compositions(wt.%) are shown as follows: 5.8Mg, 0.4Mn, 0.25Sc, 0.1Zr, :Bal. Al. After homogenization at 460°C for 24h, the ingots are cut head and milled surface to 25mm thickness. Then a hot-rolling process is applied to 6mm after heat preservation for 3h at 470 °C, which is followed by intermediate annealing at 400 °C for 2h. Subsequently, the hot-rolled sheets are cold-rolled to a thickness of 2.0 mm. The total deformation rate is up to 92%. The cold-rolled sheets are annealed at 340 °C or 150°C for 1h. Homogenization; intermediate annealing and stabilizing annealing processing are made in SPC box-type resistance furnace, and the error is ±2℃.

2.1 Optical observation

The optical specimens were prepared firstly by mechanical polishing, then electrolytic polishing and anodizing, and at the end were examined using a POLYVER-Met optical microscope. The compositions of electro-polished solution were shown as follows: 10%$HClO_3$+90%C_2H_6O(vol.%). Electrolytic polishing voltage is 25V, about 30 second. The compositions of anodizing solution were shown as follows: 30mlHF+11gH_3BO_3+970mlH_2O, and anodizing voltage is 20V, about 3min.

2.2 Measurement of the plate texture

The samples of texture test were prepared firstly by mechanical polishing, and then electrolytic polishing to remove stress layer on the surface of the sample. Testing was done using a BRUKER D8 DISCOVER automatic X-ray diffractometer. And three incomplete pole figure of {220}, {200}, {111} were measured($0\leq\alpha\leq75°$, $0\leq\beta\leq360°$). The calculation of orientation distribution function need to the defocusing, basal correction and then Bunge series expansion, the results were showed with constant $\phi 2$ ODF section diagram. When measuring the rake was Cu Kα, measuring step length was 5 °, tube current was 40mA, tube voltage was 40kV.

3 Results

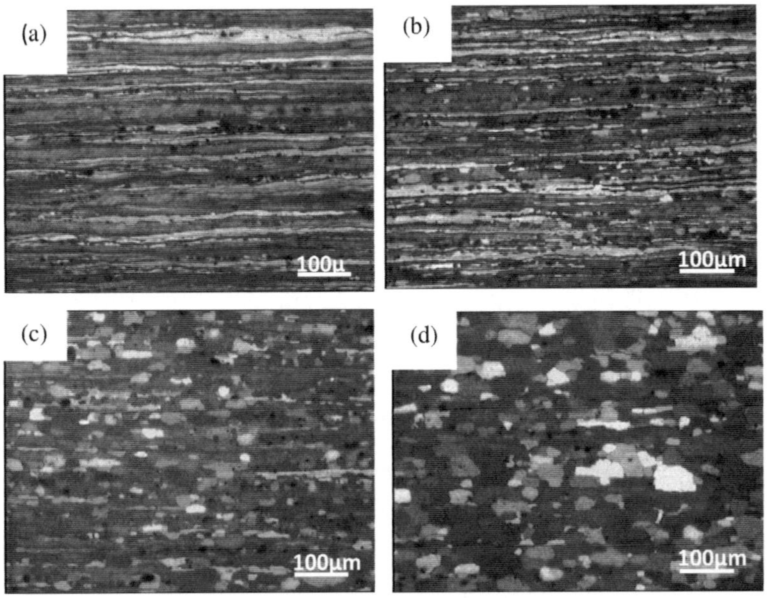

Fig.1 Optical microstructure of 01570 alloy in different annealing temperature

(a) 450℃/1h; (b) 550℃ 1h; (c) 570℃/1h; (d) 590℃/1h

3.1 Microstructure and texture of 01570 alloy

Fig.1 gives the optical microstructure of 01570 alloy in different annealing temperature. When annealing temperature is below 450℃, the microstructure keeps of the fibrous microstructure formed in the process of rolling deformation, no recrystallization phenomenon is found in the alloy(Fig.1a). When annealing temperature reaches 550℃, after annealing an hour, there are a small amount of recrystallization grains in the fibrous microstructure (Fig.b). Continuing to rise the annealing temperature to 570℃, the recrystallization phenomenon have become obvious in the alloy internal, and the fibrous microstructures become discontinuous(Fig.c). When the temperature is risen to 590℃ and the annealing time reaches an hour, the recrystallization process basically completes(Fig.d). The microstructure observations shows that when the annealing temperature is below 450℃, the deformation texture is maintained, and when the recrystallized grains appear in the alloy, the texture configurations come up with evolution, and when the temperature reaches 590℃, the deformation texture will be replaced by recrystallization texture.

3.2 ODF analysis of texture

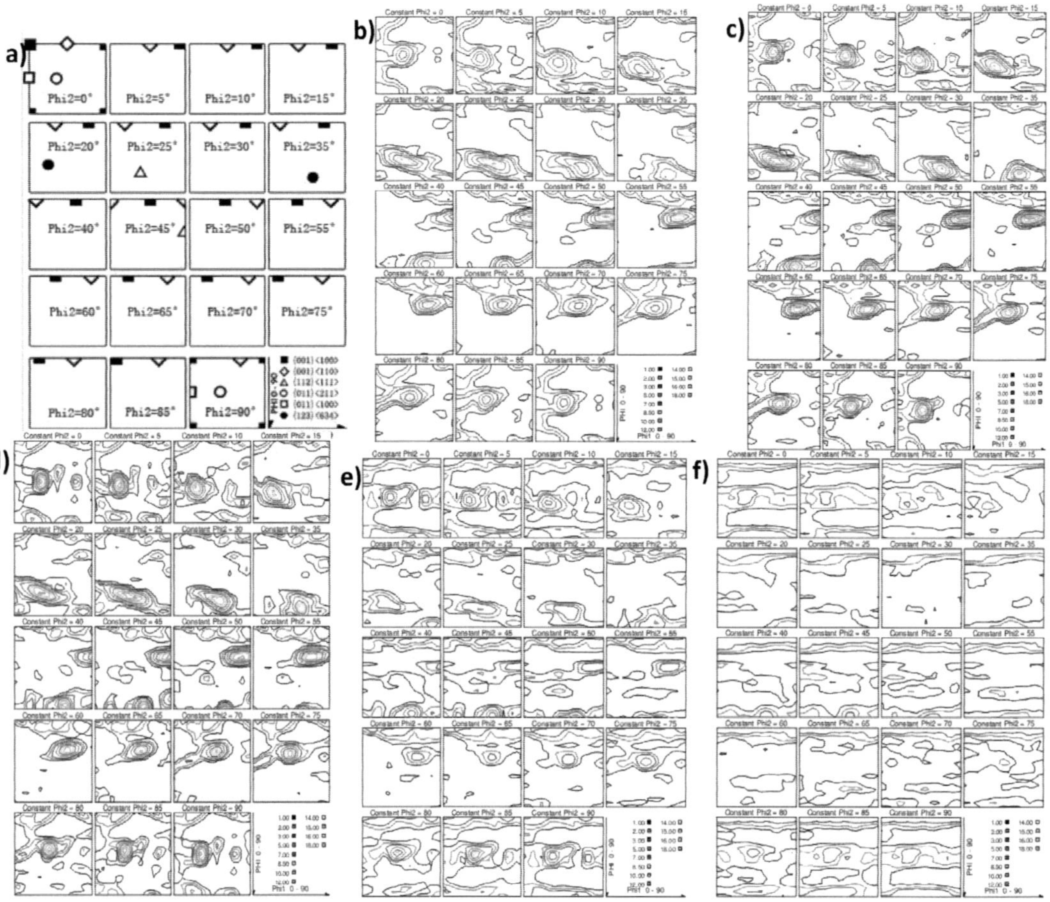

Fig.2 ODF of 01570 alloy sheet in different annealing temperature

a) the important orientation of cubic crystal; b) cold-rolling sheet

c) 450℃/1h; d)550℃/1h; e) 570℃1h; f) 590℃/1h

Figure 2 gives the orientation distribution of 01570 aluminum alloy under different annealing temperature(constant φ_2), and the distribution in Euler space of some common texture in cubic crystal is shown in Fig.2a. They are illustrate that the textures in 01570 aluminum alloy cold-rolling sheet are mainly composed of Brass texture{011}<211>, Copper texture {112}<111> and S texture{123}<634>, and the strength of Brass texture is significantly higher than the other two(Fig.2b). These sheets were annealed for an hour at 450℃ or 550℃, the texture components didn't change obviously, they were still mainly Brass texture, Copper texture and S texture, these results were shown in Fig2c,d. After 570℃/1h stabilizing annealing, because the partial recrystallization have taken place in the alloy, so the deformation texture strength decreases obviously, begins to slowly disappear, and a few cubic textures appear and are loose distribution in the plates, but their strength is low(Fig.2e). When annealing temperature rises to 590℃, after an hour annealing, the microstructure have changed into fully recrystallization microstructure, the deformation textures almost disappear, and the recrystallization texture of rotating cube texture and cubic texture become main texture, their strength increases than 570℃(Fig.2f).

3.3 Change of the texture strength with the annealing temperature

The changes of texture strength along the line of orientation in different annealing temperature are showed in figure 3. It can be seen that Brass texture{011}<211> distributes in the alpha orientation of face-centered cubic metal, and with the increase of annealing temperature. At the initial stage the strength of Brass texture changes not obviously, but in some stage drops significantly. Namely when annealing temperature is below 570℃, with the increase of annealing temperature, Brass texture orientation density decreases slightly. But when annealing temperature is higher than 570℃, such as annealing for an hour at 570℃ or 590℃, Brass texture orientation density decreases sharply. Fig.3b is illustrate that there are deformation textures of Brass texture{011}<211>, Copper texture {112}<111> and S texture{123}<634> distributing in the beta orientation of face-centered cubic metal, their orientations begin from Brass texture, via S texture and then transfer to Copper texture. In 01570 alloy cold-rolling sheet the orientation density of Brass texture{011}<211> is highest, and the orientation density of S texture{123}<634> and Copper texture {112}<111> are also higher. With the increasing of annealing temperature, the orientation densities of these three kinds of textures have a tendency to decrease. When annealing below 570℃, orientation density slightly reduced, but annealing above 570℃, the orientation densities of these three kinds of texture decrease sharply, and annealing at 570℃ the textures have

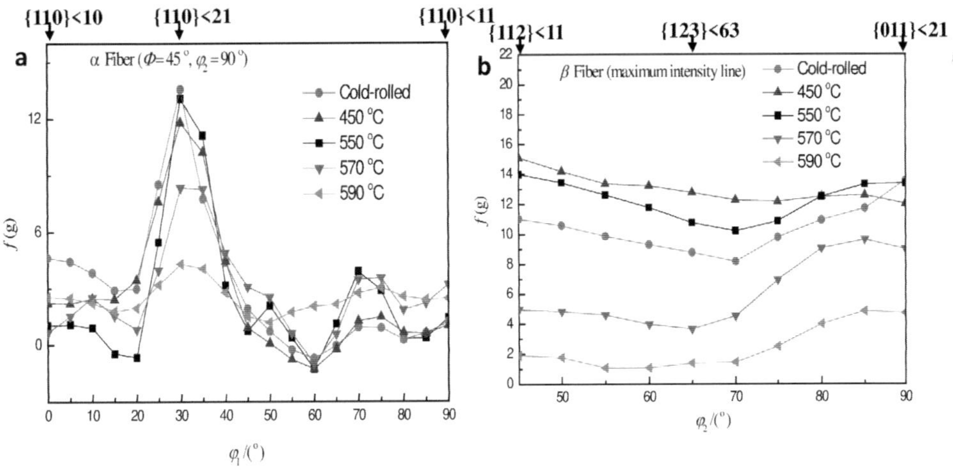

Fig.3 the orientation distribution of 01570 alloy sheet in annealing process
a) α orientation line; b) β orientation line

been almost disappear.

3.4 Texture changes of the volume fraction

Figure 4 gives the changes of texture configuration volume fraction with the change of annealing temperature. The figure shows that when the annealing temperature is below 400℃, the texture changes in volume fraction is very small, the texture configuration is composed of Brass texture{011}<211>, Copper texture {112}<111> and S texture{123}<634>, rotating cube texture almost does not exist, cubic texture is very few. When annealing temperature is over 450℃, the speed of change in volume fraction will increase. And when the annealing temperature is over 550℃, the speed will mutate, at this time the rotating cubic texture and cubic texture are in the leadership place. At 590℃, Brass texture, Copper texture and S texture almost disappear. These are mainly because when the temperature is above 450℃, the partial recrystallization phenomenon occurs in the alloy, and with the temperature rising the degree of recrystallization increases. As the temperature reaches to 590℃, the microstructure has been fully recrystallization microstructure, thus the deformation texture transfers into recrystallization texture.

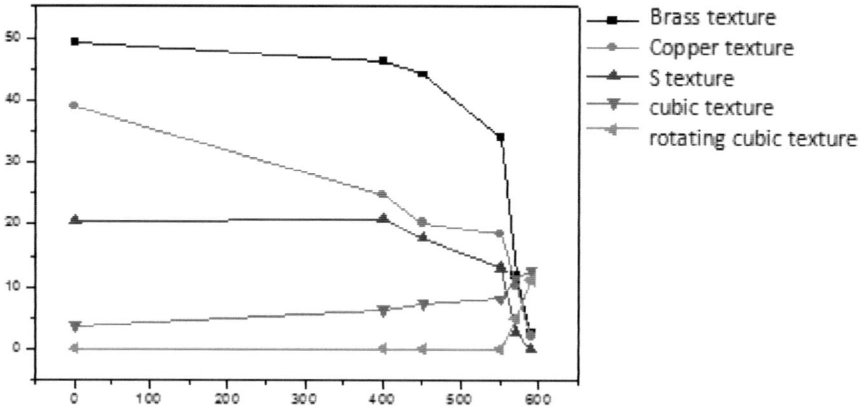

Fig.4 Texture changes of the volume fraction of 01570 alloy sheet in annealing process

3.5 Grain orientation difference distribution

The grain orientation difference distributions of 01570 aluminum in different annealing temperatures are showed in figure 5. As can be seen from the Fig.5a, the orientation difference in cold-rolling condition mainly concentrated in the vicinity of 2°, thus the most part in the grain boundary is less than 15° and belongs to small angle grain boundary, and the large angle grain boundary that the orientation difference is over 15° is account for only a little part of it. The reasons are that after cold rolling deformation, the grains bear the extrusion and appear serious breakage, the dislocations form cellular substructure by tangles. These substructures and the dislocation slip line producing by slip pile up together in the grain interior, thus make the alloy internal appear a large number of small angle grain boundaries. Fig.5b gives the orientation difference distribution of the alloy after 350℃/ 1 h annealing. It shows that the distribution density of small angle grain boundary is in reducing, but the large angle grain boundary is on the rise. This is mainly because there is reply phenomenon appearing in the alloy when annealing at 350℃, under the effect of thermal the dislocation will move, and the substructures also can migrate and merge, so can form more large angle grain boundaries. When the annealing temperature rises to 550℃, the partial recrystallization appears in the alloy, and some isometrics are found in the fibrous microstructure. At this time the small angle grain boundary content is further reduced, and the large angle grain boundary content increases significantly(Fig.5c). Continuing to increase annealing temperature to 570℃, the content of small angle grain boundary is already very low, and the large angle boundary takes leadership in the alloy. The boundary orientation differences are mainly concentrated in 25°~60° and take normal distribution, and the value of large angle boundary which the orientation difference is about 45° is most. These are mainly because with the increase of annealing temperature the degree of recrystallization increases, the isometric number is on the rise correspondingly, and large angle grain boundary density also increases. By above observation the results already show that when the annealing temperature reaches to 590℃ the alloy will be fully recrystallization microstructure and almost large angle grain boundaries,

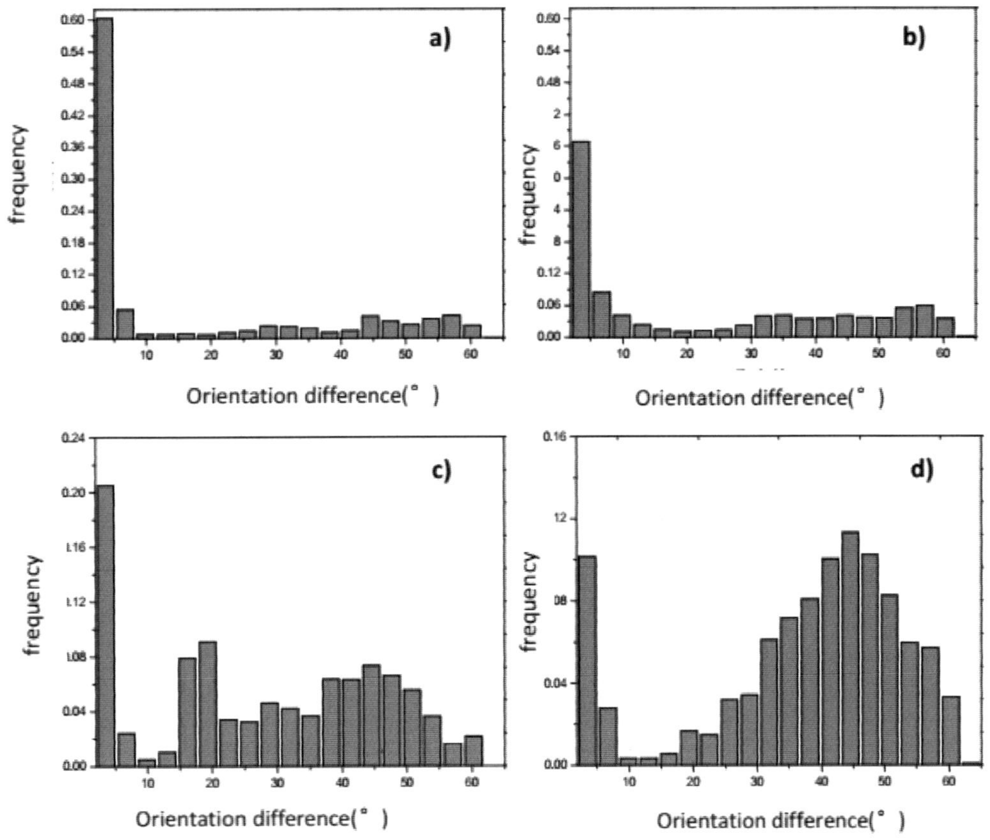

Fig.5 Interface orientation difference distribution of 01570 alloy sheet in different annealing temperature

a) cold-rolling condition; b) 350℃/ 1 h; c) 550℃/ 1 h; d) 570℃/ 1 h

4 Discuss

4.1 Texture forming mechanism of 01570 cold-rolling sheet

Brass texture{011}<211>, Copper texture {112}<111>, and so on, which belong to deformation textures, have appeared in 01570 cold-rolling sheets. These textures often appear in the metal of face-centered cubic crystal structure. Figure 6 is the orientation triangle of polar stereographic projection in face-centered cubic metal. When the face-centered cubic metals are rolled, their primary slip system is, and the secondary slip system is. Under the action of tensile stress, the crystals take rotation, and the action line rotates to the direction of . When it passes through the attachment line of [001]-[111], the primary slip system is replaced by the second slip system, and the tensile stress axis direction will turn to the new slip direction of [011], when it crosses the attachment line of [001]-[111], the primary slip system is reactivated. This phenomenon repeats for many times, at the end the action line direction falls on the attachment line of [001]-[111], and on the [112] direction which is symmetric with [101] and [011]. Point C in figure 6 is the pressure axis direction, at the beginning it is in the orthogonal state with Under the action of compression force along the normal direction on the rolling surface, point C will also rotate to direction. When point T crosses the attachment line of [001]-[111] the original slip system is replaced by the new slip system. And point C will turn to again, after turning back and forth, point C falls on direction. So in 01570 alloy can form Brass texture{ 011 }< 211 > which has high strength. In addition to the above sliding systems, there will be other sliding systems involved in sliding during the process of rolling. So in 01570 alloy can also form Copper texture{ 112 }< 111 >, Gauss texture, S texture and the other.

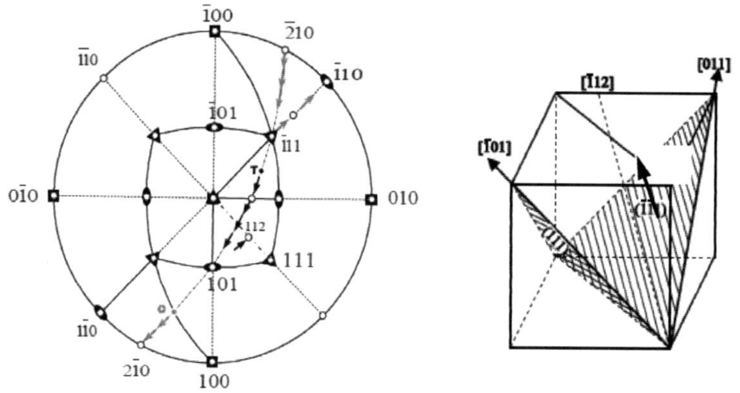

Fig.6 Face-centered cubic crystal orientation change in the rolling process

4.2 Texture evolution of 01570 aluminum during the process of stabilizing annealing

The texture configurations of 01570 aluminum alloy will vary with the change of stabilizing annealing. When the annealing temperature is below 550℃, the areas of Brass texture, Copper texture and S texture decrease as the temperature increases and discontinuously distribute along the rolling direction. The type of copper texture and R texture make their occupying areas widen by expanding them into the areas occupying by Brass texture and Copper texture. So it can be judged that the formation of R texture transforms from S texture and Copper texture. When annealing temperature reaches to 570℃, the recrystallization has taken place in the alloys, Copper texture and the type of copper texture gradually reduce, the cubic texture obviously increases, but there is a small amount of Brass texture preserved. In addition, there exist R texture and the grains which orientation close to {110}<111>. Work [12] had shown that the orientation formation of {110}<111> deflected from Brass texture and Copper texture, it is the transition texture from deformation texture transforming into cubic texture, and is particularly unstable texture type. Under high temperature annealing, it can transform into the stable cubic texture via rotating around <110> and <111> axes. When annealing temperature reaches to 590℃, the alloy has been fully recrystallization microstructure, this time the deformation texture will be replaced by recrystallization texture.

Figure 7 is the space schematic diagram which shows Brass texture how to transform into annealing texture in the annealing process in 01570 alloy sheet. With the annealing temperature rising, the grains with the orientation of (011) [2$\bar{1}$1] will do 30° rotation around <111> axis[11], and make {011}<2$\bar{1}$1> orientation gather in {112}<1$\bar{1}$0> orientation area. The non-ideal orientation grains which clustering near the brass texture orientation do 30°~40° rotation and transform into {011}<111> unstable texture. {112}<1$\bar{1}$0> and {011}<111> belong to unstable orientation, under the high temperature annealing, the grains with the two orientations will rotate around <100> axis and transform into stable cubic texture orientation[13], thus make the transformation process from deformation texture into recrystallization texture complete.

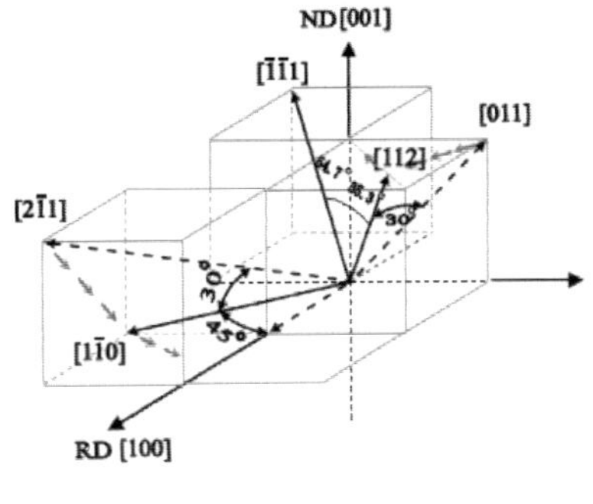

Fig.7 Texture evolution during the process of annealing

Through the rotation of grains and the migration of grain boundaries, Copper texture and S texture will gradually form the transition orientation zone which are composed of R texture orientation and the type of copper texture orientation[14]. R texture orientation during the subsequent annealing process will gradually turn to cubic texture orientation. Al_3(Sc,Zr) particles which have the role that can strongly hinder the grain boundary migration impede the process that makes R texture orientation transform into recrystallization texture orientation, so that R texture orientations preserves a lot of in 01570 alloy. When continuing to improve the annealing temperature, it makes the role of Al_3(Sc,Zr) particles to hinder the grain boundary migration reduce, thus R texture orientation is able to transform into the stable cubic texture orientation. So when the annealing temperature reaches to 550℃ the density of deformation texture in 01570 alloy still keeps in high level. But when the annealing temperature reaches to 570℃, the recrystallization has taken place obviously in the alloy, thus lead to the density of deformation texture sharply reduce and the strength of recrystallization texture begin to increase significantly.

5 Conclusions

(1) The textures in cold-rolling sheet of 01570 aluminum alloy are mainly composed of Brass texture{011}<211> , Copper texture{112}<111> and S texture{123}<634>, and the strength of Brass texture is highest.

(2) 01570 aluminum alloy is annealed for an hour below 550℃, the texture components have no obvious changes, still are mainly Brass texture, Copper texture and S texture.

(3) When annealing temperature reaches to 570℃, partial recrystallization microstructure appears in the alloy sheet, and deformation texture intensity drops, some of deformation textures transform into recrystallization texture. When annealing temperature reaches to 590℃, there are full recrystallization microstructure existing in the alloy, the deformation texture almost disappearance, and the recrystallization textures, such as rotating cube texture and cubic texture, become main textures.

Acknowledements

This project was supported by the post doctoral foundation of Central South University(126226), the national key basic research development plan(2012CB619503), the Guangdong Province science and technology plan（2012B011000038） and the Zhaoqing city science and technology plan plan(2013C011)。

References:

[1] Yin Z M, Zhu D P and Jiang F 2004 Recrystallization o f Al-Mg-Mn and Al-Mg-Mn-Sc-Zr Alloys[J]. *J. Mater. Eng.*, 6:3-6.

[2] Zhu D P, Yin Z M, Teng H, etal 2004 Effect of trace Sc and Zr on microstructure and properties of Al-Mg-Mn alloy[J]. *Aero. Mater. Techno.*, 6:45-49.

[3] Wang Y, Pan Q L, Song Y F, etal 2013 Recrystallization of Al-5.8Mg-Mn-Sc-Zr alloy[J]. *Trans. Nonferrous Met. Soc.*, 11:3235-3241.

[4] Peng Y Y, Yin Z M, Nie B, etal 2007 Effect of minor Sc and Zr on superplasticity of Al-Mg-Mn alloys[J]. *Trans. Nonferrous Met. Soc.* 4:744-750.

[5] Chen Q, Pan Q L, Wang Y, etal 2012 Microstructure and mechanical properties of Al-5.8Mg-Mn-Sc-Zr alloy after annealing treatment[J]. *J. Cent. South Univ.,* 7: 1785-1790.

[6] Wang G J 2004 Deviation of Random Orientation Distribution of Aluminium and Aluminium Alloy Sheets Strips[J]. *Light Alloy Fabric. Techno.*, 6:28-33.

[7] Zhang X M, Han N M, Liu S D, etal 2010 Inhomogeneity of texture, tensile property and fracture toughness of 7050 aluminum alloy thick plate[J]. *Chin. J. Nonferrous Met.*, 2: 202-208.

[8] Zhang Y H, Yao Z Y, Huang G J, etal 2009 EBSD investigation on microstructure and texture in rolling aluminium alloys[J]. *J. Chin. Elect. Micro. Soc.,* 1: 43-45.

[9] Merriman C C, Field D P and Trivedi P 2008 Orientation dependence of dislocation structure evolution during cold rolling of aluminum[J]. *Mater. Sci. Eng. A*, 1:28-35.

[10] Chen Y, Tian N, Zhao G, etal 2006 Effect of pre2heat treatments on cold rolling and recrystallization textures in Al alloy 6111[J]. *Chin. J. Nonferrous Met.*, 8: 1411-1415.

[11] Engler O and Hirsch J 1996 Recrystallization textures and plastic anisotropy in Al2Mg2Si sheet alloys[J]. *Mater. Sci. Forum*, 217-222: 479-486.

[12] Yao Z Y, Liu Q, Godfrey A, etal 2009 Microstructure and texture evolutions of AA1050 aluminum alloy cold rolled to high strains[J]. *ACTA Metal. Sinica*, 6: 647-651.

[13] Hua F A, Di H S, LI J P, etal 2009 Simulation of effects of crystallographic textures on the macroscopic anisotropy of metal sheet[J]. *ACTA Metal. Sinica*, 6: 657-662.

[14] Chen Z Y, Cai H N and Chang Y Z 2008 Texture evolution of polycrystalline aluminum during rolling deformation[J]. *ACTA Metal. Sinica*, 11:1316-1321.

Effects of loading methods on microstructure of diffusion welded joint of AZ31B/Cu with Ni interlayer

S M Du[1,a], Y Q Zhang[1,b], C Du[1,c], J Hu[1,d]
[1]College of Materials Science and Engineering, Xi'an University of Science and Technology, Xi'an 710054, China)

E-mail:[a]shuangmingdu@163.com, [b]18710980374@163.com, [c]980513781@qq.com, dclhujie2009@163.com

Abstract: Diffusion brazing was carried out to weld AZ31B magnesium alloy and copper with Ni foil interlayer under different loading methods that are divided into intermittent gradient pressure, gradient pressure and constant pressure. The microstructure and element diffusion of welded joint were analyzed by SEM and EDS. The results show that the AZ31B/Cu can achieve good bonded joint composed of brazing seam zone and magnesium substrate grain boundary penetration zone at 500 ℃ for 20 min. The loading methods have great influence on the thickness of brazing seam zone, microstructure and α-Mg grain size of the welded joint. At intermittent gradient pressure, the brazing seam zone reaches the biggest width of 0.18 mm, and the microstructure of brazing seam zone is composed of $Cu_{11}Mg_{10}Ni_9$, (α-Mg+Mg_2Cu+Mg_2Ni)eutectic structure and α-Mg. Under gradient pressure, the width of brazing seam zone reduces, and the microstructure of brazing seam zone is mainly dominated by (α-Mg+Mg_2Cu+Mg_2Ni)eutectic structure and $Cu_{11}Mg_{10}Ni_9$ compounds closed magnesium matrix. When using constant pressure, the microstructure of joint is similar to that under intermittent gradient pressure, the brazing seam zone reaches the smallest width of 0.11 mm, and the size of α-Mg grains reduces.

1 Introduction

Diffusion brazing is a kind of connection method that heats the heterogeneous metals to a temperature below the melting point and then hold for a period of time, low melting liquid phase forms by mutual diffusion between heterogeneous metals, liquid phase dissolves and diffuses with both sides of parent metal, and then a connection joint forms after cooling solidification. This process has the characteristics of brazing and diffusion welding, which filler is not needed and the deformation of workpiece is small after welding, so it is suitable for the connection of dissimilar materials with poor welding ability or large melting point difference[1,2]. At present, the hot topics in study of diffusion brazing mainly focus on the optimization of process parameters which include heating temperature, holding time, the value of the pressure, and the middle layer[3-5], but research about the effects of pressure methods on joint microstructure and properties is rarely reported. According to the theory of diffusion welding, at constant pressure, the smaller welding pressure is difficult to ensure physical contact between parent metal so that atomic diffusion is influenced, however, larger welding pressure makes the formation of low melting liquid extrusion effect, which leads the welding workpiece serious thinned[6]. Therefore, it is necessary to study the effects of pressure methods on dissimilar metal diffusion brazing behaviors.

Magnesium alloy is the lightest engineering material with the characteristics of high specific strength and stiffness, good thermal conductivity and electrical conductivity, strong vibration damping and electromagnetic shielding performance, easy processing and forming performance, and easy recycling performance of waste, etc, which makes it widely used for aerospace, automobile manufacturing, 3C products and other fields[7]. Copper that has the performance of good conductivity, thermal conductivity, corrosion resistant and wear-resisting is mainly used as the device with conductivity, thermal conductivity and corrosion resistance. Mg-Cu dissimilar metal composite component can fully play the performance advantage of two kinds of parents, so it has a broad application in chemical industry, electronics, Marine, military and other fields[8]. Adopting the traditional welding can't obtain reliable Mg/Cu welded joint because of their poor welding ability and large thermal physical property differences[9].

It is easy to form $Mg_{17}(Cu/Al)_{12}$ and Cu_2Mg intermetallic compounds that affect the joint properties if Mg/Cu are welded directly[10], in order to improve this situation, appropriate interlayer is used for welding materials to form a gradient bonded joint, which reduces the thermal stress and hard brittle phase of joint[11].It can be seen from phase diagram that copper and nickel can form continuous solid solution, Mg-Ni-Cu can form ternary eutectic reaction at 480 ℃, which is beneficial to improve the performance of the joint. Therefore, this paper choose Ni foil as the interlayer to weld AZ31B magnesium alloy and Cu under different pressure condition, and then analyze the effects of loading methods on the microstructure and element diffusion of welded joint.

2 Experimental materials and methods

The size of experimental materials for AZ31B magnesium alloy and industrial pure copper T2 is 10 mm×10 mm×2 mm and Ni foil is 10 mm×10 mm×0.02 mm, the chemical composition of parent metal is shown in Tables 1 and 2.

Table 1 Chemical compositions of AZ31B Mg alloy(wt, %)

Al	Zn	Mn	Si	Mg
2.5-3.5	0.5-1.5	0.2-0.5	≤0.1	Bal

Table 2 Chemical compositions of T2 Pure Copper(wt, %)

Fe	Ti	Ni	Cu
0.005	0.006	0.005	Bal

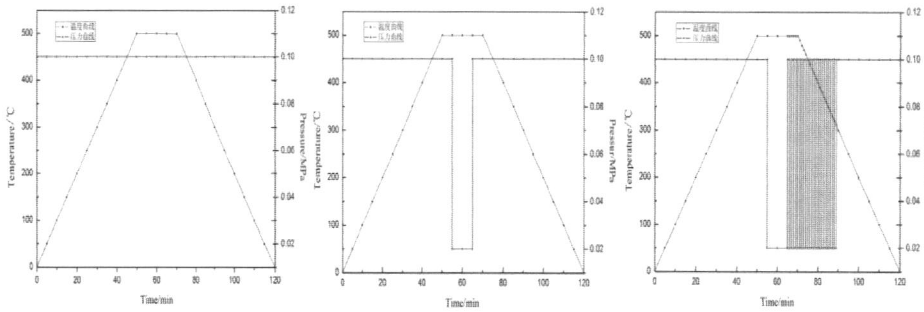

(a) (b) (c)

Fig.1 Technological parameters curve of diffusion brazing under different loading methods
(a) Constant pressure (b) Gradient pressure (c) Intermittent gradient pressure

Before welding, the intended bonding surface of specimens were ground using SiC papers until the contact surface was smooth, and then put them in acetone solution to clean 3-5 min by ultrasonic cleaner and dry, at last, speciments were then assembled in the vacuum furnace. Adopting the bonding parameters of temperature is 500 ℃ and the bonding time is 20min. The technological parameters curve of diffusion brazing under different loading methods that are divided into constant pressure, gradient pressure, intermittent gradient pressure are shown in Figure.1, the gradient pressure means the pressure reduces to 0.02 MPa in the middle of bonding time for 10 min and in the rest of the stage is 0.1 MPa; intermittent gradient pressure means the pressure is 0.02 MPa in the middle of the bonding time for 10min, changes with a frequency of 0.1 MPa to 0.02 MPa per minute during the last 5 minutes bonding time and the stage of cooling phase of 500-300 ℃, and in the rest of the stage is 0.1 MPa. After welding, the joint samples were grounded and polished and then the microstructure of welded joint was analyzed by modern analysis methods.

3 Results and discussions

3.1 The microstructure of welded joint under intermittent gradient pressure

Fig.2(a) and Fig.2(b) show the microstucture and the elemental line scan of diffusion brazing joint under intermittent gradient pressure. Fig.2(c) and Fig.2(d) are respectively the enlarged drawings of the diffusion region closed to copper substrate(A area) and magnesium matrix diffusion region(B area) in Fig.2(a).

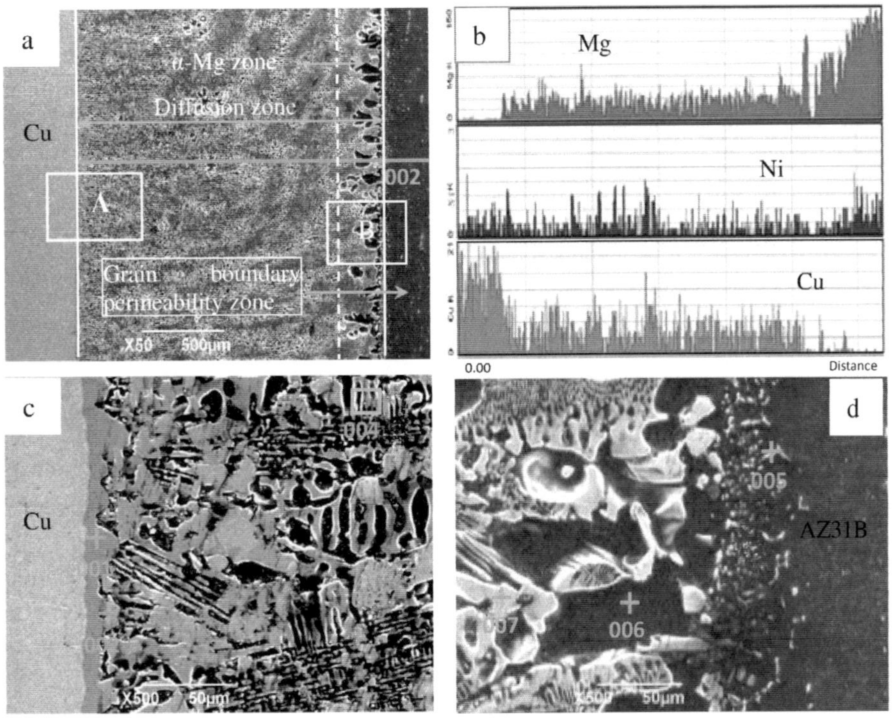

Figure.2 Microstructure and element line scan of AZ31B/Cu under intermittent gradient pressure
(a) Overall microstructure of the joint interface
(b) element line scan (c) enlarged view of the interface near copper side
(d) enlarged view of the interface near magnesium side

Fig.2(a) shows that the interlayer of nickel foil has already dissolved and the bonded joint achieves good metallurgical combination. The joint is mainly dominated by the uniform diffusion brazing seam of 1.8 mm width and magnesium based grain boundary permeable zone. The microstucture of brazing seam is dense, and found no pores, cracks, welding defects. The brazing seam is mainly composed of the copper substrate diffusion zone, eutectic zone and the magnesium matrix diffusion zone. The large black columnar crystals and dendrites growing to brazing seam zone appear at the interface between brazing seam zone and magnesium matrix permeable zone, the formation of this grain shape is due to the part of solid/liquid interface growth protruding to the front of the eutectic liquid region, a greater undercooling increases the growth rate of the protruding part and then makes it further extend to the depths of the liquid, in this condition, the solid/liquid interface is not likely to remain planar but forms many branches extending to the liquid, and these branches eventually form columnar crystals and dendrites. The decrease of the pressure in the holding stage makes more liquid fill weld zone, more atoms are activated and the diffusion reaction is more intense in the cooling stage using intermittent gradient pressure, so the width of the brazing seam is larger. It can be seen from Fig.2(c) that a small amount of lumped compounds are mainly distributed near the copper side. From Fig.2(d), it can be seen that the lamellar eutectic structure is the main structure near the magnesium substrate. The aggregation of a small amount of compounds at the copper side is mainly because the shock effect by the pressure in the cooling stages is more significant to the magnesium side atoms. In order to determine the phase of each layer, EDS is used to analyze the feature points, the results are shown in Table 3.

Table 3 The test result in different regions of the elemental composition near the weld zone(At %)

Measured point	Mg	Ni	Cu
001	3.12	1.18	95.7
002	31.46	0.93	67.61
003	33.68	29.7	36.62
004	79.43	1.32	19.25
005	96.72	1.74	1.54
006	98.24	0.23	1.53
007	50.56	2.14	47.3

It can be seen from Table 3, the gray transition layer near the Cu matrix (Fig. 2(c) 001 point) is γ-Cu. The atomic ratio of Cu/Mg of the block structure(002 point) distributed at the gray transition layer is closed to 2:1, it can be deduced as Cu_2Mg. The block structure (003 point) at central area has a chemical composition of 52.68% magnesium, 8.7% nickel and 38.62% copper, so the compound is considered as a ternary compound of $Cu_{11}Mg_{10}Ni_9$ based on shape appearance figure. According to the atomic ratio, the lamellar eutectic structure (004 point) in the central region of the joint interface is (α-Mg+Mg_2Cu+Mg_2Ni) eutectic structure. Fig.2(d) shows the diffusion layer structure closed to magnesium substrate, the color of black columnar crystals and dendrites (006 point) is similar to magnesium matrix, according to the atomic content, the layer is considered as magnesium based solid solution layer (α-Mg zone).The bright white block compound (007 point) at the edge of α-Mg grain region has a chemical composition of 50.56% Mg, 2.14%Ni and 47.3%Cu, the atomic ratio of Mg/Cu is closed to 1:1, which is considered as a mixture of Mg_2Cu and Cu_2Mg. The white grain boundary precipitation stripe (005 point) near the magnesium matrix is controlled by the interdiffusion among the elements. Under the condition of heating and loading, the atoms are activated in turn, due to the high initial concentration gradient between heterogeneous atoms, then atoms diffuse mutually, grain boundary has more defects and lower energy, so it is easy to precipitate α-Mg. With the increase of holding time, magnesium, copper and nickel will produce ternary eutectic reaction when reaching the appropriate condition.

From above analysis, it can be concluded that the microstructure of bonded joint under intermittent gradient pressure are γ-Cu, Cu_2Mg, $Cu_{11}Mg_{10}Ni_9$, (Mg_2Ni+α-Mg+$CuMg_2$) eutectic structure and α-Mg.

3.2 The microstructure of welded joint under gradient pressure

The microstucture and EDS line scan of welded joint under gradient pressure are shown in Fig.3, Similar to bonded joint under intermittent pressure gradient, the interface diffusion zone of bonded joint under gradient pressure is composed of diffusion brazing seam zone and magnesium substrate grain boundary permeability zone. Differently, the brazing seam width is reduced to 1.3 mm and no obvious α-Mg zone appears. The EDS line scan shows the middle reaction layer elements diffuse relatively uniform, so the main microstructure are eutectic structure and a little block compounds, the position of the block compounds has been changed, which is mainly distributed in the diffusion reaction layer near the magnesium matrix. Magnesium, nickel and copper elements appear peaks at the diffusion reaction layer near the magnesium matrix, so the existence of Mg-Ni-Cu ternary compounds is further confirmed.

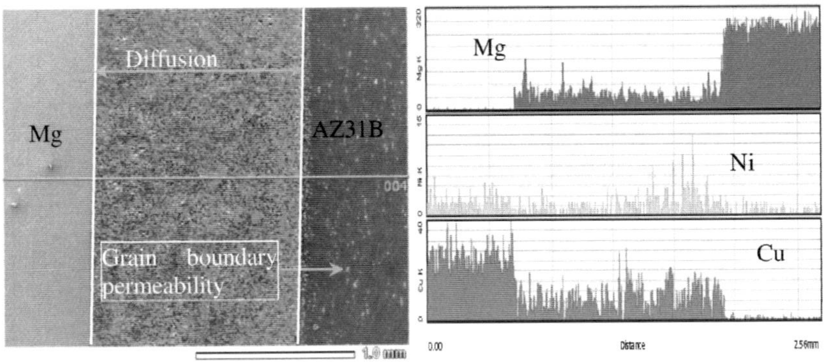

Fig.3 Microstructure and element line scan of AZ31B/Cu with Ni interlayer under gradient pressure

3.3 The microstructure of welded joint under constant pressure

The microstucture and EDS line scan of welded joint under constant pressure are shown in Fig.4.

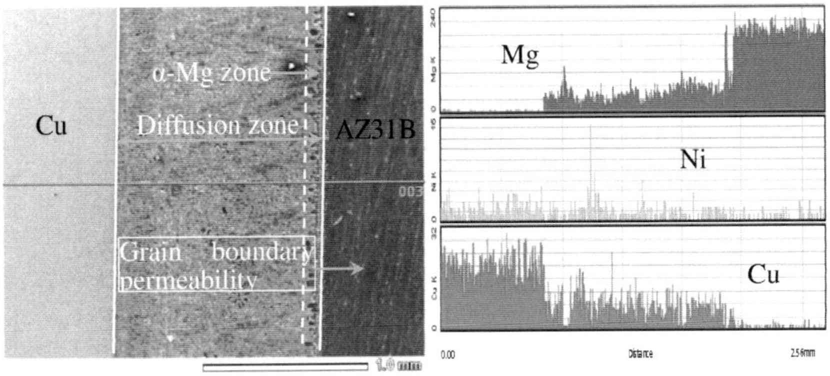

Fig.4 Microstructure and element line scan of AZ31B/Cu with Ni interlayer under constant pressure

Compared with Fig.2 and Fig.3, the microstructure of welded joint is extremely similar to that under intermittent gradient pressure, but the brazing seam zone width significantly reduces to 1.1 mm. The grain size of α-Mg presenting columnar crystals and cellular crystals decreases; At constant pressure, constant high pressure makes a large amount of eutectic liquid phase extrude, the atomic diffusion rate in liquid is greater than that in solid, the reduction of eutectic liquid phase will slow atomic diffusion velocity, which leads to the decrease of diffusion layer width. A small amount of liquid phase makes the temperature gradient on the front edge of the solid/liquid interface more uniform, compared with the vast liquid phase under intermittent gradient pressure, the cooling rate decreases and a limited amount of liquid phase will also prevent the growth of crystal nucleus, therefore, all these factors have contributed to the decrease of α-Mg size.

4 Conclusions

(1)AZ31B/Cu with Ni interlayer under different pressure methods can achieve metallurgical bonding joints composed of brazing seam zone crystals and magnesium substrate grain boundary penetration zone, and the loading methods have great influence on the thickness of brazing seam zone, microstructure and α-Mg grain size of the welded joint.

(2)At intermittent gradient pressure, the width of brazing seam zone is 0.18 mm, and the microstructure of brazing seam zone is composed of $Cu_{11}Mg_{10}Ni_9$, (α-Mg+Mg_2Cu+Mg_2Ni)eutectic structure and α-Mg. Under gradient pressure, the width of brazing seam zone reducesto 0.13 mm, and the microstructure of brazing seam zone is mainly dominated by (α-Mg+Mg_2Cu+Mg_2Ni) eutectic structure and $Cu_{11}Mg_{10}Ni_9$ compounds closed magnesium matrix. When using constant pressure, the microstructure of bonded joint is similar with that under intermittent gradient pressure, but the brazing seam zone width is minimal, which is only 0.11 mm, and the size of α-Mg reduces.

Reference

[1] Liu Z G, Guo J and Yang J M 2004 The Application of Pervasion Weld Technology[J]. *Aeronautical Science and Technology,* **(02)** 38-40.

[2] Du S M, Zhao W T, Liu G and Guo W 2012 Microstructure and Properties of Interfacial Region of AZ31 Alloy by Diffusion Brazing[J]. *Casting Forging Welding,* **(01)** 140-144.

[3] Liu G 2013 Microstructure and element diffusion in welding interface of AZ31B/Cu dissimilar materials[D]. Xi'an, Xi'an university of science and technology.

[4] Dang P P 2014 Study on the diffusion brazing of AZ31B magnesium alloy and copper of property and microstructure of the joint[D]. Xi'an, Xi'an university of science and technology.

[5] Lan T 2013 Study on the microstructure and properties of Mg-Cu dissimilar metals[D]. Xi'an, Xi'an university of science and technology.

[6] Wang F S, Gao Z and Liu S G 2008 Model of Muti-Pressure Craft Model based on Transient Liquid-Phase Bonding[J]. *Journal of Lanzhou petrochemical College of Technology,* **8 (4)** 25-27.

[7] Li W X 2007 Introduction to non-ferrous metal material engineering[M]. Beijing: Metallurgical industry press.

[8] MAHENDRAN G, BALASUBRAMANIAN V and SENTHILVELAN T 2009 Developing diffusion bonding windows for joining AZ31B magnesium and copper alloys[J]. *The International Journal of Advanced Manufacturing Technology,* **42 (7/8)** 689-695.

[9] Li C X, Yang Y and Zhao X 2016 Interfacial Microstructure and Mechanical Properties of Magnesium Alloy/Cu Ultrasonic Spot Welded Joints[J]. *Hot Working Technology,* **45 (17)** 192-196.

[10] Du S M, Liu G and Wang M J 2013 Microstructure and properties of transient liquid-phase diffusion bonded joint of AZ31B/Cu dissimilar metal[J]. *The Chinese Journal of Nonferrous Metals,* **23 (5)** 1255-1260.

[11] Fen Z, Ma T F and Cao R 2015 Research progress of Mg-Cu dissimilar metals welding technology[J]. *Welding & Joining,* **(08)** 8-12.

Adsorption and Exchange Kinetics of Hydrophilic and Hydrophobic Phosphorus Ligands on Gold Surface

X Q Zhuge[1,a], Z C Bian[1,b], Z H Luo[1,c], Y Y Mu[1,d] and K Luo[1,e]

[1] *Guangxi Key Laboratory of Universities for Clean Metallurgy Comprehensive Utilization of Nonferrous Metal Resource,* Collaborative Innovation Center for Exploration of Hidden Nonferrous Metal Deposits and Development of New Materials in Guangxi, College of Materials Science and Engineering, Guilin University of Technology, Guilin, 541004, PR China;

[2] College of Chemistry and Chemical Engineering, Huazhong University of Science and Technology, Wuhan 430075, PR China

E-mail:[a]843343874@qq.com ,[b]765952245@qq.com ,[c]luozhihong615@163.com ,[d]1083354578@qq.com ,[e]luokun998@gmail.com

Abstract. The adsorption kinetics process of hydrophobic ligand (triphenylphosphine, PPh_3) and hydrophilic ligand (tris(hydroxymethyl)phosphine oxide, THPO) on the surface of gold electrode were estimated by using electrical double layer capacitance (EDLC). Results showed that the adsorption process of both ligands included fast and slow adsorption processes, and the fast adsorption process could fit the first order kinetic equation of Langmuir adsorption isotherm. During the slow adsorption process, the surface coverage (θ) of PPh_3 was higher than that of THPO due to the larger adsorption kinetic constant of PPh_3 than that of THPO, which implied that PPh_3 could replace THPO on the gold electrode. The exchange process of both ligands on the surface of gold electrode proved that PPh_3 take the place of THPO by testing the variation of EDLC which promote the preparation of Janus gold, and the theoretic simulation explained the reason of ligands exchange from the respect of energy..

1. Introduction

Surface modified gold is the most frequently used in the fields of sensors [1-4], corrosion protection [5-7], molecular electronic devices [8-9], and other nanoscale devices. During which, using self-assembling monolayers (SAMs) to modified gold substrates was a powerful method with high orderliness, orientation, thermodynamic stability and low defect. Thiol modified gold is widely applied in fundamental and practical issue due to the strong interaction between gold and sulfer[10-13].The adsorption kinetics of organic molecules on the surface of gold electrode was investigated by quartz crystal microbalance (QCM)[14-16], atomic force microscopy(AMF)[17], electrochemical methods[18,19] and surface plasmon resonance (SPR). The results showed that adsorption kinetics had two process, the first process was a fast adsorption finished in a short time, the second was a slow adsorption or reorganizing process needed 10 h to 20 h[20]. Compared to thiol, phosphorus ligands can also adsorb onto gold surfaces due to their electrostatic interactions. However, few reports investigated the adsorption process of phosphorus ligands.

For the preparation of modified nanoparticles, ligand exchange reaction is an approach to produce amphiphilic nanoparticles[21]. Recently, various techniques have been applied to investigate the process of the ligand exchange reaction. The previous reports considered that the electrostatic interaction between phosphorus ligands and gold substrate was weak, however, the replacement of phosphorus ligands on gold substrate was feasible. We inferred that Janus gold particles with THPO and PPh_3 ligands can be obtained. In this work, we studied the adsorption process of two kinds of ligands (PPh_3 and THPO) on the surface of gold electrode by using EDLC, as well as the adsorption kinetics. And the ligand exchange process of THPO with PPh_3 was investigated by means of XPS and contact angle measurement, theoretical simulation was used to describe the possibility of Janus gold formation.

2. EXPERIMENTAL

2.1 Chemicals

Tris(hydroxymethyl)phosphine oxide (THPO, AR, 95.5%) was purchased from Wuhan Jinlin Chemical Technology Co., Ltd. (China). Triphenylphosphine ($C_{18}H_{15}P$, PPh_3, CP), toluene (AR, 99.5%), acetone (AR, 99.5%), sulfuric acid (H_2SO_4, AR, 95-99%), cupric sulfate (AR, 99%) ethanol (AR, 99.7%) and hydrogen peroxide (H_2O_2, AR, 30%) were purchased from Sinopharm Chemical Reagent Co., Ltd. (China). Deionized water was used in all experiments.

2.2 Adsorption and exchange process of ligands on gold electrodes

A gold disk electrode (2 mm in diameter) was employed as working electrode, which was carefully polished by alumina powder (0.05 μm). After rinsing in deionized water, the electrode was immersed in a ''pirhana'' solution (H_2SO_4/ 30% H_2O_2 ,v/v, 3:1) at 50 ℃ for 3 min, followed by ultrasonic in ethanol and deionized water for 5 min, respectively. To cleaning the gold electrode further, cyclic voltammetry was carried out for 20 times in 0.5 M H_2SO_4 solution with a scan rate of 100 mV/s and the potential range from -0.6 to 1.1 V, during which Hg/Hg_2SO_4 was used as reference electrode. Then, the gold electrode was rinsed subsequently with acetone, ethanol and deionized water, and dried in ambient temperature.

The surface modification of gold by THPO molecule followed a method described in previous literature [20], typically the gold electrode was immersed in 20 μM THPO aqueous solution of at 25 ℃, at the different time intervals (from 0 s to 24 h), as-modified gold electrode was measured by cyclic voltammetry after rinsing with acetone, ethanol and deionized water in sequence. The adsorption process of PPh_3 was the same as THPO except by using 20 M PPh_3 toluene solution. To investigate the ligand exchange reaction, the THPO-modified gold electrode (adsorbed in THPO solution for 24 h) was used, which was soaked and sampled at different intervals in 20 M PPh_3 toluene solution for further characterizations.

2.3 Characterizations

The adsorption of phosphorus molecules on gold surface was mainly investigated by electrochemical analysis, which was performed on a potentiostat (CHI 660, Shanghai Chenhua Instruments Co., Ltd. China) by a three-electrode system, including gold disk (2 mm in diameter), platinum plate and Hg/Hg_2SO_4 electrode as working electrode, counter electrode and reference electrode, respectively. The surface property of the gold electrode was characterized by a contact angle goniometer (JWA-360, Xiamen Chongda Intelligent Technology Co., Ltd., China) and an X-ray photoelectron spectroscopy (ESCALAB250Xi, Thermo Fisher Scientific).

2.4 Theoretic caculation

Using Discover of Materials Studio package, molecular dynamics was employed to calculate the adsorption of phosphorus ligands on the Au (111), (110) and (100). The dynamics was performed in NVT ensemble, with the time step of 0.1 fs, and the total dynamics time is 100 ps.

3. Results and Discussion

3.1 Adsorption process of ligands on gold surface

The adsorption of the THPO and PPh_3 was accompanied by the decrease of the real surface area of the gold electrode, and the reduction degree of the surface area was related to the adsorption capacity which can be measured by cyclic voltammogram. Fig.1 showed the CV plots of the gold electrode in 0.5 M H_2SO_4 solution after 20 times scanning at a rate of 100 mV/s, in which the bands in the area B were assigned to the formation of AuO monolayer according to previous reports [37], and the peak A is the reduction of AuO monolayer with a charge of 31.11 μC (integral area of peak A). So, the real area of gold is determined as 0.085 cm^2, calculated by the ratio of the reduction charge (31.11 μC) to the theoretical value (400 μC/cm^2). The roughness factor R_f (2.72) was then obtained according to ratio of the real area (0.085 cm^2) to the geometrical area (0.0314 cm^2).

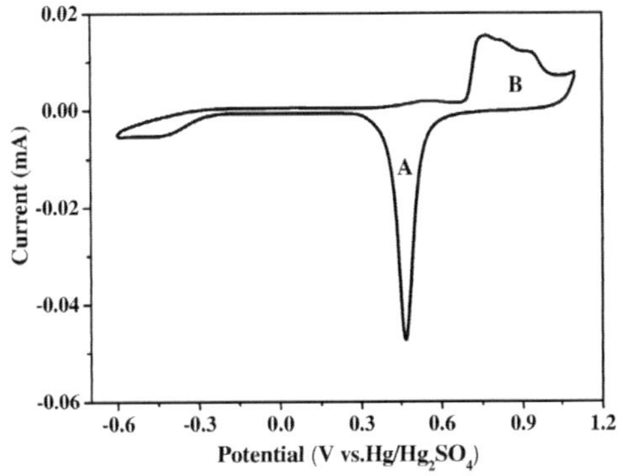

Fig.1 Cyclic voltammetry of gold electrode.

3.2 Adsorption kinetics of ligands on gold surface

Fig.2a showed the CV plots of the gold electrode adsorbed with PPh_3 molecules at variant periods. It was noticed that the difference of charge and discharge current (Δi, μA) gradually reduced with increasing the adsorption time, which indicated that the adsorption capacity of PPh_3 on the surface of the gold electrode was increased. Within 45 s, the Δi reduced rapidly. However, when the adsorption time exceeded 45 s, the Δi reduced slowly. Fig.2b showed the CV plots of the THPO adsorbed gold electrode at variant adsorption periods. Within 60s, the Δi was reduced fastly(rapidly). But the Δi reduced slowly when the adsorption time exceeded 60s. At the initial stage, the Δi of PPh_3 reduced faster than that of the THPO.

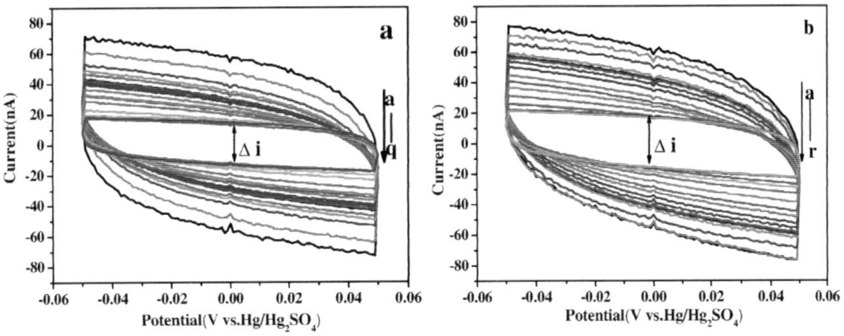

Fig.2 CV plots of gold electrode after soaking in PPh_3 (a) and THPO (b) for different adsorption time (a-q:Bare gold, 5 s, 30 s, 45 s, 1 min, 2 min, 3 min, 4 min, 5 min, 10 min, 20 min, 30 min, 1 h, 2 h, 4 h, 20 h, 24 h. a-r: Bare gold, 5 s, 30 s, 1 min, 2 min, 3 min, 4 min, 5 min, 10 min, 20 min, 30 min, 1 h, 2 h, 4 h, 5 h, 17 h, 20 h, 24 h)

The electrochemical capacitance measurement was useful to probe the packing structure of the ligand adsorbed on the electrode. The EDLC on the gold electrode could be calculated according to the following equation

$$C_d = \Delta i / 2\upsilon A \qquad (1)$$

Where υ was scanning rate (V/s), A was the real area of the electrode (cm^2). Fig. 3 showed that trend of C_d varying with adsorption time. The differential capacitance was inversely proportional to the surface coverage, with the time increasing C_d obtained from PPh_3 adsorbed gold decreased at the earlier stage ($t < 1800$ s) and then reached its final value (ca. 16.72 μF/cm^2). C_d obtained from THPO adsorbed gold needed at least 7200 s to reach a constant value (ca. 18.87μF/cm^2). It was obvious that the final value of C_d in THPO solution was lower than that of PPh_3, which indicated that the film formed by PPh_3 on the gold electrode surface is compact and less defects.

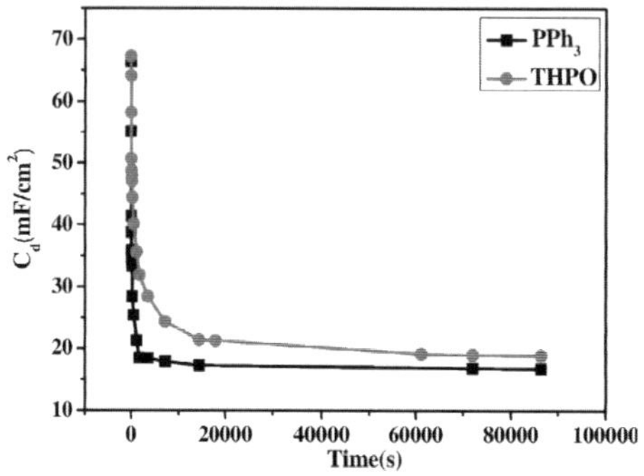

Fig.3 Curve of C_d changing with time (t).

According to C_d, the value of surface coverage (θ) during the adsorption process could be calculated by the following formula,

$$\theta_t = \frac{C_{d(0)} - C_{d(t)}}{C_{d(0)} - C_{d(film)}} \tag{2}$$

Where $C_{d(0)}$ and $C_{d(t)}$ was the differential capacitance of the ligand adsorbed gold electrode at the time of 0, t , respectively. $C_{d(film)}$ was the differential capacitance at equilibrium adsorption.

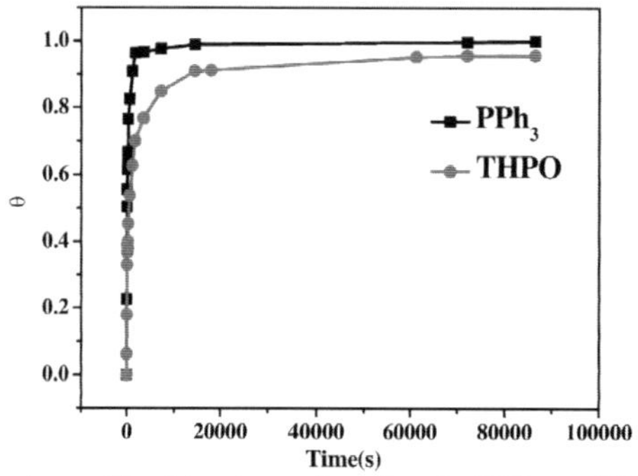

Fig.4 The variation of θ with adsorption time in PPh_3 and THPO solution.

The θ was presented in Fig. 4 obtained from the gold electrode after soaking in PPh_3 and THPO solution, respectively. For PPh_3, at the initial adsorption step, the surface coverage (θ) was increased from 0.226 to 0.908 as the adsorption time increasing from 5 s to 20 min, which approached to 1 finally. While for THPO, when deposition time t was 5 s, the surface coverage (θ) was only 0.062, θ approached to 0.9 needed at least 2 h, and the final value of θ was 0.958. This indicated that adsorption rate of PPh_3 was faster than that of THPO. At the fast increasing range of θ, the time dependence of coverage could be described with Langmuir adsorption model:

$$\ln(1-\theta) = -k_{ad}ct \tag{3}$$

Where k_{ad} was the adsorption rate constant, c was the concentration of ligands.

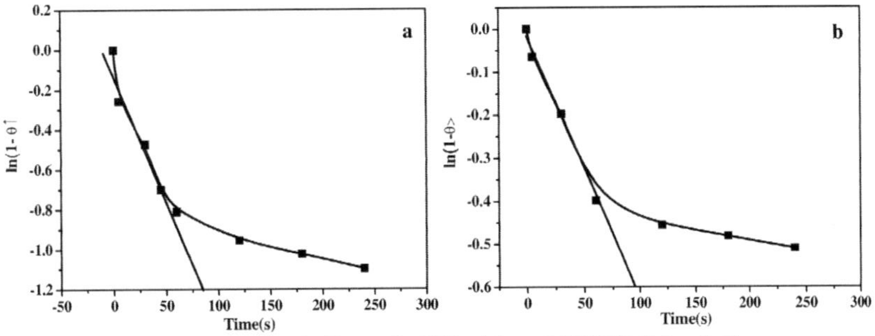

Fig.5 curve of ln(1-θ) to t in PPh_3(a) and THPO (b) solution.

As showed in Fig.5a, plot of ln(1-θ) versus t showed good linearity in the initial adsorption process ranging from 0 to 60 s due to the formation of PPh_3 film on the gold electrode surface. And the fitting coefficient r was 0.97451. The adsorption rate constant (k_{ad}) calculated by formula (3) was 625 $M^{-1}s^{-1}$. The k_{ad} of THPO was 319 $M^{-1}s^{-1}$. It was obvious that k_{ad} of PPh_3 was much higher than that of THPO indicating that the adsorption of PPh_3 was much easier than THPO.

3.3 Ligands exchange processes on gold surface

To investigate the exchange possibility of PPh_3 and THPO, gold electrode was immersed in the both ligand solution successively. As showed in Fig.6, curve A showed the C_d of the THPO modified Au electrode soaked in 20 μM PPh_3 toluene solution for the different time to investigate the kinetic of PPh_3 replacing THPO. As the dipping time increasing, variation of C_d can be dividing into four stages: fast increasing stage (t < 300 s), fast decreasing stage (300 ≤ t < 1 h), slow decreasing stage (1 h ≤ t < 5 h), and equilibrium stage (t ≥ 5 h).

To further probe the effect of toluene on the exchange process of THPO, PPh_3, THPO modified Au electrode soaked in toluene solution for the different time, the variation tendency of C_d (curve B) was similar to that of curve A except the mission of slowly decreasing stage and the variation time: C_d increased fast when t < 120s, then decreased quickly when 120 s < t < 30 min, and then reached equilibrium finally.

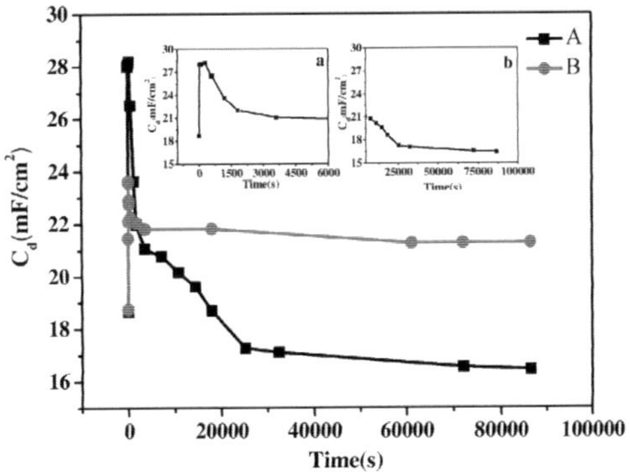

Fig.6 EDLC at different adsorption time in THPO (A) and PPh_3 (B) solution, inset: enlarged image of curve A (a) and curve B (b)

3.4 XPS analysis

XPS measurements were used to investigate the composition of the capping ligands in the replacement processes on the surface of gold electrode. For the PPh_3 modified gold electrode, the chemical states of phosphorous with the binding energies of 130.7 eV, 132.83 eV and 134.32 eV were attributed to PPh_3, PPh_3O (oxidation form of PPh_3) and the adsorption of phosphorus on the gold surface. For the C 1s spectrum, the binding energy of 284.85 eV, 286.36 eV and 288.12 eV are attributed to C-C, C-P and -C=O groups.

For the different replacement time (30 min, 2 h, 12 h, 24 h) of the samples, there were not only THPO ligand, but also PPh_3 and PPh_3O ligands. Considering the data analysis of the P 2p spectrum, the fixed ration of THPO or PPh_3 and PPh_3O for 30 min, 2 h, 12 h, 24 h samples were about 0.39:1, 0.48:1, 0.93:1, 1:1, respectively. For the C 1s spectrum, the ratio was 0.37: 1, 0.46:1, 0.91:1, 0.95:1, respectively. With the increasing of the replacement time, the content of PPh_3 and PPh_3O molecules were increasing. Finally, the fixed ratio of THPO or PPh_3 and PPh_3O on the surface of gold was close to 1:1, which showed that the substitution reaction of phosphorus ligands on the surface of gold electrode has occurred, about half the THPO molecules replaced by PPh_3 molecules.

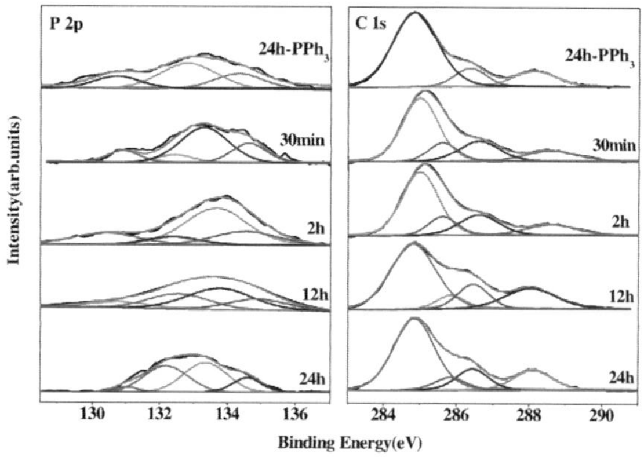

Fig 7 C 1s and P 2p spectra of the samples at different ligands exchange process

3.5 Amphiphilic property

Static contact angles were further characterized to investigate the amphipathicity of the self-assembled films. The bare gold electrode surface turned to be a hydrophilic interface after soaking in THPO toluene solution due to the adsorption of THPO (b), and which became a hydrophobic surface after soaking in PPh_3 toluene solution (c). It is worthy to noted that the hydrophobic content of THPO-modified gold surface increased after soaking in PPh_3 toluene solution for 24 h (d). The contact angle of water on surface of bare (a), THPO- modified (b), PPh_3 -modified (c) and THPO/PPh_3 modified (d) gold were 77.1°, 40.1°, 95.3° and 63.3°, respectively. Which demonstrated that the co-adsorption of THPO and PPh_3 contributed to the amphipathicity of the gold surface.

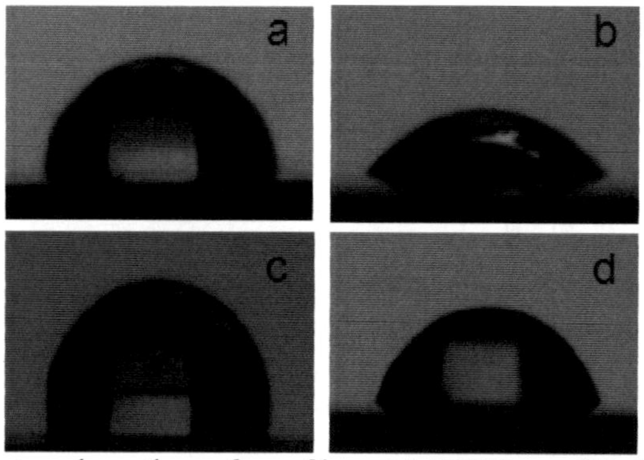

Fig.8 Static water contact angle on the surface of bare (a), THPO- modified (b), PPh_3 -modified (c) and THPO/PPh_3 -modified gold (d).

3.6 Theoretic simulation

The optimized structure of PPh_3 and THPO on Au (111), (110) and (100) were showed in Fig.9. The adsorption energy (E_{ad}) of ligands was calculated through equation (4)

$$E_{ad} = E_{molecule} + E_{surface} - E_{total} \quad (4)$$

Where $E_{molecule}$, $E_{surface}$ and E_{total} was the energy of ligands, facet of Au and the optimized structure. The adsorption energy of PPh_3 on Au (111), (110) and (100) was 108.474 kcal/mol, 55.440 kcal/mol and 70.080 kcal/mol, respectively. As for THPO, the adsorption energy was 43.250 kcal/mol, 26.683 kcal/mol and 25.000 kcal/mol, respectively. It was obvious that the adsorption energy of PPh_3 was more than twice of THPO. The bond length between P of PPh_3 and Au of (111), (110) and (100) was 3.705 Å, 4.001 Å and 3.769 Å, and the bond length between P of THPO and Au of (111), (110) and (100) was 3.745 Å, 4.012 Å and 3.982 Å, respectively. The higher adsorption energy of PPh_3 and shorter bond length between P and Au demonstrated that the adsorption of PPh3 on each facet of Au was much easier than THPO, which explained the exchange process from the respect of theoretic study.

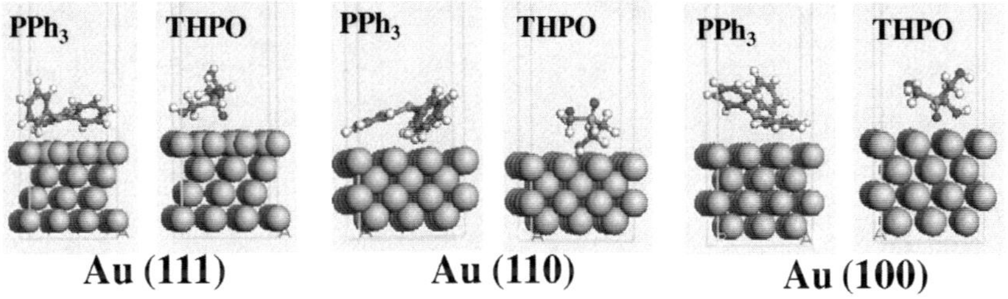

Fig.9 Side view of optimized structures of PPh_3 and THPO on Au (111), (110) and (100).

4. Conclusions

In this study, the adsorption kinetics of THPO and PPh_3 on gold electrode was studied by electrical double layer capacitance, which was an effective method to investigated dynamic process of ligand adsorption on the surface of gold electrode. The results showed that the adsorption processes of both ligands included fast and slow adsorption process, and the fast process followed the langmuir adsorption isotherm. The surface coverage (θ) for PPh_3 was higher than that of THPO, and the adsorption of PPh_3 was quicker compared to THPO due to the larger adsorption kinetic constant of PPh_3. In addition, the exchange process of the ligands was studied on the surface of gold electrode by electrical double layer capacitance indicating that the ligand exchange was a complex process, and about half of THPO was replaced by PPh_3 on the surface of THPO-Au resulting Janus Au with amphiphilic property.

Acknowledgment

This work was supported by Guangxi Natural Science Foundation (No. 2015GXNSFBA139220 and 2016GXNSFAA380107) and the funding from Guangxi Ministry-Province Jointly-Constructed Cultivation Base for State Key Laboratory of Processing for Non-ferrous Metal and Featured Materials.

References

[1] Campuzano S and Pedrero M 2005 *Talanta*, **66**, 1310-1319.
[2] Boozer C and Ladd J 2004 *Anal. Chem.*, **76**, 6967-6972.
[3] Chaki N K and Vijayamohanan K 2002 *Biosens. & Bioelectron.*, **17**, 1-12.
[4] Motesharei K and Ghadiri M R 1987 *J. Am. Chem. Soc.*, **119**, 11306-11312.
[5] Zamborini F P, Campbell J K and Crooks R M 1998 *Langmuir*, **14**, 640-647.
[6] Jennings G K and Laibinis P E, J 1997 *Am. Chem. Soc.*, **119**, 5208-5214.
[7] Ramachandran S, Tsai B L, Blanco M 1996 *Langmuir*, **12**, 6419-6428.
[8] Kitagawa K, Morita T and Kimura S 2005 *Langmuir*, **21**, 10624-10631.
[9] Chen J, Reed M A and Rawlett A M 1999 *Science*, **286**, 1550-1552.
[10] Zheng G, Chen M and Liu X Y 2014 *Electrochim. Acta*, **136**, 301–309.
[11] Motaghedifard M, Behpour M and Ghoreishi SM 2014 *Sens. and Actuator B: Chem.*, **203**, 802–811.
[12] Sunil K, Park M K 2014 *Biosens. & Bioelectron.*, **61**, 260–265.
[13] Du X, Miao Z Y, Zhang D 2014 *Biosens. & Bioelectron.*, **62**, 73–78.
[14] Wu B, Wu K and Wang P 2007 *J. Phys. Chem. C*, **111**, 1131-1135.
[15] Liao S, Shnidman Y and Ulman A 2000 *J. Am. Chem. Soc.*, **122**, 3688-3694.
[16] Dilushan R, Ronan J and Laura S 2011 *Langmuir*, **127**, 13029–13036.
[17] Hu K and Bard A 1998 *Langmuir*, **14**, 4790-4794.

[18] Shrikrishnan S, Sankaran K and Lakshminarayanan V 2012 *J. Phys. Chem. C,* **116** 16030−16037.
[19] Pesika N S, Stebe K J and Searson P C 2006 *Langmuir*, **22**, 3474-3476.
[20] Zuo G F, Liu X H and Yang J D 2007 *J. Electroanal. Chem.*, **605**, 81-88.
[21] Reyes E, Madueñõ R and Blázquez M 2010 *J. Phys. Chem. C*, **114**, 15955–15962.

Influences of MWCNTS Concentration on the Friction and Anti-Wear Behavior of Ni-P Coating

J M Xu, W C Sun, Y Wang, Z W Jia, F Guo

College of Materials Science and Engineering, Xi'an University of Science and Technology, Xi'an, Shaanxi 710054, P.R.China

E-mail: sunwanchang@tsinghua.org.cn

Abstract. Ni-P alloy and Multi-walled carbon nanotubes were codeposited on medium carbon steel by electroless plating. The effect of MWCNTS concentration on the friction and anti-wear resistance of the coating was researched. The results showed that the introduction of MWCNTS into the composite coating enhanced obviously the wear resistance of Ni-P-MWCNTS coating. It was found that the Ni-P-MWCNTS composite coating exhibited a minimum coefficient of friction, with a MWCNTS concentration of 0.3g/L, under 5N applied load. The wear rate of Ni-P-MWCNTS coating was lowest when the MWCNTS concentration in the bath was 0.3g/L.

Introduction

Corrosion and anti-wear resistance of carbon steels are often improved using surface coatings or hot working technologies, such as chemical conversion, anodization, electrolytic deposition, electroless plating, PVD, CVD and so on. Compared with other surface technologies, the electroless plating can deposite various coatings which possesses competitive corrosion and anti-wear resistance [1,2]. Carbon nanotubes are beneficial in the deposition of composite coatings owing to their nanometer size and wonderful properties [1, 3-5]. For the past few years, the corrosion and abrasion resistance of Ni-P coating with carbon nanotubes dispersed phase were researched by Liu X.J [6] and Zhi Yang et al. [7]. M.M.Tian and W.C.Sun researched the effect of CNTs concentration on the morphologies and electrochemical properties of Ni-P-MWCNTS coating [8]. In view of this, in the present research, Ni-P alloy and Multi-walled carbon nanotubes were codeposited on medium carbon steel by electroless plating, and then the influences of MWCNTS concentration on the friction and anti-wear behavior of Ni-P-MWCNTS coating were discussed.

Experimental

The composition of the bath and process parameters were described previously [8]. The wear tests were measured under dry sliding using a rotational tribology tester. Si3N4 balls were adopted as the counter body. The tests were carried out on a track radius of 5 mm at 560 rpm with a load of 5 N, 10 N, 15 N for 10 min and a sliding speed of 0.293 m/s. The friction coefficients and sliding time were recorded automatically. Each sample was cleaned by ultrasonic washing and then weghed before and after each test. The worn morphologies were characterized by scanning electron microscope.

Results and Discussion

Fig.1 shows the relationship between the the wear rate of Ni-P-MWCNTS coating and the MWCNTS concentration in the bath. It was shown that the wear rate of MWCNTS coating is greatly reduced, with the increase of the MWCNTS concentration in the bath. This could be attributed to two reasons. For one thing, the micro-hardness of Ni-P-MWCNTS coating is increased by the MWCNTS addition in the composite coating. For the other thing, the shear friction of the coating and the wear of nickel substrate are weakened, which is because that the carbon nanotubes could carry load. Besides, it can be indicated that the wear rate of Ni-P-MWCNTS coating rises with increasing the load. This is because the friction of Ni-P-MWCNTS coating elevates the shear force, with increasing the load, resulting in nickel substrate being easily peeled off, which finally leads to the increased wear rate of Ni-P-MWCNTS coating.

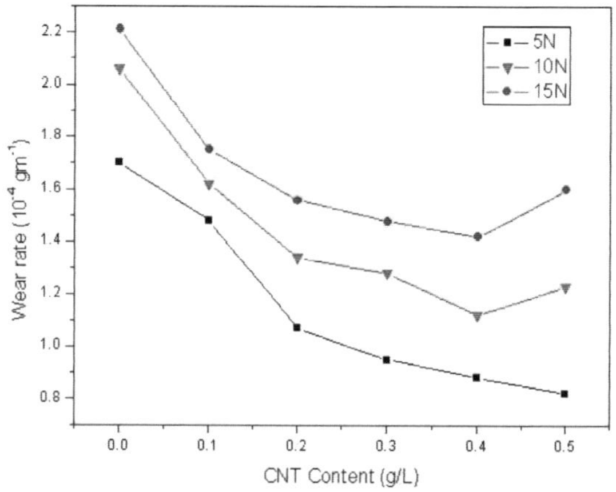

Fig.1 Effect of MWCNTS concentration on the wear rate of Ni-P-MWCNTS coatings

The relationship between the friction coefficient of Ni-P-MWCNTS coating and MWCNTS concentration under different loads was shown in Fig.2. It was shown that the friction coefficient of the composite coating has been significantly reduced, due to the co-deposition of carbon nanotubes and the nickel substrate. The friction coefficient showed a downward trend, with the increase of the concentration of MWCNTS. It could be put down to two factors. One side, the increase of MWCNTS concentration in the composite coating, which is increased with the MWCNTS concentration in plating bath, reduced direct friction with the substrate. On the other side, the carbon nanotubes acted as lubricant due to the

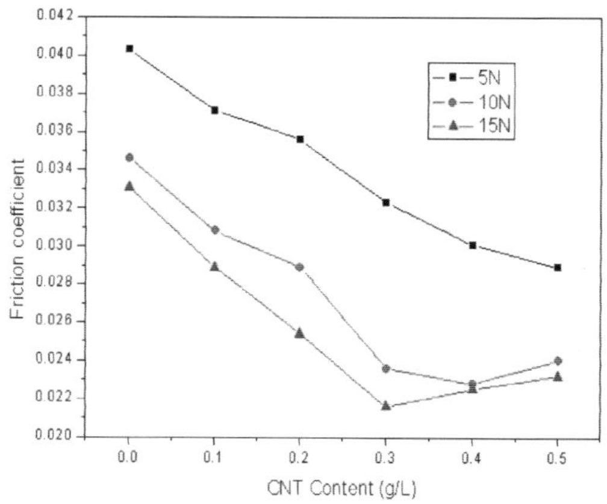

Fig.2 Influence of MWCNTS concentration on the friction coefficient of Ni-P-MWCNTS coatings

intrinsic tubular structure and high toughness. Furthermore, the conclusion could be drawn that the friction coefficient of Ni-P-MWCNTS coating rises as the load elevated.

Fig.3 shows the abrasive surface morphologies of Ni-P-MWCNTS coating under 10N load with various MWCNTS concentration. Compared with that of Ni-P alloy coating, the wear scar of Ni-P-MWCNTS coating became narrowed and faded significantly, when the MWCNTS were added into the bath, the worn surface is relatively smooth, shedding and plastic deformation is not obvious. This suggests that carbon nanotubes, which dispersed uniformly in the composite coating, worked well in resisting the plastic flow, which in turn effectively inhibited the composite coating adhesion and plowing wear, enhancing the anti-wear resistance of Ni-P-MWCNTS coatings.

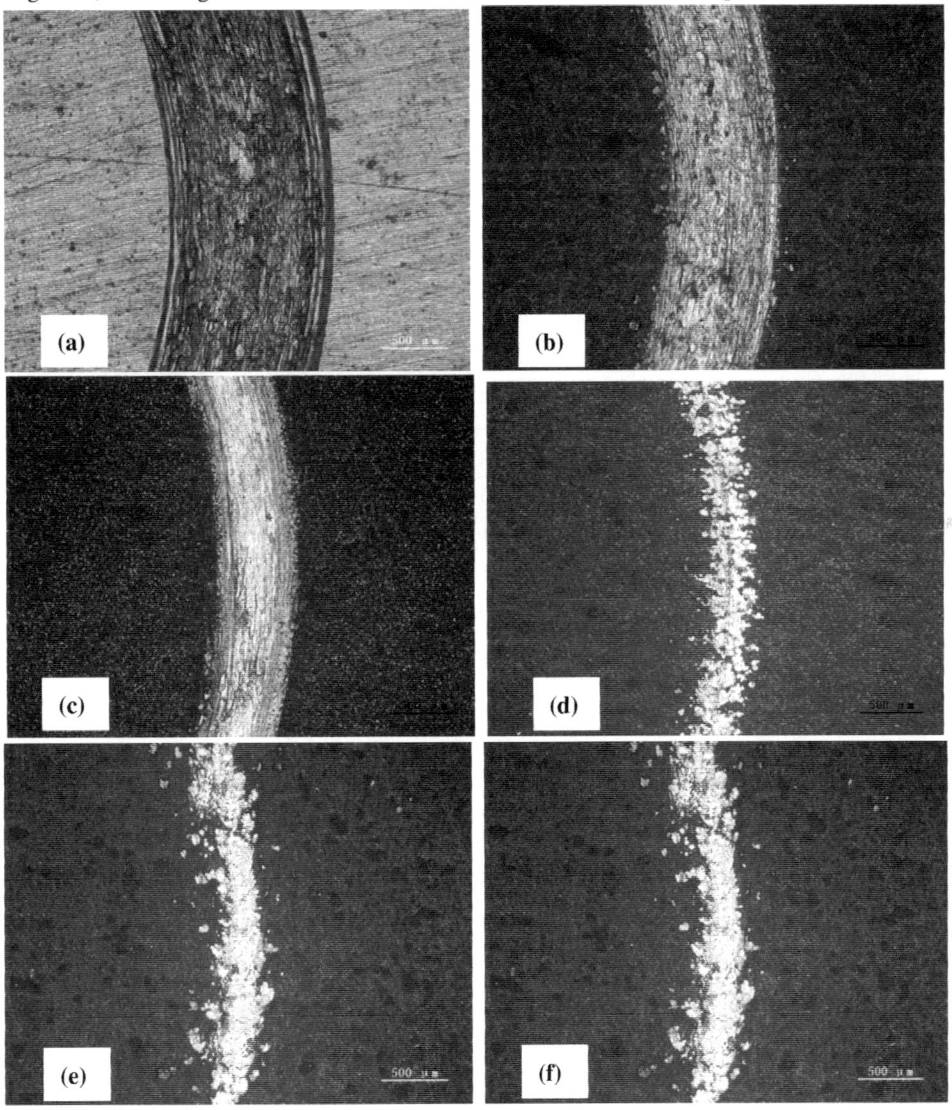

Fig.3 Influence of MWCNTS concentration on the worn morphologies of Ni-P-MWCNTS coatings. (a) Ni-P, (b) 0.1g/L, (c) 0.2 g/L, (d) 0.3 g/L, (e) 0.4 g/L, (f) 0.5 g/L

It was shown also that with the increase of MWCNTS concentration in the bath, the wear scars of Ni-P-MWCNTS coating also become shallower and narrower. When the MWCNTS concentration in the bath is 0.3g/L (Fig.3(d)), the worn signs of the composite coating is lighter than that of other coatings, the plastic deformation of the surface is not obvious and wear tracks are shallow. When the concentration of carbon nanotubes is 0.5g / L (Fig.3(f)), the wear scar of Ni-P-MWCNTS coating becomes wider, and the furrow-shaped wear tracks become deeper. Therefore, with the increasing of carbon nanotubes in the bath, the friction and anti-wear resistance of Ni-P-MWCNTS coating get improved. When the concentration of MWCNTS in the coating is moderate, the uniformly distributed carbon nanotubes could efficiently disperse the strengthening effect, improving the friction and anti-wear resistance of Ni-P-MWCNTS coatings. However, When the amounts of carbon nanotubes is excessive, carbon nanotubes would distribute as the reunited state in the coating, due to the carbon nanotubes prone to agglomerating during the deposition process, which consequently weaken the dispersion strengthening effect, making the friction and anti-wear resistance of Ni-P-MWCNTS coating degrade. Therefore, MWCNTS can substantially improve the micro-hardness and toughness and also contribute to a higher anti-wear resistance of Ni-P-MWCNTS coating.

Summary

1) MWCNTS dispersed uniformly in Ni-P coating effectively improved the anti-wear resistance of Ni-P-MWCNTS coatings. The wear scars on Ni-P-MWCNTS coating with 0.4g/L concentration of MWCNTS in the bath is shallower and narrower than that of other composite coatings, and the wear rate reached the minimum value, when the MWCNTS concentration in the bath is 0.4g/L.
2) Compared with Ni-P coating, Ni-P-MWCNTS composite coatings improve greatly friction reduction and anti-wear performance. The friction coefficient of Ni-P-MWCNTS coating decreases gradually with adding MWCNTS concentration from 0.1g/L to 0.5g/L, and the friction coefficient is only 0.021, when the addition of MWCNTS is 0.3 g/L.

Acknowledgments

This work is supported by Shaanxi Industrial Science and Technology Research（2014K08-09）.

References

[1] Gu C , Lian J S and Li G Y 2005 *Journal of Alloys and Compounds*. **Vol.391,** p. 104-110.
[2] Ambat R and Zhou W 2002 *Surface & Coating Technology*, **Vol.179**, p.124-131.
[3] Zhao G G 2010 *Journal of Heilongjiang Institute of Science &Technology,* **Vol.20,** p.1-6.
[4] Rudd A L, Breslin C B and Mansfeld F 2000 *Corros. Sci.* **Vol.42 No.2,** p.275-281.
[5] Zhang W X, He J G, Jiang Z H, Jiang Q and Lian J S 2007 *Surface & Coating Technology*, **Vol.201 No.8**, 4594.
[6] Li G, Tong Y, Wang Y and Liu G 2003 *Electrochim. Acta,* **Vol.48**, p.4061-4069.
[7] Wang K L, Zhang Q B, Sun M L, Wei X G and Zhu Y M 2001 *Appl. Surf. Sci.*, **Vol.174**, p.191-197.
[8] Tian M M, Sun W C and Shi Q, et al 2015 *Materials Science Forum,* **Vols.809-810**, p.610-614.

Strengthening in a copper composite containing graphene nanofillers

Y Song[a], W W Liu[b] and Y Chen[c]
School of Mechanical and Electric Engineering & Collaborative Innovation Center of Suzhou Nano Science and Technology, Soochow University, 215123 Suzhou, China

E-mail:[a]songying55555@163.com, [b]liuweiwei@suda.edu.cn, [c]chenyao@suda.edu.cn

Abstract: Multi-layer graphenes (MLGs) reinforced Cu matrix composites were fabricated using spark plasma sintering in this research. Microstructure observation indicated that MLGs homogeneously disperse in the matrix to result in grain refinement. Instrumented spherical micro-indentation tests were conducted to evaluate micro-mechanical properties of the composites, and the elastic modulus and indentation yield strength of the 1.5 wt.% MLG/Cu composite increase up to ~ 36% and ~ 41% than those of the unreinforced Cu, respectively. Grain refinement, the enhanced dislocation density (EDD) due to the thermal expansion mismatch between Cu matrix and MLGs and Orowan are thought to be the main strengthening mechanisms in this research.

1.Introduction

Copper (Cu) and its alloys with excellent electrical and thermal conductivities have been widely used in machinery industries including the electric train overhead trolley frame, the blast furnace of metallurgical industry and the large capacity contactor switch[1]. However, relatively poor mechanical properties such as tensile strength and hardness limit their further applications. It is expected that addition of suit second phases might be an effective way to improve their mechanical properties, and considerable research efforts have been devoted to achieve copper matrix composites reinforced with various ceramic particles/fibers in the past decades[2-4]. However, these ceramic reinforcements used in the Cu composites are nonconductive, resulting in improved mechanical properties with sacrifice of good electrical and thermal conductivities to some extent.

Carbonaceous materials have emerged as important reinforcements in recent years because of their good mechanical properties and excellent electrical conductivity. Graphene has excellent mechanical properties and electronic and thermal conductivity [5], and very high specific surface area (~ 2630 m^2 g^{-1}). It is noted that multi-layer graphenes (MLGs) consisting few graphene layers have similar properties to the single layered graphene. Hence, MLGs reinforced metal composites have attracted a lot of attentions [6-10]. Kim et al [7] reported that yield strength and tensile strength of the 1.0 vol.% MLG/Cu composite increase ~ 14.6% and ~ 10.9%. Also, Li et al [8]found that a good dispersion and strong interfacial bonding between graphene and Cu matrix using Ni nanoparticles decorated MLGs, a significant improvement in ultimate tensile strength (~ 42%) of 0.8 vol% Ni-GPL composite was achieved. However, above research mainly focused on macro-mechanical properties, however, there is a significant difference between macro and micro-mechanical properties , the later is mainly dependent upon the localised microstrusture.

The purpose of this research is to evaluate micro-mechanical properties of the sintered composites using instrumented micro-indentation technique, and the effects of MLGs on microscopic mechanical properties of the copper composites and their strengthening mechanisms are also discussed in this research.

2.Experimentals

Copper (purity 99.5%, diameter: 60 nm) and MLGs with a thickness of ~ 1-5 nm and a length of 1-5 μm from Nanjing Xian Feng Nano Material company (China) were employed as the starting materials. MLGs were well dispersed using ultrasonic treating in distilled-water with sodium dodecyl-benzenes sulfonate (SDBS) as dispersant, and then nanosized copper powders were added into the MLG suspension for further ultrasonication. The as-obtained composite powders were mixed in the ethanol using Si_3N_4 ball milling with a speed of 350 rpm for 12 h. After fully dried, the composite powders were sintered using spark plasma sintering system (SPS, Dr. Sinter 1050, Sumitomo Coal Co. Ltd., Japan) at a temperature 900°C under a uniaxial pressure of 40 MPa for 6 min with a heating rate of 150°C/min. The weight fractions of MLGs in Cu matrix composite were 0.5, 1.0 and 1.5 wt.%.

X-ray diffraction (XRD, X'Pert-Pro MPD, The Netherlands) used Cu Ka radiation with a scanning rate of 2°/min to analysis phase constituents of the sintered samples. The additive MLGs into copper composites and as-received MLGs were analyzed by Raman spectroscopy (JR HR800, France) using an Argon ion laser of wave length 514 nm and acquisition time of 20 s. Microstructures of the sintered samples were analyzed using scanning electron micrography (SEM, Hitachi S-4700, Japan). Hardness and elastic modulus of the sintered samples were tested using instrumented micro-indentation by MCT tester (CSM, Switzerland) with a load of 500 mN and a dwell time of 10 s. Spherical indenter radius was 20 μm and load-uploading rate was 1000 mN/min. Furthermore, Continuous Multi-Cycle (CMC) load-unloading technique was also conducted to achieve the indentation stress-strain curves of these sintered samples with spherical indenter radius 20 μm. The CMC process consisted of 12 cycles of load-partial unloading indentation performed on the same spot, and the applied load increased quadratically from 100 mN to 1000 mN. Ten tests were conducted at different locations for each sample.

3.Results

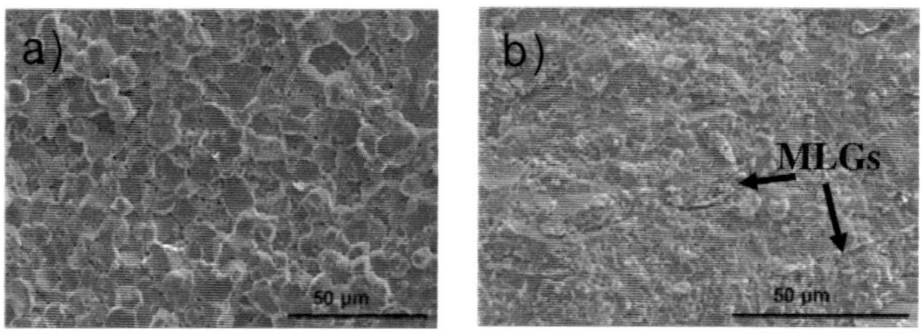

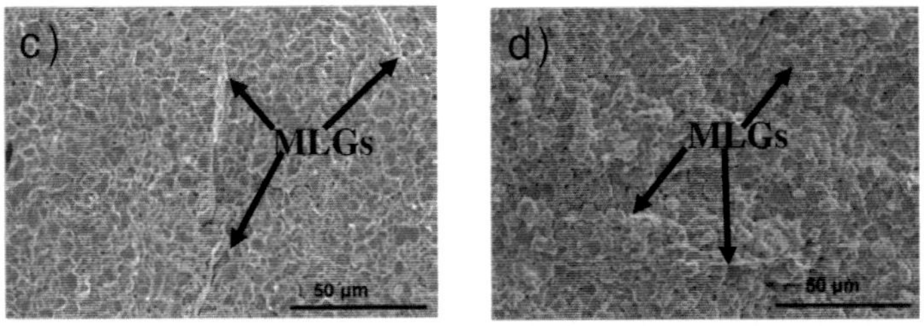

Fig. 1 SEM images of the fracture surfaces of unreinforced sample (a), and sintered composites with 0.5wt.% MLG (b), 1.0 wt.% MLG (c) and 1.5 wt.% MLG (d).

Fig.1 shows SEM images of the fracture surfaces of sintered samples, it is clear that the average grain sizes of the unreinforced sample are ~ 8 μm (Fig. 1a), while those of the samples with MLGs are all ~ 5 μm (Fig. 2b-d). The grain refinement might ascribe to MLGs dispersed in the grain boundaries restrict grain growth. Moreover, MLGs disperse homogeneously in the copper matrix (Fig. 1). As shown in Fig. 2a, it is clear the main phases of the sintered copper and MLG/Cu composites are Cu and some oxide particles, implying Cu powders underwent oxidation in the ball milling and SPS. Furthermore, the content of oxide particles decreases with increasing of MLG content in light of the peak intensity, this is due to rapid transfer of thermal energy from surface to the inner of the composites in the SPS process by means of excellent thermal conductivity associated with MLGs. Fig. 2b shows the Raman spectra of as-received MLGs and the sintered 1.5 wt% MLG/Cu composite, it is clear that D (1347 cm^{-1}) and G (1595 cm^{-1}) peaks which are related to defects in the structure and plane C-C bond stretching in graphene, respectively. The results sufficiently confirm that MLGs survive in the composites after harsh SPS process conditions.

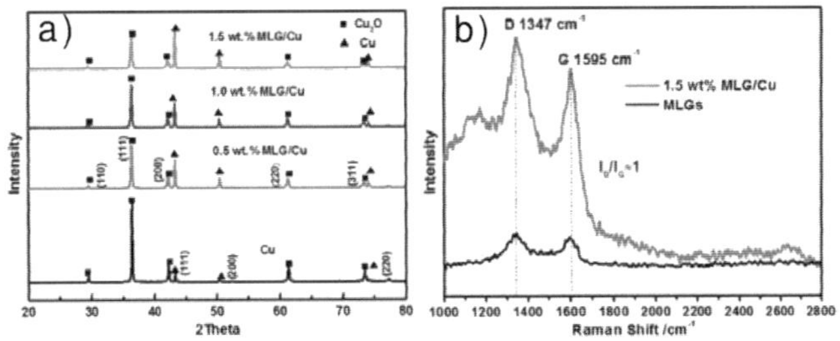

Fig. 2 XRD results of the milled powders and sintered samples (a) and Raman spectroscopy of the 1.5 wt.% MLG/Cu Composite powders and as -received MLGs (b).

Table 1 Micro-mechanical properties of sintered samples

Samples	*Hardness(GPa)*	*Elastic Modulus(GPa)*	*Hardening Exponents*	*Yield strength (MPa)*
Cu	1.73±0.06	31.85±1.67	0.298±0.16	196±9.33
0.5 wt.%MLG/Cu	1.66±0.03	33.92±2.54	0.274±0.08	218±7.93
1.0 wt.%MLG/Cu	1.45±0.03	36.45±1.37	0.168±0.09	266±6.77
1.5 wt.%MLG/Cu	1.37±0.10	43.30±2.00	0.166±0.08	276±7.82

Fig. 3 depicts the representative load-displacement curves and the reconstructed indentation stress-strain curves of the tested samples using Field and Swain method [11] , in which indentation yield strength was defined as the minimum stress at 0.2% plastic deformation[12]. As summarized in Table 1, elastic modulus and indentation yield strength of the sintered samples show an increase with MLG content. It is clear that elastic modulus of the 1.5 wt.% MLG/Cu composite exhibits an improvement by up to ~ 36% as compared with that of monolithic copper. High elastic modulus of the MLGs gives rise to the significant increase in the elastic modulus of the composites. Also, hardness of the 1.5wt.% MLG/Cu composite decreases to ~ 1.37 GPa while that of unreinforced Cu is ~ 1.73 GPa. As mentioned above, the content of oxide particles decreases with increasing of MLGs content, consequently resulting in the decrease of the hardness. Most importantly, the indentation yield strength of the 1.5 wt.%MLG/Cu composite significantly increases up to ~ 41%. Therefore, the addition of the MLGs to the Cu matrix plays a significant role in the improvement of the mechanical properties. It should be noted that the indentation yield strength increases ~ 22% when the content of added MLGs increases from 0.5 wt.% to 1.0 wt.% , but only ~ 4% improment occurs with further increase in MLG content (1.5 wt.%). These mainly because the MLG length of the 1.0 wt.% MLG/Cu composite is ~ 6 µm as depicted in Fig. 4a when that of the 1.5 wt.% MLG/Cu composites is apparent agglomeration (~ 20 µm, Fig. 4b). Also there are some micro-voids existing around the MLGs due to their agglomeration as shown in Fig. 4, which would decrease the stress transfer efficiency from Cu matrix to MLGs. These results show that the reinforced effect would decrease when the MLG content exceed 1.0 wt.%.

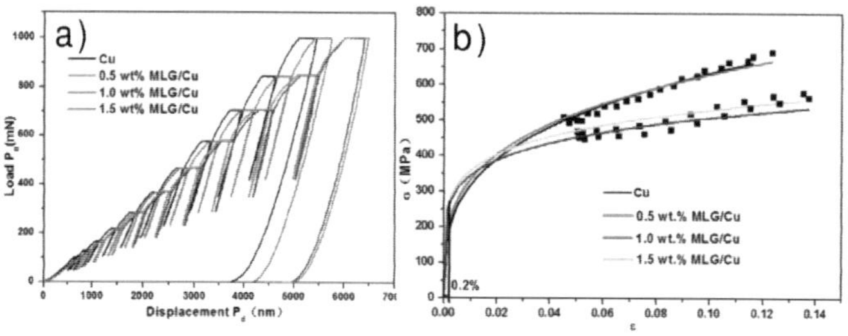

Fig.3 Typical load-displacement curves of spherical indentation Continuous Multi-Cycle (CMC) load-unloading of the sintered samples (a), and reconstructed indentation stress-strain curves of the sintered samples (b).

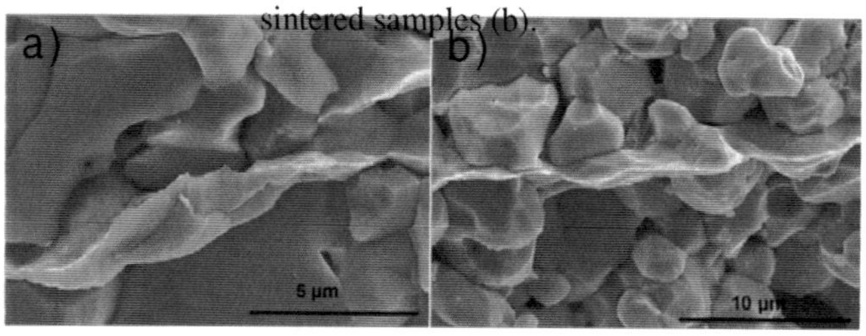

Fig. 4 SEM images of fracture surfaces of sintered composites containing of 1.0wt.% (a) and 1.5 wt.% MLG(b)s.

Hihger residual stress should exist in the sintered composites due to thermal expantion mismatches between Cu ($17\times10^{-6}K^{-1}$) [1] and MLGs ($(-8.0\pm0.7)\times10^{-6}K^{-1}$) [13] during cooling from the processing temperature to ambient temperature. Carbonaceous materials reinforced composites usually lead to enough residual thermal stresses to induce dislocation around the carbonaceous materials [14]. Therefore, the dislocation density would be increased duing the SPS to produce higher dislocation interaction and a shorter traveling distance of dislocations, subsequentely leading to an increased resistance to further plastic deformation, which will lead to the decrease of the hardening exponent. This phenomenon is similar to the fact that the strain hardening exponent of copper is controlled by dislocation density and grain size because smaller grain sizes and larger dislocation densities[15].

As for the nanocomposites, it is well known that grain refinement, enhanced dislocation density (EDD) and Orowan mechanism have significant effects on the mechanical properties. (i) Grain refinment. The grain refinement would lead to the strength improvement according to the Hall-Petch relationship. As depicted in Fig. 1, the grain size of the composites is ~ 5 μm, which is much smaller than that of the unreinforced Cu (~ 8 μm). Therefore, grain refinement plays a significant role in the improvement of the mechanical properties. (ii) Enhanced dislocation density mechanism (EDD), which results from the increase of the dialocation destiny due to the thermal expanation mismatches between MLGs and Cu. (iii) Orowan mechanism. MLGs are capable of restricting dislocation movement and forming loop-like configuration around the MLGs when the dislocations by-pass these MLGs on the slip plane to increase the strength. Therefore, the improvement of yield strength is ascribed to the grain refinement, EDD and Orowan mechanism.

4.Summary

Cu composites reinforced with MLGs were fabricated by SPS, in which MLGs with their pristine structure could homogeneously disperse in the composites and lead to grain refinement of the composites. Instrumented indentation results indicate that mechanical properties of the MLG/Cu composites such as elastic modulus and yield strength are improved. Analysis indicates that the grain refinement, the enhanced dislocation density (EDD) due to the thermal mismatch between copper and MLGs, and Orowan mechanism are attributed to the strengthening in this research.

5.References

[1] Tjong S C 2013 Recent progress in the development and properties of novel metal matrix nanocomposites reinforced with carbon nanotubes and graphene nanosheets, *Mater. Sci. Eng, R.* **74** 281-350.

[2] Wang T M, Zou C L, Chen Z N, Li M Y, Wang W, Li R G, et al. 2015 In situ synthesis of TiB_2 particulate reinforced copper matrix composite with a rotating magnetic field, *Mater. Des.* **65** 280-288.

[3] Li C, Feng X, Shen Y and Chen W 2016 Preparation of Al_2O_3/TiO_2 particle-reinforced copper through plasma spraying and friction stir processing, *Mater. Des.* **90** 922-930.

[4] Yin J, Yao D, Hu H, Xia Y, Zuo K and Zeng Y P 2014 Improved mechanical properties of Cu matrix composites reinforced with β-Si3N4 whiskers, *Mater. Sci. Eng, A.* **607** 287-293.

[5] Alexander S G, Balandin A, Bao W, Calizo I, Teweldebrhan D, Miao F and Lau C N 2008 Superior Thermal Conductivity of Single-Layer Graphene, *Nano Lett.* **8** 902-907.

[6] Chu K and Jia C 2014 Enhanced strength in bulk graphene-copper composites, *Physica status solidi. A.* **211** 184-190.

[7] Kim W J, Lee T J and Han S H 2014 Multi-layer graphene/copper composites: Preparation using high-ratio differential speed rolling, microstructure and mechanical properties, *Carbon.* **69** 55-65.

[8] Li M, Che H, Liu X, Liang S and Xie H 2014 Highly enhanced mechanical properties in Cu matrix composites reinforced with graphene decorated metallic nanoparticles, *J. Mater. Sci.* **49** 3725-3731.

[9] Rashad M, Pan F, Hu H, Asif M, Hussain S and She J 2015 Enhanced tensile properties of magnesium composites reinforced with graphene nanoplatelets, *Mater. Sci. Eng, A.* **630** 36-44.

[10] Wang J, Li Z, Fan G, Pan H, Chen Z and Zhang D 2012 Reinforcement with graphene nanosheets in aluminum matrix composites, *Scripta Mater.* **66** 594-597.

[11] Field J S and Swain M V 1994 Determining the mechanical properties of small volumes of material of submicrometer spherical indentation, *J. Mater. Res.* **10** 101-112.

[12] Huang Q Q, Song Y, Liu W W, Chen Y, Qi F, Zhao D, et al. 2015 Spherical indentation with multiple partial unloading for assessing the mechanical properties of ZrB_2–SiC composites, *Ceram. Int.* **41** 12349-12354.

[13] Yoon D, Son Y W and Cheong H 2011 Negative Thermal Expansion Coefficient of Graphene Measured by Raman Spectroscopy, *Nano Lett.* **11** 3227-3231.

[14] Chen Y, Balani K and Agarwal A 2012 Do thermal residual stresses contribute to the improved fracture toughness of carbon nanotube/alumina nanocomposites?, *Scripta Mater.* **66** 347-350.

[15] Fattah-alhosseini A, Imantalab O, Mazaheri Y and Keshavarz M K 2016 Microstructural evolution, mechanical properties, and strain hardening behavior of ultrafine grained commercial pure copper during the accumulative roll bonding process, *Mater. Sci. Eng, A.* **650** 8-14.

The effect of gas assisted length on polymer melt extrusion based on the gas-assisted extrusion technique

B Wan[1], Z Ren[2,3], G D Liu[2], X Y Huang[3]

[1] President's Office, Jiangxi Science and Technology Normal University, Nan Chang, Jiangxi, China, 330038
[2] Key Laboratory of Optic-Electronic and Communication, Jiangxi Science and Technology Normal University, Nanchang, 330038, China
[3] School of Mechanical and Electrical Engineering, Nanchang University, Nanchang, 330031, China

E-mail: renzhong0921@163.com

Abstract. In this study, the gas-assisted technique was used into the process of polymer melt extrusion to overcome the extrudate swell problem. The gas length is an important factors in the gas-assisted extrusion technique. To ascertain the mechanism of the gas-assisted extrusion technique, and to determine the optimal gas length, the effect of gas length on the extrudate swell ratio of melt was numerically investigated. In finite element numerical simulation, PTT constitutive model and full slip boundary condition were used to achieve the gas-assisted mode. Compared with the traditional no gas-assisted extrusion, numerical results showed that the extrudate swell problem was well eliminated by the gas-assisted method. Moreover, the extrudate swell of melt decreased with the increasing of the gas length because the pressure and shear stress of melt were greatly decreased. Moreover, the flow velocity of melt is uniform at the die outlet.

1. Introduction

In the process of polymer processing, the extrudate swell and sharkskin phenomena usually occur due to some factors, such as high elasticity of polymer melt, high shear stress and tensile stress. Many methods have been proposed to eliminate these extrusion problems, for example, fluoropolymer coating [1], fluoropolymeradditive [2], lubrication oil coating [3], vibration extrusion [4] and gas assisted extrusion [5], etc. The gas-assisted extrusion (GAE) is a promising and environmental method. Brzoskowski [6] firstly proposed the GAE technique. Liang [7] firstly applied the GAE technique into the polymer melt extrusion. Then, the GAE technique was studied by Arda [8], Huang[9], and Liu [10], etc. Experimental and numerical researches have fully demonstrated that the GAE can well overcome the extrudate swell of melt, melt fracture and extrusion distortion phenomena. However, in GAE, the gas layer length is one of the most important factors of impacting the extrusion effect of GAE. Huang [11], Huang [12], and Deng [13] thought that, if the gas length was short, the extrudate problems were not eliminated because the elastic energy effect of melt was not completely diminished, whereas the extrusion of melt was also influenced by gas flow if the gas length was too long. Therefore, the gas length is a key factor in the designing of the gas extrusion die of polymer melt. In order to determine the optimal gas length, the influence of the gas length on the flow behavior and extrudate effect of polymer melt was studied in this paper.

2. Numerical model and method

2.1 Numerical model

The geometric model and finite element mesh of GAE are shown in Fig.1(a) and (b), respectively.In Fig.1(a), OA is the inlet, DG is the exit, AB is wall boundary, BC is gas layer, CD is free boundary, OG is the symmetry axis, where the length of inlet OA and exit DG are 5mm, the length of AB is 5mm, as well as the CD, the length of BC can be changeable from the range of 1 to 20mm.

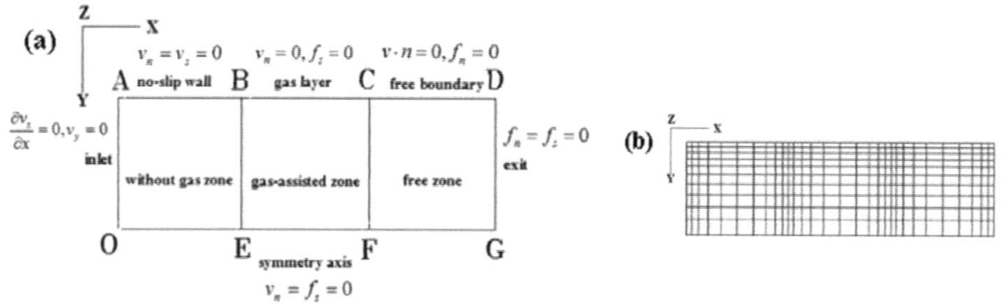

Figure 1. Numerical model. (a) geometric model; (b) finite element mesh

2.2 Governing equations

In the simulations, the polymer melt is regarded as the non-compressible, iso-thermal, steady, laminar and viscoelastic Non-Newtonian fluid. The inertia, gravity force, viscous dissipation effects are neglected.

The mass and momentum equations are given as follows:

$$\nabla \cdot v = 0 \quad (1)$$

$$-\nabla p + \nabla \cdot \tau = 0 \quad (2)$$

where ∇ is Hamilton operator, v is velocity of melt, p is the pressure drop, τ is the extra stress tension. In this study, Phan-Thien-Tanner (PTT) constitutive model [14] was used because it not only describe the characteristic of shear-thinning, but also represent the elastic properties of viscoelastic fluid, PTT constitutive equation is given as follows:

$$\exp\left[\frac{\varepsilon\lambda}{(1-\eta_r)\eta} tr(\tau_1)\right]\tau_1 + \lambda\left[\left(1-\frac{\xi}{2}\right)\overset{\nabla}{\tau_1} + \frac{\xi}{2}\overset{\Delta}{\tau_1}\right] = 2(1-\eta_r)\eta D \quad (3)$$

where $\eta_r = \eta_2 / \eta$ is viscosity ratio, η is the total viscosity of melt, η_2 is Newtonian viscosity component of melt, λ is the relaxation time, ε and ξ are the parameters of melt correlated with material tensile and shear characteristics, respectively. $\overset{\nabla}{\tau_1}$ is the upper convected derivative of the extra stress tensor τ_1, η_1 is the Non-Newtonian component viscosity of melt, D is the strain-rate of tensor.

In this simulations, for PTT construction equation, its parameters are given in Table 1.

Table 1 The parameters of PTT construction equation

Parameters	η(Pa.s)	λ(s)	ε	ξ	ηr
Values	2700	0.2	0.18	0.23	0.12

2.3 Boundary conditions
(1)For the inlet face, the flow of melt is full-developed, steady and laminar flow. The following equations should be satisfied, i.e., $\partial v_x / \partial x = 0, v_y = 0$;(2) For no gas-assisted wall, no slip wall is suitable for the traditional no gas-assisted wall, the following equation is satisfied, i.e., $v_n = v_s = 0$;(3)For the gas-assisted wall, the full-slip boundary condition is used, i.e., $v_n = 0, f_s = 0$;(4)For the free surface, no normal stress and entangle velocity are imposed. $v \cdot n = 0, f_n = 0$;(5) For the symmetric axis, the following equations are satisfied, i.e., $v_n = f_s = 0$;(6)For the exit of melt, no trance force and entangle velocity are imposed. $f_n = f_s = 0$.

3. Numerical results and discussion

3.1 Effect of gas length on the swell ratio
To investigate the effect of gas length on extrudate swell ratio of melt, the different gas length from 0 to 20mm were test, and the extrudate swell ratio were obtained, which are shown in Figure 2. From Figure 2, it can be seen that the extrudate swell ratio was decreased with increasing gas length. However, the extrudate swell ratio is nearly equal to zero when the gas length is about 15mm, and the change of extrudate swell ratio is not obvious when the gas length is larger than 15mm, which demonstrates that the gas length of 15mm is the optimal length.

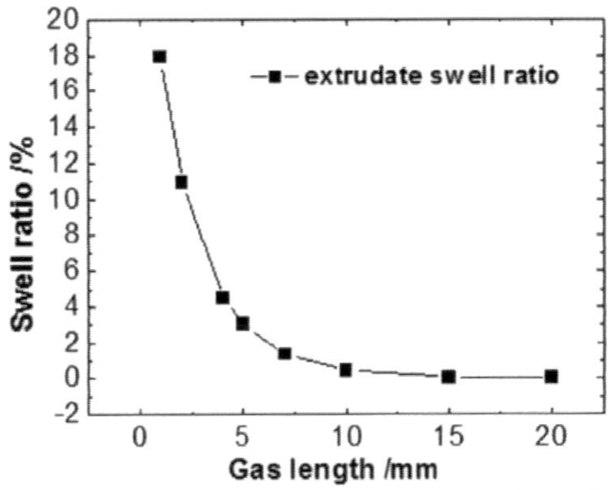

Figure 2. The extrudate swell ratios of polymer melt under different gas lengths

3.2 Effect of gas length on the flow velocity of melt
Figure 3 (a) and (b) are the X velocity distributions of polymer melt at the axial and radial direction, respectively. From Figure 3, it can be seen that the change of X velocity at the axial direction is not obvious for the different gas lengths, but for the X velocity at the radial direction, the X velocity amplitude of melt decreases with increasing the gas length, which demonstrates that the extrudate swell effect is decreased with the gas length.

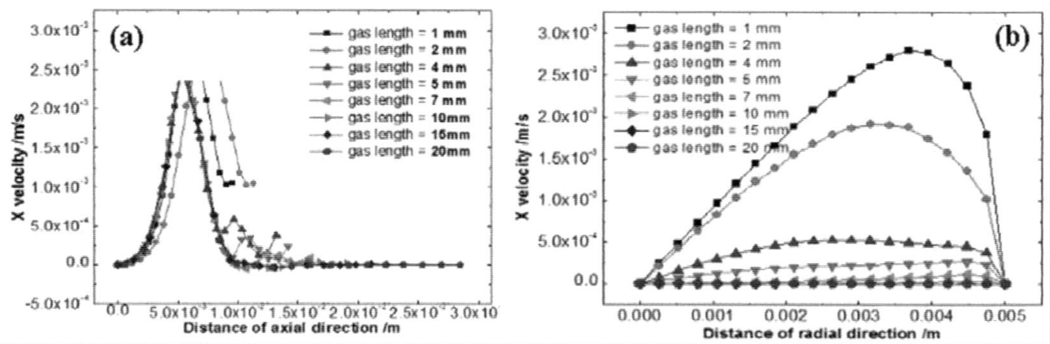

Figure 3. X velocity distribution. (a) axial direction; (b) radial direction.

3.3 Effect of gas length on the pressure drop of melt

Figure 4 is the pressure drop of polymer melt at the axial direction under the different gas lengths. In general, the pressure drop of melt increases with increasing the change of flow channel. However, from Figure 4, it can be seen that the change pressure drop of melt is not large with increasing the gas length or flow channel, which demonstrates that the increasing of gas layer decreases the pressure drop of melt.

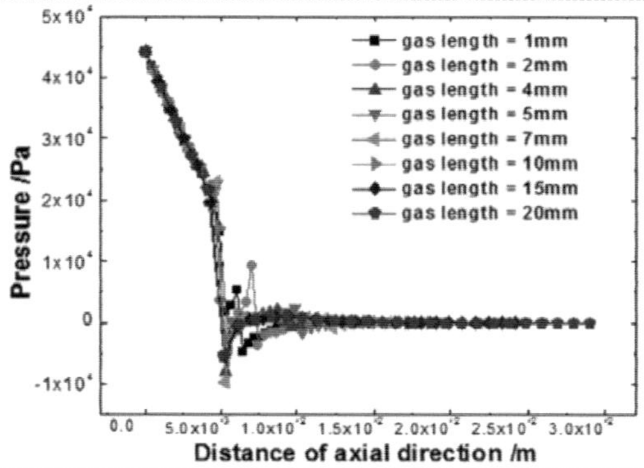

Figure 4. The pressure distribution of polymer melt under different gas lengths

3.4 Effect of gas length on the shear stress of melt

Figure 5 is the shear stress of melt under the different gas lengths. From Figure 5, it can be seen that the shear stress of melt decrease with the gas length at the outlet of die, which demonstrates that the extrudate swell effect and the melt fracture phenomena can be well eliminated with increasing the gas length.

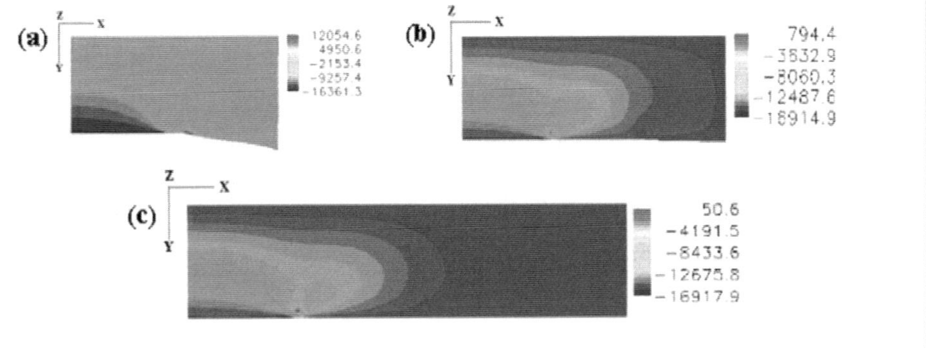

Figure 5. The shear stress of melt under the different gas lengths. (a) gas length is 1mm, (b) gas length is 5mm, (c)gas length is 10mm

4. Conclusions

The gas length is an important factor in the gas-assisted extrusion process because it can impact the effect of overcoming extrudate swell of plastic products extrusion. To ascertain the mechanics of gas-assisted extrusion technique, and to determine the optimal gas-length, the effect of gas length on the extrudate swell of plastic products was numerically studied by using the finite element computation method. In the numerical simulation, the gas-assisted extrusion model based on the full slip boundary condition was used, and PTT constitutive model was used to characterize the viscoelastic properties of plastics. Numerical results showed that the gas-assisted technique well eliminated the extrudate swell problem. Moreover, the extrudate swell decreased with the increasing of the gas length. The reason was that the flow velocity of melt became uniform when the melt was extruded out the die, the pressure and the shear stress of melt was greatly decreased due to the gas-assisted mode. Therefore, the gas-assisted extrusion technique has the good prospect in the field of plastics extrusion process.

Acknowledgements

This work is supported by the Natural Science Foundation of Jiangxi Province (20151BAB202011), the Science and Technology Pillar Program of Jiangxi Province (20132BBG70103), Top-notch talent foundation project of JXSTNU (2014QNBJRC004), the Scientific Research Foundation of Jiangxi Provincial Education Bureau (No. GJJ150814), and 2011 Cooperative Innovation Center Project of Jiangxi Province (2014XTCX005).

References

[1] Piau J M, El Kissi N, Tousssaint F and Mezghant A 1995 *Rheol. Acta* **34** 40
[2] Schwetz M, Merten A and Munstedt H 2001 *Kunststoffe-Plast Europe* **91** 86
[3] Kulikov O and Hornung K 2004 *J. Non-Newton. Fluid Mech.* **124(1-3)** 103
[4] Intawong N T and Sombatsompop N 2004 *Polym. Eng. Sci.* **44** 2298
[5] Ren Z, Huang X Y, Liu H S, Deng X Z and He J T 2015 *J. Appl. Polym.Sci.* **132** 12365
[6] Brzoskowski R, White J L, Szydlowski W, Weissert F C, Nakajima N and Min K 1987 *Rubber Chem. Technol.* **60** 945
[7] Liang R F and Mackley M R 2001 *J. Rheol.* **45** 211
[8] Arda D and Mackley M 2005 Rheol. Acta **44** 352
[9] Huang X Y 2006 Theoretical and experimental study of polymer gas-assisted extrusion from dies (Nanchang: Nanchang University) (in Chinese)
[10] Liu H S, He J T, Huang X Y, Deng X Z and Wan Q F 2014 *Polym.Mater. Sci. Eng.* **30** 113 (in Chinese)
[11] Huang X Y, Liu H S, Zhou G F, et al. 2005 *J. Plast. Eng.* **12** 101 (in Chinese)

[12] Huang Y B 2011 Theoretical and experimental study on polymer gas-assisted co-extrusion (Nanchang: Nanchang University) (in Chinese)
[13] Deng X Z 2014 Experimental and theoretical study on gas-assisted co-extrusion of plastic profile with an irregular cross-section (Nanchang: Nanchang University) (in Chinese)
[14] Phan T and Tanner R 1977 *J. Non-Newton. Fluid Mech.* **2** 353

Microstructure, texture and mechanical properties of a Mg–Gd–Nd alloy under different thermo-mechanical treatments

X L Hou[a], Y X Zhai, C L Zhang, P Zhang[b], Q F Guan

School of Material Science and Engineering, Jiangsu University, Zhenjiang Jiangsu, 212013, China

E-mail:[a]houxiuli@ujs.edu.cn, [b]zhangpjlxb@ujs.edu.cn

Abstract. In current work, the influences of hot extrusion and rolling treatments on the microstructure and texture developments, as well as mechanical properties of Mg–1.5Gd–1.5Nd (wt.%) alloy were investigated. It was shown that the alloy underwent completely dynamic recrystallization during thermal-mechanical processing, resulting in fine equiaxial grain structure. Especially for the hot rolling treatment that heavy deformation led to even smaller grain size of the alloy. Unlike the strong basal textures obtained in conventional Mg alloy sheets, the extruded and rolled alloy sheets both revealed modified basal texture with tilt of basal poles towards extrusion/rolling direction. Through orientation distribution function (ODF) analysis, the textures were deeply analyzed and two types of rare earth textures were obtained in the extruded and rolled alloy sheets. The variations in microstructure and texture played an important role in the mechanical properties, and also planar anisotropies of the alloy sheets.

1. Introduction

Currently, massive CO_2 emissions have seriously influenced the environment. As one of the major responsible sectors, transportation industry is in constant pursuit of lightweight resources to reduce the fuel consumption. Mg alloys are undoubtedly promising for structural applications in automotive and aerospace industries because of their low densities [1]. However, the use of Mg alloys as a structural material has not been so competitive that a few challenges still exist. The relatively low ductility and unsatisfactory formability at ambient temperatures are key factors limiting the practical application of Mg alloys. This can be attributed to the limited active slip systems of hexagonal close-packed (hcp) crystal structure and thereby the development of strong crystallographic texture during sheet processing. It corresponds to the basal planes of most grains being nearly parallel to the extrusion/rolling direction, which severely restricts component manufacturing from the sheet products. Recent studies have shown that the texture of Mg alloy sheets is more dependent on alloy composition than sheet processing parameters (i.e. temperatures or strain path) [2-4]. Thus, one effective way to improve the formability of Mg alloys is engineering the alloy composition. Alloying Mg with rare earth (RE) elements is known to can significantly weaken the texture of wrought alloys, both by reduce the texture sharpness and change the orientation of texture intensity peak [3-10], introducing an additional "RE texture" component [3-5]. The modified alignment of grains for basal slip leads to higher ductility and improved formability of wrought alloys.

In this study, the Mg−1.5Gd−1.5Nd (wt.%) alloy was processed by hot extrusion and hot rolling technique, respectively. The purpose of this study is to identify the effects of thermal-mechanical treatment and RE elements on the microstructural and texture features of the alloy, and to analyze the related mechanical behavior.

2. Experimental procedure

The Mg−1.5Gd−1.5Nd (wt.%) alloy ingot was prepared from high purity Mg (99.9 wt.%), Mg−20Gd and Mg−30Nd (wt.%) master alloys by semi-continuous casting technique. Then the as-cast ingot was performed on homogenization treatment at 500 °C for 16 h, followed by hot water quenching at 60 °C. The homogenized ingot was machined into cylindrical samples with dimensions of Φ120 mm × 200 mm. Hot extrusion was carried out at 380 °C with a ram rare of 1 mm/s. The cross dimensions of extruded alloy sheet was 80 mm × 12 mm, which corresponds to an extrusion ratio of 8:1. After extrusion treatment, hot rolling was further performed on the alloy sheet at the temperature of 400 °C with the rolling rate of 20 mm/s. The extrusion direction is parallel to the rolling direction. The reduction in thickness per pass was 15 %, and between two passes the sheets were reheated at 400 °C for 5 min. After four passes the sheets were hot rolled from 12 mm to 6.2 mm in thickness.

For microstructural observations, Olympus GX71 optical microscopy, and ESEMFEG scanning electron microscopy (SEM) equipped with an X-ray energy-dispersive spectrometer (EDS) were adopted. Texture analysis was performed on the PANalytical X-ray diffractometer using Cu Kα radiation by measuring incomplete pole figures between Ψ = 0–75°. Specimens for measurements were obtained from the mid-layer of the extruded/rolled alloy sheets. Orientation distribution function (ODF) was calculated by MTEX software. Bunge notations of the Euler angles (φ_1, Φ, φ_2) were used to represent the ODF. Tensile specimens with the gauge length of 15 mm and gauge width of 5 mm were cut along the extrusion/rolling direction (ED/RD) and transverse direction (TD) of the alloy sheets, respectively. Uniaxial tensile tests were carried out using WSM-50KB universal testing machine under a strain rate of 1.0×10^{-3} s^{-1}.

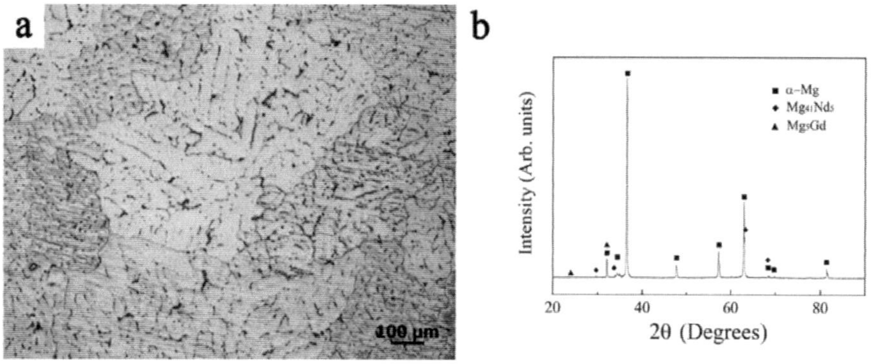

Figure 1. (a) Optical micrograph and (b) X-ray diffraction pattern of the as-cast alloy

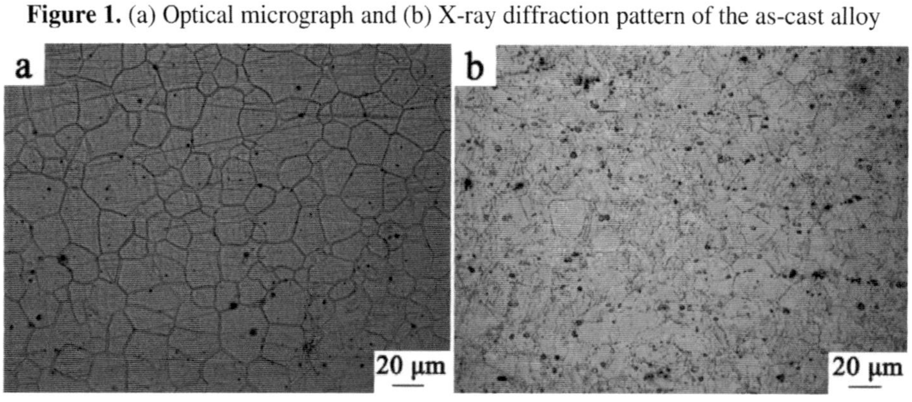

Figure 2. Optical micrographs of the (a) as-extruded and (b) as-rolled alloy

3. Results and discussion

3.1. Microstructure

Figure 1a shows optical micrograph of the as-cast alloy. Obviously it reveals a typical eutectic microstructure consisting of dendrite-like α-Mg matrix and semi-continuous eutectic compounds crystallizing at grain boundaries. According to the X-ray diffraction (XRD) analysis, as indicated in Figure 1b, the eutectic compounds involve $Mg_{41}Nd_5$ and Mg_5Gd phases. The two binary phases are stable intermediate phases in the Mg–Gd–Nd ternary system that can exist at both room and elevated temperatures [11,12]. Hot extrusion and hot rolling treatments took completely dynamic recrystallization for the alloy as shown in Figure 2. The as-extruded alloy (Figure 2a) exhibits fine equiaxial grain structure, while the microstructure of as-rolled alloy (Figure 2b) is more complex that the heavily deformed, twinned and very small grains all emerged in it. This can be attributed to the heavy and rapid deformation condition during rolling processing. Overall, the grain size of as-rolled alloy is definitely smaller than that of as-extruded alloy. Additionally, the eutectic phases have been crushed into small particles in the as-extruded and as-rolled alloys, as indicated in Figures 2a and b.

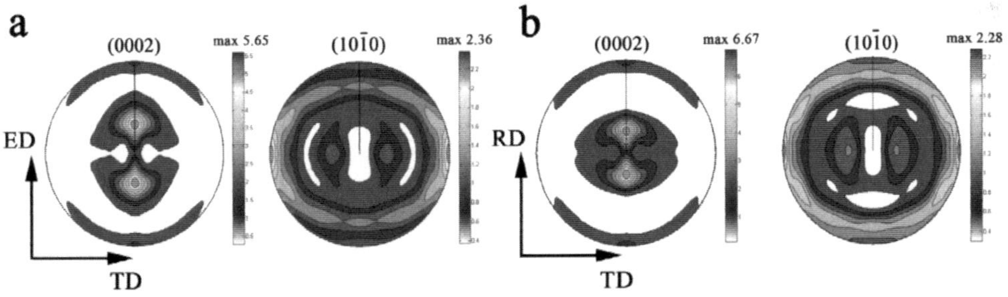

Figure 3. (0002) and (10$\bar{1}$0) pole figures of the (a) as-extruded alloy and (b) as-rolled alloy

3.2. Texture evolution

Figure 3 displays the {0002} and $\{10\bar{1}0\}$ pole figures of the as-extruded and as-rolled alloys. It can been seen that the as-extruded alloy reveals a double peak non-basal texture (Figure 3a), in which the {0002} basal poles tilt about ± 25° from the center (normal direction) towards extrusion direction (ED). While for the $\{10\bar{1}0\}$ prismatic poles the intensity maximum mainly located at the transverse direction (TD). After hot rolling treatment, the texture of as-rolled alloy does not show obviously change in nature, as indicated in Figure 3b. But in detail, the intensity distribution becomes even denser and simultaneously the intensity maximum becomes higher. Additionally, the tilting angle of basal poles towards rolling direction (RD) is shrunk to about ± 20°.

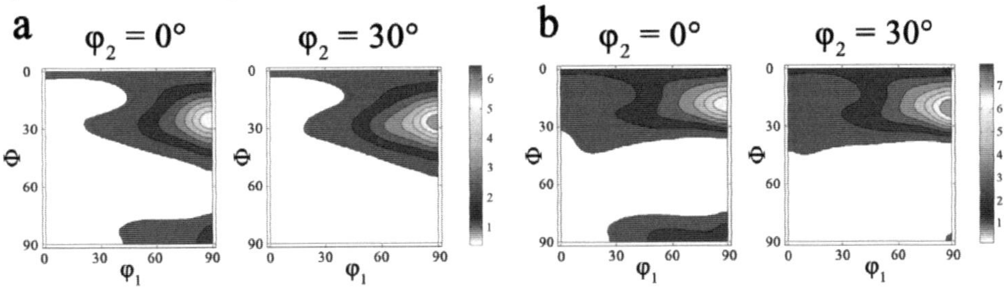

Figure 4. ODF sections at $\varphi_2 = 0°$ and $\varphi_2 = 30°$ for the (a) as-extruded alloy and (b) as-rolled alloy

To better explore the texture components of two alloy sheets, ODF sections at φ_2 = 0° and 30° are presented in Figure 4, where more detailed information about the texture of Mg alloys can be achieved [1]. As seen in Figure 4, the major texture components of two alloy sheets both emerge at the section of φ_2 = 30°. For the as-extruded alloy the strongest texture component appears at the orientation (90°, 27°, 30°) and for the as-rolled alloy the tilt angle of c-axes from the sheet normal (i.e. angle Φ) is obviously reduced that the strongest texture component distributes at orientation (90°, 20°, 30°). Simultaneously, the intensity peaks of major texture components increase from 6.36 multiples of random distribution (mrd) for the as-extruded alloy to 7.82 mrd for the as-rolled alloy. According to the relationship between the orientation g{φ_1, Φ, φ_2} and the Miller-Bravais indices{*h k i l*}<*u v t w*> in Bunge system [13], the two major orientations corresponding to the Miller-Bravais indices {01$\bar{1}$3}<0223> and {01$\bar{1}$4}<01$\bar{1}$2>, respectively. Generally, the basal planes of Mg alloy sheets are mainly aligned parallel to the extrusion/rolling direction [1,13]. In this work, the two major texture components reveal a rotation of conventional basal texture component {0001}<10$\bar{1}$0> (90°, 0°, 30°) to the ED/RD. At the section of φ_2 = 0°, there are two minor texture components, including the component {1$\bar{2}$17}<1$\bar{2}$12> (90°, 26°, 0°) for the as-extruded alloy and the component {1$\bar{2}$19}<1$\bar{2}$12> (90°, 19°, 0°) for the as-rolled alloy, respectively. Also, the two minor texture components show the transition of basal texture component {0001}<11$\bar{2}$0> (90°, 0°, 0°) towards ED/RD. In recent years, many investigations have been put on the study of textures in RE-containing Mg wrought alloys [4, 14-16]. It was found that except the weakened texture strength, another interesting phenomenon is the texture transition from conventional crystallographic orientations, which led to the occurrence of new "RE textures" [4]. It is essentially consistent with the observations in current work.

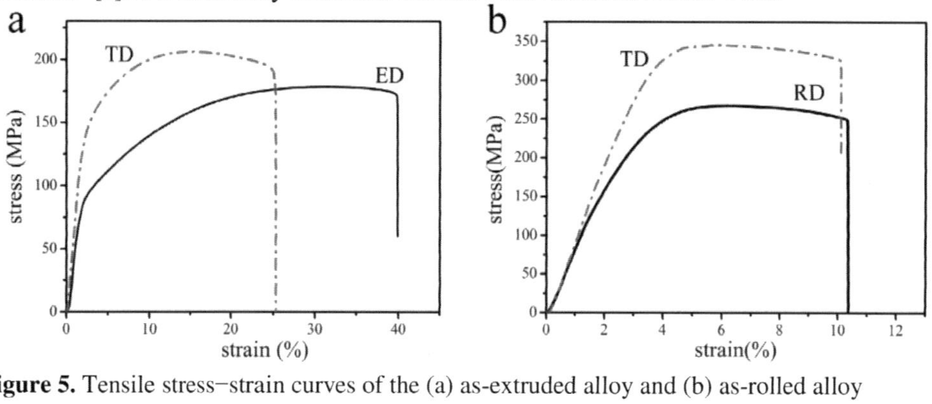

Figure 5. Tensile stress–strain curves of the (a) as-extruded alloy and (b) as-rolled alloy

3.3. Mechanical properties

Figure 5 exhibits the tensile stress–strain (s–s) curves of the as-extruded and as-rolled alloy sheets in different directions. It is found that regardless of specimen directionality, the as-extruded alloy shows superior ductility (Figure 5a) while the as-rolled alloy reveals higher strength (Figure 5b). For the as-extruded alloy it can be attributed to the homogenous microstructure and weaker texture intensity which facilitate the uniform plastic deformation and allow it to accommodate larger strains prior to failure [17]. That the elongation in the ED is as high as 37.5 % and in the TD is 23.2 %. While for the as-rolled alloy, the smaller grain size and higher texture intensity greatly raise its tensile strength that the tensile yield strength in the RD is 245.6 MPa and in the TD is 326.8 MPa. But on the other hand, the strong texture and non uniform grain size distribution impede the continuous plastic deformation, leading to lower elongation. This is reflected by the little strain hardening behavior in the s–s curves of the as-rolled alloy (Figure 5b). For Mg and its alloys which possess the hexagonal close-packed (hcp) crystal structure, the intensity distribution of {0002} basal poles is critical for the planar mechanical properties because the room temperature deformation is dominated by basal slip system [13]. For the as-extruded and as-rolled alloys, the {0002} basal poles are mainly inclined to the ED/RD, as indicated in Figure 3, which means the grains are oriented favorably for basal slip during tension in ED/RD. It definitely reduces the deformation resistance, causing lower tensile yield strength and partly ensures the higher elongation. While in the TD of two alloy sheets, the grains are in hard orientations for basal slip or $\{10\bar{1}2\}$ extension twinning [18]. More non-basal slip systems with higher critical resolved shear stress (CRSS) must be activated to accommodate the plastic deformation, which contributes to the higher tensile yield strength in TD. Overall, the planar anisotropy of the as-rolled alloy is weaker than the as-extruded alloy due to the smaller rotation angle of its {0002} basal poles.

4. Summary

In this work, the Mg−1.5Gd−1.5Nd (wt.%) alloy was processed by hot extrusion and hot rolling treatments, and the microstructures, textures and mechanical properties were studied and compared. The results demonstrate that the as-cast alloy reveals a typical eutectic microstructure consisting α-Mg solid solution and semi-continuous eutectic compounds. Through XRD analysis the eutectic compounds including $Mg_{41}Nd_5$ and Mg_5Gd phases were observed. Thermo-mechanical treatments have a significant effect on the grain refinement of the alloy, and the as-rolled alloy shows a smaller grain size than the as-extruded alloy. Additionally, textures characterized by splitting basal poles towards ED/RD were obtained. Compared with as-extruded alloy, the tilt angle of basal poles in as-rolled alloy was obviously reduced. According to the ODF, the major RE texture components for the as-extruded alloy and as-rolled alloy are $\{01\bar{1}3\}\langle 0\bar{2}23\rangle$ and $\{01\bar{1}4\}\langle 01\bar{1}2\rangle$, respectively. The as-rolled alloy processes higher texture strength. Both the microstructure and texture can influence mechanical properties of the alloy, and the planar anisotropy of the alloy sheets is mainly affected by the texture. In detail, the as-extruded alloy shows superior ductility while the as-rolled alloy reveals higher tensile strength. The planar anisotropy of the as-rolled alloy is weaker than the as-extruded alloy due to the smaller rotation angle of its {0002} basal poles.

Acknowledgements

This study was financially supported by the Natural Science Foundation of Jiangsu Province for Youths (Grants No. BK20130519) and Foundation of Jiangsu University (Grants No. 12JDG094).

References

[1] Hirsch J and Al-Samman T 2013 *Acta Mater.* **61** 818
[2] Zeng Z R, Zhu Y M, Xu S W, Bian M Z, Davies C H J, Birbilis N and Nie J F 2016 *Acta Mater.* **105** 479
[3] Stanford N and Barnett M R 2008 *Mater. Sci. Eng. A* **496** 399
[4] Stanford N and Barnett M R 2008 *Scr. Mater.* **58** 179
[5] Stanford N 2010 *Mater. Sci. Eng. A* **527** 2669
[6] Hantzsche K, Bohlen J, Wendt J, Kainer K U, Yi S B and Letzig D 2010 *Scr. Mater.* **63** 725
[7] Mackenzie L W F and Pekguleryuz M O 2008 *Scr. Mater.* **59** 665
[8] Lentz M, Nissen J, Fahrenson C, Vogel S C and Reimers W 2016 *Mater. Sci. Eng. A* **655** 17
[9] Jung I H, Sanjari M, Kim J and Yue S 2015 *Scr. Mater.* **102** 1
[10] Al-Samman T and Li X 2011 *Mater. Sci. Eng. A* **528** 3809
[11] Chen X H, Geng Y X and Pan F S 2016 *Rare Met. Mater. Eng.* **45** 13
[12] Xiao L, Zhong Y, and Chen C 2014 *Trans. Nonferrous Met. Soc. China* **24** 777
[13] Wang Y N and Huang J C 2003 *Mater. Chem. Phys.* **81** 11
[14] Stanford N, Atwell D, Davies A beer C and Barnett M R 2008 *Scr. Mater.* **59** 772
[15] Huppmann M, Gall S, Muller S and Reimers W 2010 *Mater. Sci. Eng. A* **528** 342
[16] Bohlen J, Yi S, Letzig D and Kainer K U 2010 *Mater. Sci. Eng. A* **527** 7092

[17] Steiner M A, Bhattacharyya J J and Agnew S R 2015 *Acta Mater.* **95** 443
[18] Mu S, Tang F and Gottstein G 2014 *Acta Mater.* **68** 310

Phase field modelling on the growth dynamics of double voids of different sizes during czochralski silicon crystal growth

X J Guan[1,2,3] **and J Wang**[1,2]

[1] School of Materials Science and Engineering, Shandong University, Jinan 250061, China
[2] State Key Laboratory of Crystal Materials, Shandong University, Jinan 250100, China

Abstract. To investigate their dynamics and interaction mechanisms, the growth process of the two voids with different sizes during Czochralski silicon crystal growth were simulated by use of an established phase field model and its corresponding program code. On the basis of the several phase field numerical simulation cases, the evolution laws of the double voids were acquired as follows: the phase field model is capable to simulate the growth process of double voids with different sizes; there are two modes of their growth, that is, either mutual integration or competitive growth; the exact moment of their fusion can be also captured, and it is τ of 7.078 (simulation time step of 14156) for the initial vacancy concentration of 0.02 and the initial space between two void centers of $44\Delta x$.

1. Introduction

In the past decades, along with the development of computer technology, the simulation research on microdefect evolution in Czochralski silicon crystal develops so rapidly that numerical modelling becomes one of the important research approaches to understanding microdefect dynamics thoroughly as well as predicting the microdefect size and distribution. Phase field modelling is a burgeoning numerical simulation method to be usually applied to microstructure evolution in materials in recent years. Unlike traditional sharp interface model, phase field model is capable of modelling the evolution process of microstructures without real-time tracking of the interface movement. Thus, it is widely used in the simulation research of the material microstructure evolution at mesoscopic level.

For the sake of investigating the dynamics of microdefect evolution and capturing the essential features of the formation and growth of the grown-in microdefects, different theoretically analytical models and simulation tools were developed [1-3]. In recent years, researchers [4-5] in the field of nuclear materials are carrying out relative researches about the void-type microdefect evolution in irradiated materials using phase field modelling, which provides us the inspiration of applying phase field modelling on the microdefect evolution in single crystal growth.

The growth dynamics of double voids of different sizes during CZ silicon crystal growth are concerned using phase field modelling in this paper, which is the follow-up work and extension of our previous study [6], in which phase field simulations of the growth dynamics of double voids of the same size during CZ silicon crystal growth were reported. Based on the constructed phase field model, the phase field modelling experiments for the growth dynamics of double voids of different sizes were conducted to illustrate the evolution mechanism of the concerned voids and study the influence laws of relative factors.

2. Simulation model and conditions

2.1. Phase field model

The simulation system concerned is consisted of two phases, the matrix phase containing vacancies of supersaturated concentration and the void phase. The vacancy concentration field, c_v, and order parameter field, ϕ, were introduced as phase variables to describe the vacancy diffusion and void evolution, where $\boldsymbol{r}$ is spatial position and t is time. In the matrix phase, ϕ is 0. While in the void phase, both c_v and ϕ are set to be 1. The total free energy functional of the system includes the chemical free energy and gradient energy as follows:

$$E = N\int_V (F(c_v,\phi,T) + wg(\phi) + k_v|\nabla c_v|^2 + k_\phi|\nabla\phi|^2)\mathrm{d}V \quad (1)$$

where N is the number of the lattice sites per unit volume of the crystal; $F(c_{v},\phi,T)$ is the total chemical free energy; T is the absolute temperature; $g(\phi)$ is a double-well function related to phase transition barrier; w is a positive constant; $\kappa_v|\nabla c_v|^2$ and $\kappa_\phi|\nabla\phi|^2$ represent the interfacial energy contributed by the gradients of composition and chemical ordering, repetitively, where κ_v and κ_ϕ are gradient energy coefficients. The total chemical free energy and the double-well function are:

$$F(c_v,\phi,T) = (1-h(\phi))f_m(c_v,T) + h(\phi)f_v(c_v) \quad (2)$$

$$g(\phi) = \phi^2 - 2\phi^3 + \phi^4 \quad (3)$$

where $h(\phi)$ is a interpolating function; $f_m(c_v,T)$ and $f_v(c_v)$ are chemical free energy of matrix phase and void phase, repetitively:

$$h(\phi) = 3\phi^2 - 2\phi^3 \quad (4)$$

$$f_m(c_v,T) = (E_{f,v} - k_B T\ln(c_{v,0}))c_v + k_B T(c_v\ln(c_v) + (1-c_v)\ln(1-c_v)) \quad (5)$$

$$f_v(c_v) = (c_v - 1)^2 \quad (6)$$

where $E_{f,v}$ is the vacancy formation energy; k_B is the Boltzmann constant; $c_{v,0}$ is the pre-exponential factor of the thermal equilibrium vacancy concentration in the matrix, $c_{v,e} = c_{v,0}\exp(-E_{f,v}/k_B T)$.

The process of vacancy diffusion and void evolution were described by Cahn-Hilliard conservative field equation [7] and Allen-Cahn non-conservative field equation [8], repetitively:

$$\frac{\partial c_v}{\partial t} = \nabla\cdot M_v\nabla(\frac{1}{N}\frac{\delta E(c_v,\phi,T)}{\delta c_v}) \quad (7)$$

$$\frac{\partial\phi}{\partial t} = -L(\frac{\delta E(c_v,\phi,T)}{\delta\phi}) \quad (8)$$

where $M_v = D_v c_v/k_B T$ is the vacancy mobility with D_v being the vacancy diffusivity; L is the free surface mobility.

2.2. Initial conditions and parameters

The two-dimensional simulation region was divided into isometric grids of 256Δx×256Δx, where Δx was grid length. There were two isolated voids with the radiuses of 10Δx and 30Δx, respectively, at the center of the solution domain, surrounded by the matrix containing uniform vacancies of supersaturated concentration, and the initial temperature was set to be 1385K. In the simulation, periodic boundary condition was applied. The simulation parameters were listed in Table 1, and the physical meanings can be referred in the Reference [6].

Table 1. The parameters used in the phase-field simulation.

Parameters	Value
$E_{f,v}$ [eV]	4
$\Delta\tau$	5.0×10^{-4}
q [$K\cdot\Delta\tau^{-1}$]	0.5
$\tilde{D}_v$	2
$\tilde{L}$	100
κ_v [$eV\cdot l^{*2}$]	1
κ_ϕ [$eV\cdot l^{*2}$]	1
w	1
$c_{v,0}$	10^{11}

3. Results and Discussion

Figure 1 shows the evolution of order variable field for the initial vacancy concentration, $c_{v,i}$, of 0.02 and the initial space between two void centers of $50\Delta x$ at τ = 0, 50, 100, 150, respectively. With the pulling of the silicon single crystal and the cooling of the simulation system, the supersaturated vacancy in the matrix enhances the instability of this thermodynamic system, which prompts the vacancy' s inward diffusion to void and void growth. As seen in Figure 1, the two voids grow discretely in the whole simulation. The left void grows into an irregular one with edges and corners, while the right void grows into an oval shape one. The irregular left void can be interpreted by the nine-point method applied on the calculation of Laplace operator, and the oval-shaped right void attributes to the more affluent vacancies in its upper and lower matrix, compared with the vacancies in the matrix of the two voids gap which are absorbed by the bigger left void.

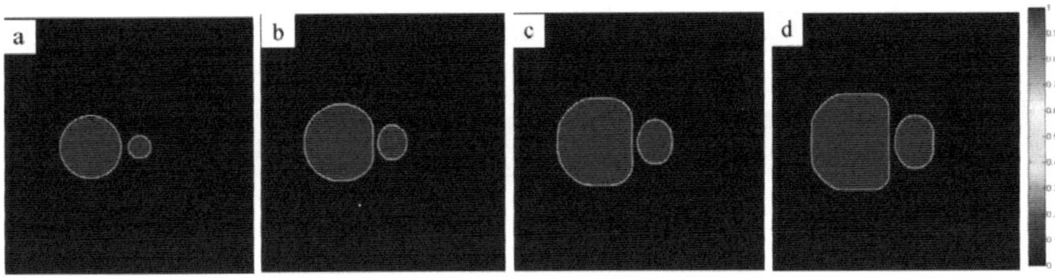

Figure 1. The evolution of ϕ field for $c_{v,i}$ = 0.02 and the initial space between two void centers of $50\Delta x$ (a) τ = 0; (b) τ = 50; (c) τ = 100; (d) τ = 150.

Figure 2 shows the evolution process of order variable field for the initial vacancy concentration, $c_{v,i}$, of 0.02 and the initial space between two void centers of $44\Delta x$ at τ = 0, 50, 100, 150, respectively. Unlike the independent growth mode shown in Figure 1, the initial isolated voids in Figure 2 merge together at some moment and grow into an irregular void sequentially. The two processes demonstrate that the initial space between void centers affects the growth modes of the double voids, that is, either mutual integration or competitive growth.

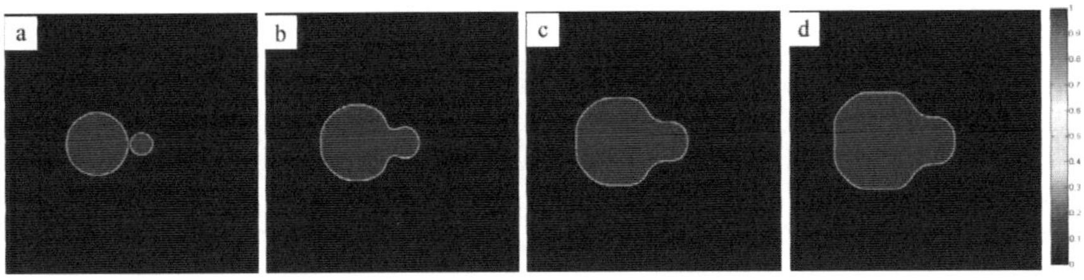

Figure 2. The evolution of ϕ field for $c_{v,i}$ = 0.02 and the initial space between two void centers of $44\Delta x$ (a) τ = 0; (b) τ = 50; (c) τ = 100; (d) τ = 150.

To capture the exact fusion moment of the two voids, a result- processing program was written and their fusion process for the initial vacancy concentration, $c_{v,i}$, of 0.02 and the initial space between two void centers of 44Δx is simulated in Figure 3. The exact moment of fusion, τ_e, is 7.078 in the phase field simulation, which corresponds to simulation the time step of 14156. From this moment, the two isolated voids merge into a void, and the single void will go on its growth by absorbing vacancies in the matrix until the simulation system reaches thermodynamic equilibrium state, where the vacancy concentration in the matrix is equilibrium concentration at some temperature.

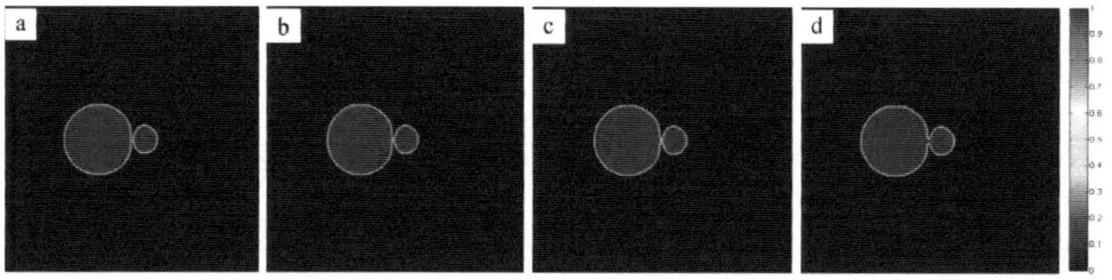

Figure 3. The fusion process of voids for $c_{v,i}$ = 0.02 and the initial space between two void centers of 44Δx (a) τ = 6.5; (b) τ =6.75 ; (c) τ = 7.0; (d) τ = 7.078.

4. Conclusion

The growth dynamics of double voids of different sizes were studied by using an established phase field model and related program codes for simulating the evolution behavior of void growth during Czochralski silicon single crystal growth. Series of simulation cases aimed to illustrating the growth dynamic laws of the double voids of different sizes were conducted. The results show that the phase field model can simulate the growth process of the two voids, and the initial space between the two void centers has a significant effect on their growth modes, and the related program codes have also the ability to capture the process and exact moment of void fusion.

Acknowledgments

We greatly acknowledge for the financial support from the Opening Foundation Project granted by State Key Laboratory of Crystal Materials, Shandong University (granted no. KF1303).

References

[1] Kulkarni M S, Voronkov V V and Falster R 2004 Quantification of defect dynamics in unsteady-state and steady-state czochralski growth of monocrystalline silicon *J. Electrochem. Soc.* **151** G663-78

[2] Sinno T and Brown R A 1999 Modelling microdefect formation in czochralski silicon *J. Electrochem. Soc.* **146** 2300-12

[3] Dai J, Seider W D and Sinno T 2007 A lattice kinetic monte carlo study of void morphological evolution during silicon crystal growth, *Mol. Simul.* **33** 733-45

[4] Hu S Y and Henager C H 2009 Phase-field modelling of void lattice formation under irradiation *J. Nucl. Mater.* **394** 155-59

[5] Rokkam S, El-Azab A and Millet P 2009 Phase field modelling of void nucleation and growth in irradiated metals *Modell. Simul. Mater. Sci. Eng.* **17** 064002

[6] Guan X J, Wang J and Zhang X Y 2015 Phase-field simulation for the growth dynamics of doub le voids during czochralski silicon crystal growth *J. Syn. Cryst.* **44** 1207-12

[7] Cahn J W and Hilliard J E 1958 Free energy of a nonuniform system: Ⅰ. interfacial free energy *J. Chem. Phys.* **28** 258-67

[8] Allen S M and Cahn J W 1979 A microscopic theory for antiphase boundary motion and its application to antiphase domain coarsening *Acta. Metall.* **27** 1085-95

Application and Preparation of Enteric Coating Materials

M M Zhou[1, a], L Wang[2, b], X L Zhang[1], H J Zhou[1], X Q Chen[1], Y T Li[1, c], S L Yang[1]

[1]Institute of Energy Source, Hebei Academy of Sciences, Shijiazhuang, 050000, China
[2]Hebei Institution for Product Quality Supervision and Inspection, Shijiazhuang, 050000, China

E-mail:[a]applezhoumeng@aliyun.com, [b]282932157@qq.com,
[*1,c]13903116163@163.com (corresponding author)

Abstract. In this paper, polymethacrylate enteric coated materials based on the equal mass of methyl acrylic acid and ethyl acrylate as the main raw materials were synthesized through emulsion polymerization. Omeprazole Enteric-coated Capsules were prepared by the fluidized bed coating technology using above materials as enteric layer and in vitro enteric test was considered according to standard. The results showed that the material had good coverage in the surface of omeprazole isolated pellets, excellent acid resistance in artificial gastric acid environment, and reached the disintegration effect in the buffer solution of 20min. Moreover the drug release reached 88.2% and had excellent long-term storage.

1. Introduction

Omeprazole was widely used in treatment of gastric ulcer and duodenal ulcer because of its strong acid-suppression, high specificity and long duration, which was a representative drug of first generation proton pump inhibitors (PPIs). However, its property of easy decomposition for the structure that the human body is not easy to absorb in an acid environment, decided that their oral preparations are all enteric preparation. Poly methacrylate has been widely concerned as coating materials, which could be prepared for enteric coated materials for its COOH group in the molecular structure. This material can control disintegration of drug in a specific pH environment, while in the gastric acid environment not released, so as to guarantee the effectiveness of drug treatment [1~4]. At present, there has been many studies on enteric coated polyacrylic acid resin. Yao et al[5,6] prepared enteric coating materials, according to change -COOH content, using methycl acrylic acid and methyl methacrylate as raw materials, which can be dissolved in pH>6 or pH>7 medium. However, it was necessary to add mass plasticizer before use.

This paper aimed at preparing enteric coated pharmaceutical coating materials with good solution in pH>5.5 medium, which took methacrylic acid and ethyl acrylate as raw materials. Then this material was used to coat on the omeprazole isolated pellets to form capsules by fluidize bed coating technology. And in vitro enteric test results meet the limits of "Chinese pharmacopoeia" for various index.

2. Experimental

2.1. Materials.

Methacrylic acid (MAA) (Sanyi Technology Co., Jiangsu, China) were purified by distillation under reduced pressure before use. Ethyl acrylate (EA) and potassium persulfate (C. P. grades) were supplied by Tianjin Taixing Reagent Factory. Alkyl ether sulfates, disodium laureth sulfosuccinate and fatty alcohol-polyoxyethylene ether were donated by Honesty Fine Chemical Co., Ltd. (Shanghai, China). Omeprazole isolated pellets were obtained from Hebei pharmatrade Biotechnology Co., Ltd. Triethyl citrate (99.9% of purity, Sinopharm Chemical Reagent Co. China), talcum powder (10μm of size) and distilled water were used.

2.2. Synthesis of Enteric Polymethacrylate Coating Materials.

The pre-emulsion with solid content of 46% was obtained through pre-emulsified technology. The equal mass of methyl acrylic acid and ethyl acrylate were put into a reactor. Then the emulsifier and initiator and distilled water were added into reaction system. And the amount of emulsifier (the ratio of anionic emulsifier and nonionic emulsifier was 3:1) and initiator were 3% and 0.46%, respectively. Subsequently the pre-emulsion was obtained after 1 hour's pre-emulsification under high-speed mixing. Moreover, the seeded emulsion was obtained by adding 1/5 of the pre-emulsion and remaining deionized water into a reactor, and heated to 80 ℃ for 0.5 h. Then the remaining mixture was added dropwise for 4~5 hours, and the reaction mixture was maintained for 1h at 80 ℃. After cooled to room temperature and filtered, the enteric polymethacrylate water dispersion coating materials were obtained.

2.3. Preparation of Omeprazole Enteric-coated Capsules.

1%(w/w) triethyl citrate and 2.4%(w/w) talcum powder were added to distilled water and stirred until a homogenous dispersion was obtained. And the 15%(w/w) enteric coating liquid was obtained after adding polymethacrylate emulsion into the dispersion. Fluidized bed coating technology was adopted to prepare omeprazole coated pellets under the following conditions: fan frequency of 40Hz, inlet air temperature of 45℃, feed rate of 1~1.2rpm, atomization pressure of 0.1Mpa. Then omeprazole coated capsules were obtained with coated pellets weight gain of 33% after drying for 20 min in the fluidized bed and baking for 2h in the oven of 38℃.

2.4. Characterization.

The molecular structure of polymethacrylate was analyzed by Fourier Transform Infrared Spectrometer (FT-IR). Particle size and distribution were recorded on the LS900 laser particle size analyzer. The molecular weight of the polymer was measured via gel permeation chromatography and an eluent of tetrahydrofuran, at a flow rate of 1ml/min was selected. The release of omeprazole enteric coated capsules was tested according to standard of " Chinese pharmacopoeia ".

3. Results and Discussion

3.1. Basic Physicochemical Properties of Enteric Polymethacrylate Coating Materials.

The basic physicochemical properties and molecular structure of enteric polymethacrylate coating materials were tested and analyzed, which were shown in Figure.1 and Table 1.

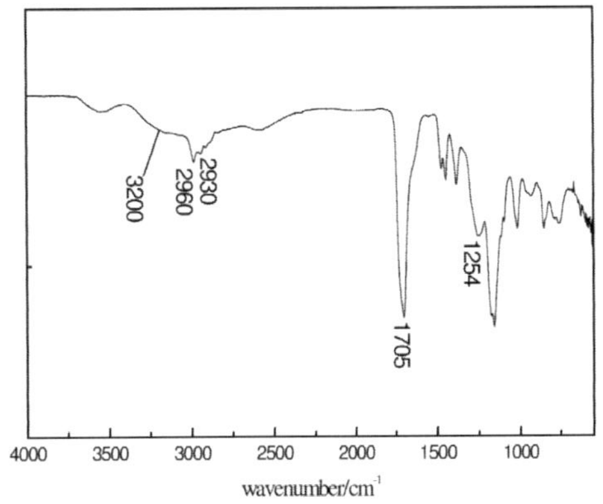

Figure.1. The FT-IR spectra of enteric polymethacrylate coating materials

The existence of carboxyl groups are the key for polymethacrylate enteric coated materials. Figure.1 depicted the FT-IR spectra of this material. There was no -C=C- stretching vibration at 1640cm-1, which indicated monomer react completely. The FT-IR showed some characteristic bands, C-O and C=O stretching vibrations were presented at 1254cm-1 and1705cm-1. And –OH stretching vibrations can also be found at 3200cm-1. These results indicated that there are COOH groups in the molecular chain and correspond to the acidity test of 2.3 in Table 1. Thus the enteric polymethacrylate coating material was prepared successfully. From Table 1 we knowed that this material has properties of low viscosity, small particle size, which benefit for the coating operation.

Table1. Basic physicochemical properties of enteric polymethacrylate coating materials

Name	Polymethacrylate	Norm
Shape	Milky white liquid	——
Structure	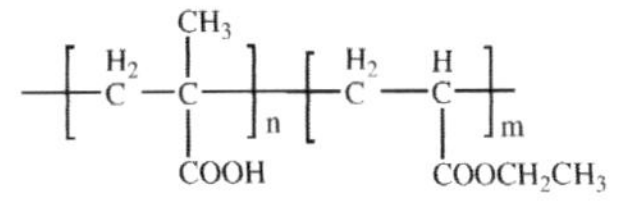	——
Molecular weight	230000	——
Particle size/nm	≤180	——
Viscosity /mPa.s	10	3~10
Acid value	320	300~330
Acidity	2.3	2.0~3.0
Amount of residue after Evaporation	29.7%	28.5%~31.5%
Residue on ignition	0.01%	0.1%<

3.2. Preparation of Enteric Coated Omeprazole Capsules.

The enteric coating material was coated on the surface of omeprazole isolation pellets by fluidized bed coating technology, and increasing weight about 33%. Then white omeprazole enteric coated pellets of smooth appearance were obtained, which were stored in condition of (25±2) ℃, RH(60±5) %. The results showed that, the acid resistance and the release rate of three batch of omeprazole enteric coated capsules were (99.19±1.2)%、(97.00±1.6)%、(96.85±2.0)% and (86.97±1.4)%、(92.14±1.8)%、(93.44±2.2)%, respectively. And the acid resistance and the release rate were no less than 90% and 80%, coinciding with standard of " Chinese pharmacopoeia " and showing excellent acid resistance and release properties. Figure.2 showed the dissolution effect of enteric coated pellets in buffer dissolution. Pellets were intact and no apparent discoloration rupture with the acid resistance of （96.85±2.0）% when put in artificial gastric juice 2h. While in buffer solution 20 min, the drug release up to 88.2% and achieve the disintegration effect.

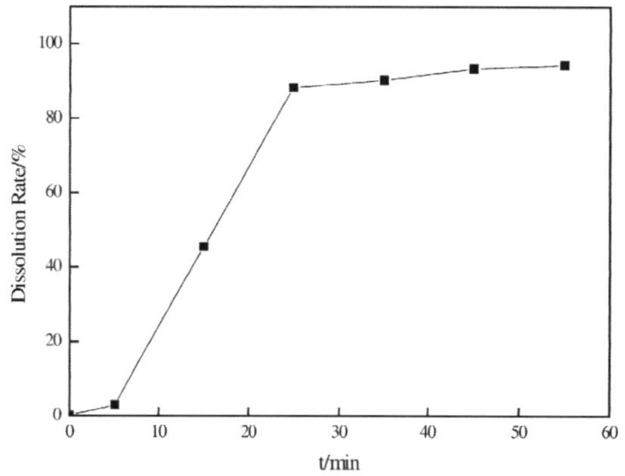

Figure.2. Dissolution curve of omeprazole enteric capsules

3.3. Stability of Long-term Storage of Enteric Coated Omeprazole Capsules.

The omeprazole enteric capsules were stored in （25±2）℃, RH （60±5）%environment. In order to observe the long-term storage stability of capsules, their acid resistance and release properties were tested every 3 month.

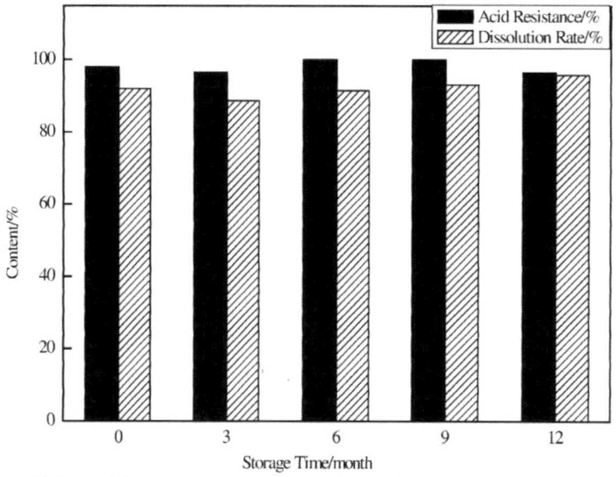

Figure.3. Stability of long-term storage of enteric coated omeprazole capsules

Figure. 3 showed that the acid resistance and release of omeprazole enteric capsules are in accordance with the limits prescribed in the standard, and have excellent long-term storage property.

4. Summary

In this paper, the enteric coating materials with low viscosity, small size were synthesized successfully through emulsion polymerization. Enteric coated omeprazole capsules were prepared by fluidized bed coating technology. The experimental results showed that taking this material as enteric layer has a good performance of excellent acid resistance, release rate, good disintegrating effect and long-term storage stability.

References

[1] Olsen K M, Bergman K L, Kaufman S S, et al. 2001 Omeprazole pharmacodynamics and gastric acid suppression in critically ill pediatric transplant patients[J]. *Pediatric Critical Care Medicine*,, **2(3):** 232-237.

[2] Zhou X, Zhang J, Wang J, et al. 2012 Preparation of omeprazole enteric capsules by encapsulating enteric-coated pellets[J]. *Pharm J Chin PLA*, **28(1)**: 32-35.

[3]Zheng L L, Huang Y, Zhang Z R, et al. 2005 Synthesis and characterization of a novel Ph-sensitive coating materials[J]. *West China Journal of Pharmaceutical Sciences.***20(3)**:235-238.

[4] Carlsson E 1985 Animal pharmacodynamics of omeprazole: A survey of its pharmacological properties in vivo[J]. *Scandinavian Journal of Gastroenterology*, **20(S108):** 23-35.

[5] Yao J S, Li M, Qiao C D, et al 2010 Study on the synthesis of aqueous acrylate copolymers using as pharma centical coatings.[J]. *Laboratory Science.* **13(02)**: 70~72.

[6] Liu L, Zhang Y N 2008 Study of the aqueous acrylic acid resin, Dissolved in intestines for pharma centical coating[J]. *Synthetic Materials Aging and Application.* **37(3)**:23~27.

Enhanced ethanol response of La_2O_3-loaded SnO_2 nanorods prepared by an improved solid state reaction

F Gao[1], Y P Zhao [1], W J Wan[1], X P Ren[2], H Y Zhao[1,3]

[1] College of Materials Science and Engineering, Yunnan University, Kunming, 650091, PR China.
[2] Yunnan Security and Technology Co. Ltd., Kunming, 650033, PR China.
[3] Yunnan Key Laboratory for Micro/Nano Materials and Technology, Kunming 650091, PR China

E-mail: hyzhao@ynu.edu.cn; rxp89@126.com

Abstract. La_2O_3-loaded SnO_2 nanorods with rutile structure were successfully synthesized by an improved solid state reaction with the surfactants in the presence of NaCl-KCl. The indirect-heating sensor was prepared using the products as sensing-material to study the response of the La_2O_3-loaded SnO_2 nanorods. The results showed that the performances of the SnO_2 nanorods to ethanol was greatly improved after loading La_2O_3. The significant improvement could be attributed to the excellent catalytic properties of loaded La_2O_3 and the high surface area.

1. Introduction

As a well-known substance used for detection of different VOC pollutants owing to chemical and thermal stability, good gas response and low cost [1], SnO_2 has been extensively studied for detection of a wide variety of inflammable and toxic VOCs gases such as ethanol, acetone, isopropanol and so on [2]. However, it is also suffering a serious problem associated with maximum sensitivity, high operating temperature, lack of long-term stability and poor selectivity and so on. In order to overcome these shortcomings, a lot of exploratory work have been devoted to the new material with enhanced gas sensing performances. Recently, one-dimensional (1-D) nanostructured SnO_2 such as nanowires [3], nanobelts [4], nanorods [5] and so on, is attracting a great deal of attention to study the applications in sensor technology due to their high surface–volume ratio and enhanced surface reactivity. Another significant effort, doping Lanthanide has been used to overcome these limitations and improve the performances [6].

In this work, La_2O_3-loaded SnO_2 nanorods synthesized by an improved solid state reaction with surfactants in the presence of NaCl-KCl. Indirect-heating sensors based on nanorods were fabricated and investigated for the response to ethanol. The gas sensing experiments of La_2O_3-loaded SnO_2 nanorods were carried out compared against the no-loaded SnO_2 nanorods to research the effect of loading La_2O_3 on the sensing properties.

2. Experimental

All utilized analytical grade chemical reagents were purchased from the Sinopharm Chemical Reagent Co. No-loaded and La_2O_3-loaded SnO_2 nanorods were synthesized via an improved solid state reaction. Typically, a mixture of the $SnCl_4 \cdot 5H_2O$ added in an amount of $LaCl_3 \cdot 6H_2O$ (Did not add for no-loaded SnO_2 nanorods) was mixed with KBH_4 in the presence of KCl-NaCl and nonyl phenyl ether (9)/(5) at room temperature. Then, the mixture was fully ground to cause the solid phase reaction. The product kept in a porcelain crucible was annealed at 680 °C for 2h. After that, the resulting product was washed with distilled water several times to remove NaCl and KCl, and was dried at 80 °C for 12 h for further characterization.

The X-ray diffraction (XRD) using a Rigaku D/MAX-3B powder diffractometer was performed to determine the crystalline structure and phase of SnO_2 nanorods. The morphology and detailed structure was characterized by an FEI Quanta 200 scanning electron microscope (SEM) and a transmission electron microscope (TEM, JEOL 2010, 200 KV). The indirect-heating sensor was elected to study the response of the SnO_2 nanorods [2]. The test of the gas-sensing properties carried out on a JF02F measurement system in a relative humidity range of 40–70%. The response was defined as the ratio of R_a/R_g, where R_a and R_g is the resistance in atmospheric air and in target gas, respectively.

3. Results and discussion

The typical XRD patterns of the as-prepared SnO_2 nanorods were shown in Figure 1(a). As seen from Figure 1(a), all the diffraction peaks of the La_2O_3-loaded SnO_2 nanorods can be indexed to the rutile structured SnO_2 (JCPDS file No. 41-1445), which is almost similar to that of the no-loaded SnO_2 nanorods. However, it is obvious that two weak additional diffraction peaks located at 28.42° and 48.56°, which could be assigned to the crystalline phases of La_2O_3 can be detected. Moreover, the relative intensities of the SnO_2 nanorods sample decreased after loading La_2O_3, suggesting the growth of SnO_2 nanorod was prevented by La_2O_3.

Figure 1(b) and (d) showed the morphologies of the SnO_2 nanorods characterized by SEM. Being compared with the no-loaded SnO_2 nanoroads (shown in Figure 1(d)), finer rod-like structured regular morphologies of the La_2O_3-loaded SnO_2 nanorods with the length in the range of 500 nm and several micrometers can be observed. It indicates that the La_2O_3 could prevent the nanorods further growth up [7]. The thinner nanorods could provide a large surface area, contributing to increase the probability of surface trapping to spur the surface reaction. The EDS spectrum of the La_2O_3-loaded SnO_2 nanorods shown in Figure 1(c) (inset) reveals the component of stannum (Sn), oxygen (O) and lanthanum (La), while the La is not observed in the EDS spectrum of the no-loaded SnO_2 nanoarods shown in Figure 1(e) (inset).

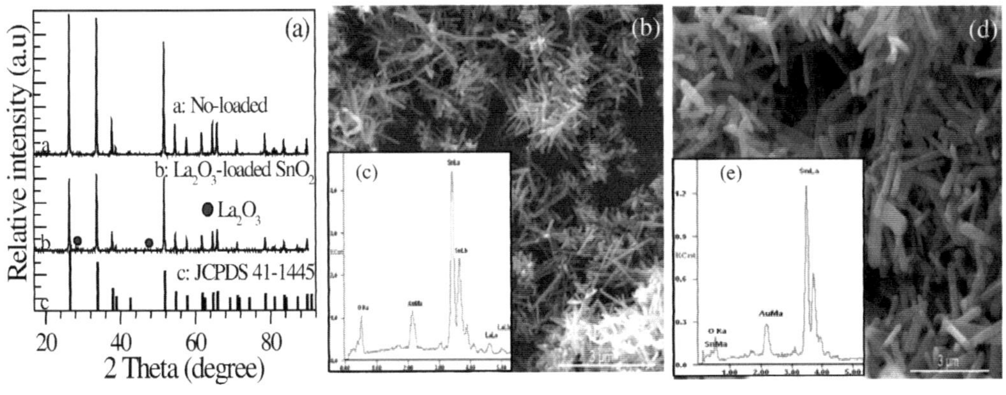

Figure 1. (a) XRD patterns of as-synthesized SnO_2 nanorods. (b) and (d) SEM images of SnO_2 nanorods. (c) and (e) EDS spectra of SnO_2 nanorods (inset).

The TEM image of the La_2O_3-loaded SnO_2 nanorods shown in Figure 2 displayed that the product mainly consists of solid rod-like structures with diameters 20 nm and length up to several micrometers. Furthermore, it is obvious that many small La_2O_3 crystalline grains are attached to the SnO_2 nanorods surface. The HRTEM of individual SnO_2 nanorod shown in Figure 2b depicted the uniform structure and clear lattice fringes. The 0.342 nm space between two adjacent lattice fringes of the La_2O_3-loaded SnO_2 nanorod corresponds to the (110) lattice plane of SnO_2 in tetragonal cassiterite structure. The

Fourier transform electron diffraction pattern (FFT) (inset) indicated the single structure of the La_2O_3-loaded SnO_2 nanorod.

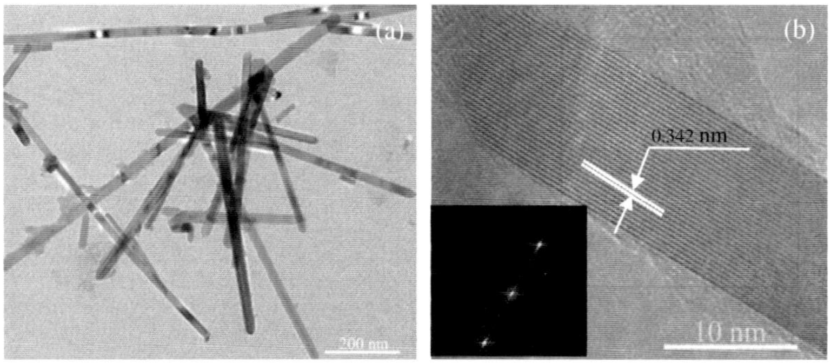

Figure 2. TEM images of La_2O_3-loaded SnO_2 nanorod. (a) General TEM image. (b) HRTEM image. Inset: Fourier transform electron diffraction pattern.

Figure 3(a) showed the response dependence of the sensors on the operating temperature for 500 ppm ethanol. It showed that the responses increase with the temperature at first, after reaching a maximum value of 38 and 116 for no-loaded and La_2O_3-loaded SnO_2 nanorods at optimum operating temperature of 246 °C, and then decreased rapidly. This performance is associated with the kinetics and mechanism of gas adsorption and desorption of semiconducting metal oxides [8]. From the Figure 3(a), it can be seen that the SnO_2 nanorods sensors have a lower optimum temperature, and the response of La_2O_3-loaded SnO_2 nanorods is much larger than that of the no-loaded SnO_2 nanorods. Figure 3(b) showed the investigated response curves of La_2O_3-loaded SnO_2 nanorods sensor as a function of different ethanol concentration at the optimum operating temperature of 246 °C. It can be observed that the response increases rapidly in the range of 5 to 1000 ppm. The response increases slightly above 1000 ppm, indicating the sensors approach to saturation.

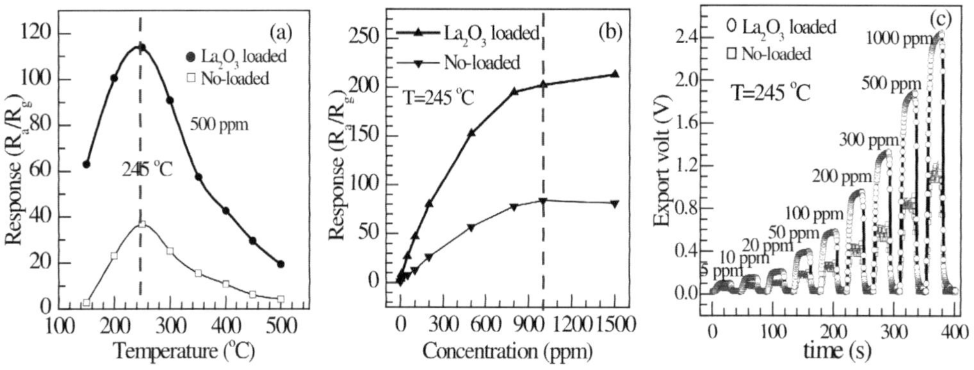

Figure 3. (a) Response depending on the operating temperature. (b) Response to concentrations at 246 °C. (c) Dynamic response and recovery behaviours.

The repeatable characteristic of the response and recovery was shown in Figure 3(c) achieved by the La_2O_3-loaded SnO_2 nanorods sensor being orderly exposed to different concentrations at the optimized temperature of 246 °C. It is evident that the response amplitude of the sensors gradationally increases with the concentration changed from 5 to 1000 ppm. The sensor presents a considerable response to low ethanol concentration of 5 ppm. It is worth noting that the characteristics of the SnO_2

nanorods were remarkably improved after loading La_2O_3 compared with the no-loaded SnO_2 nanoroads. The response (t_{Res}) and recovery (t_{Rec}) time is an important parameter to assess a gas sensor. Figure 4(a) and (b) showed the transient response–time data of the La_2O_3-loaded SnO_2 nanorods sensor to 500 ppm at 246 °C. It is easy calculated from the Figure 4(a) and (b) that the response time t_{Res} values is 9 s, while the recovery time t_{Rec} value is about 5 s. This result indicates the response (t_{Res}) and recovery (t_{Rec}) time is short enough for practical application. Furthermore, we performed a series of experiments of SnO_2 nanorods exposed to 300 ppm VOCs gas including acetone, ethanol, toluene, isopropanol, formaldehyde, petrol, carbinol and ammonia in order to explore the selectivity, and the result was shown in Figure 4(c). It showed that sensors exhibited a good selectivity to ethanol. Particularly, the response of the no-loaded SnO_2 nanorods is no greater than 36, while the response of the Ln_2O_3–loaded SnO_2 nanorods sensor achieves 84, which is more than 2 times larger than that of the no-loaded SnO_2 nanorods. The greatly improved response could be attributed to the excellent catalytic properties of loaded La_2O_3 and the restrained growth of the SnO_2 nanorods resulting in a high surface area, being beneficial to increase the amount of active sites on the SnO_2 nanorods surface [6].

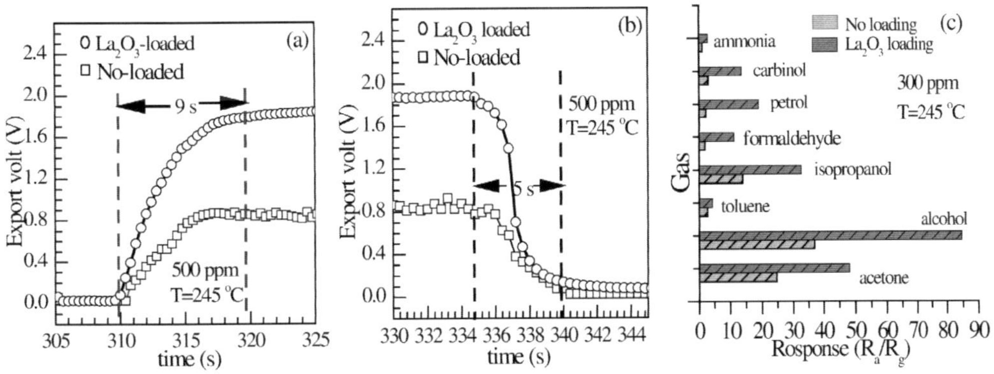

Figure 4. (a) and (b) Enlarged response and recovery transient time examined. (c) Bar chart showing the response of La_2O_3-loaded SnO_2 nanorods to different gases.

4. Conclusions

In summary, La_2O_3-loaded SnO_2 nanorods with rutile structure were successfully synthesized by an improved solid state reaction in the presence of NaCl-KCl and surfactants. The response of the La_2O_3-loaded SnO_2 nanorods to ethanol showed high response, good selectivity, low detection limit, and short response and recovery time. The response improvement of the La_2O_3-loaded SnO_2 nanorods to ethanol is remarkable. Doping Lanthanide is effective and significant route to improve the performance of the SnO_2 nanorods for applications in detecting ethanol.

Acknowledgements

This work was financially supported by the National Natural Science Foundation of China (No. 61564009 and 61161008), the Natural Science Foundation of Yunan Province (2009CD015) and the Graduate Student Innovational Scientific Research Project of Yunnan University (No.YNUY201693).

References

[1] Wang Z J, Liu, S Y, Jiang T T, Zhang J, An C H and Wang C 2015 *RSC Adv.* **5** 64582
[2] Qin G H, Gao F, Jiang Q P, Li Y H and Zhao H Y 2016 *Phys.Chem.Chem.Phys.* **18** 5537
[3] Wang B, Zhu L F, Yang Y H, Xu N S and Yang G W 2008 *J. Phys. Chem. C* **112** 6643
[4] Fields L L and Zheng J P 2006 *Appl. Phys. Lett.* **88** 263102
[5] Yu L M, Fan X H, Qi L J, Ma L H and Yan W 2011 *Appl. Surf. Sci.* **257** 3140
[6] Zhang G Z, Zhang S P, Yang L, Zeng D W and Xie C S 2013 *Sens. Actuators B* **188** 137
[7] Ruiz A M, Cornet A, and Morante J R 2004 *Sens. Actuators B* **100** 256
[8] Yamazoe N, Fuchigama J, Kishikawa M and Seiyama T 1979 *Surf. Sci.* **86** 335

IOP Conf. Series: Materials Science and Engineering **170** (2017) 012023 doi:10.1088/1757-899X/170/1/012023

Studies of Composition and performance of mountain sand concrete powders

Z H Tang[1,b], X G Li[2,a], X P Nie[3,c], J C Wang[4,d], S X Shuai[3]
[1]China Construction Forth Engineering Division Corp.Ltd.,Guanzhou,Guandong,510000.
[2]College of Civil Engineering,Guizhou University,Guiyang,Guizhou,550006,China.
[3]Guizhou Construction Science Research & Design Institute Limited Company of CSCEC,Guiyang, Guizhou,550006,China.
[4]Guizhou Ultrafine Powder Technology Co.Ltd,Guiyang,Guizhou,550006

E-mail:[a]zwlxg@126.com,[b]765123279@qq.com,[c]290740231@qq.com,[d]13908506244@163.com

Abstract: We have designed experiments to study the effect of gravel and sand on different labels of concrete. The result show that the compressive strength can be improved by increasing the proportion of mortar in low-grade concrete and increasing the proportion of stone in high-grade concrete. Tests showed that the powder of gravel and sand can significantly improve the mixing performance of concrete. The correlation of performance indicators with the strength of fresh concrete was also analyzed. The results in this research have important guiding significance for the design of concrete mix ratio and the application of powder in concrete.

1 Introduction

Concrete material was usually wasted and its compressive strength was not high when it was damaged in the potential state. Compressive destructive test of concrete found, C15-C30 low grade concrete block was crushed when the surface mortar layer and the interfacial transition zone appearing cracks,with some stone crushed, fracture occurred mostly in the mortar layer and interfacial transition zone, few occurred in the crushing gravel;C35-C60 high grade concrete block was crushed when stones were crushed, with some mortar layer and the interfacial transition zone appearing, fracture occurred mostly in crushing stone, few in layer and interfacial transition zone. We tried to change the proportion of gravel and sand in different grades of concrete composition to increase the compressive strength[1-5].

2 Test

2.1 Raw materials used in the test

Cement: P·O52.5 manufactured by Guizhou Hailuo Pangjiang Cement Factory.
Sand1,Sand2: Manufactured sand locally produced.
Stone: Limestone gravel, continuously graded in 5-25mm
powder: Produced by Ultrafine Powder company
Coal ash: Grade II coal ash in Guizhou Mingchuan, Active index of 28d 76%
Stone powder:Made in Guizhou;

Admixture: Polycarboxy late-type high performance water-reducing admixture

2.2 Test procedure

We tried to increase the compressive strength by changing the proportion of gravel and sand pf composition in different grades concrete. Each test for 30L material, designing of six modules and seven modules installed powder concrete two sets for each one, six modules and seven modules installed concrete three sets for each one(each set from C15-C60).To verify the effect of change of gravel and sand ratio on strength of different grades concrete, we formed six modules for more gravel and seven modules for more sand. The effect of powder on the performance of concrete was also tested by the contrast test of powder concrete. A total of 100 sets of concrete test was tested.

2.3 Concrete composition

The composition of powder concrete is as follows: (with powder instead of just 20% of cement for the other composition)

Concrete grade	52.5Cement(kg)	Stone powder(kg)	Coal ash (kg)	Sand1 (kg)	Sand2(kg)	Gravel5～25(kg)	Water(kg)	Admixture (kg)
C15	70-80	150-160	100-120	530-540	490-510	860-870	167-169	4-6
C20	100-110	130-140	100-120	520-530	490-510	870-880	167.5-169.5	4-6
C25	130-140	110-120	90-110	510-520	490-510	880-890	168-170	4-6
C30	180-190	90-100	80-100	500-510	490-510	890-910	168.5-170.5	4-6
C35	200-210	80-90	70-90	690-700	290-310	900-910	169-171	4-6
C40	230-250	70-80	60-80	570-580	390-410	900-910	169.5-171.5	4-6
C45	270-280	60-70	50-70	460-470	490-510	910-920	170-172	4-6
C50	300-320	50-60	40-60	350-360	590-610	920-930	170.5-172.5	4-6
C55	340-350	40-50	30-50	240-250	690-710	930-940	171-173	4-6
C60	370-390	30-40	20-40	130-140	790-810	940-950	171.5-173.5	4-6

2.4 Test data: Data was summarized as follows which reflected the various properties of concrete:

Molding type	Rewinding time((s)	Slump/degree of expansion(mm)	Wet bulk density(kg/m³)	Gas content()	56d strength MPa
C15 7 molds doped powder	3.1	240/600	2521	1.9	18
7 molds doped powder	2.1	230/620	2577	1.5	19.1
6molds doped powder	2	230/600	2554	2.0	16.6
6molds doped powder	2.5	240/610	2548	1.1	18.6
7molds non-doped powder	3.0	250/620	2542	1.5	23.5
7molds non-doped powder	2.1	240/580	2477	1.6	23
7molds non-doped powder	2.8	250/640	2572	1.1	22.8
6molds non-doped powder	3.2	230/620	2527	1.2	22.5
6molds non-doped powder	2.7	230/570	2518	1.5	22
6molds non-doped powder	2.8	250/640	2572	1.1	22.8
C20	3.0	230/620	2492	1.7	25.3
	2.6	230/590	2497	2.1	23.3
	2.3	240/600	2548	1.5	24
	2.1	240/600	2568	1.5	23
	3.5	240/640	2521	1.7	26.5
	2.3	250/640	2497	1.2	28.1
	2.8	220/610	2500	2.2	27.3
	2.7	240/640	2500	1.7	25.0
	2.7	230/590	2518	2.0	26
	2.5	230/600	2545	2.0	27
C25	2.5	220/610	2515	1.9	31.9
	2.5	240/620	2512	2.3	31.9
	3.0	230/600	2554	2.4	31.9
	2.5	230/610	2557	1.6	31
	2.5	220/620	2500	1.7	34.4
	3.5	230/580	2500	2.4	36.8
	3.2	240/600	2530	1.3	36.4
	2.3	230/600	2562	2.5	33.9
	2.4	240/650	2527	2.4	34.6
	3.1	240/610	2527	2.6	34.3
C30	2.5	220/610	2515	1.9	39.6
	2.5	240/620	2512	2.3	39.3
	3.0	230/600	2554	2.4	36.7
	2.5	230/610	2557	1.6	38.3
	2.5	220/620	2500	1.7	38.4
	3.5	230/580	2500	2.4	39.8
	3.2	240/600	2530	1.3	40.2
	2.3	230/600	2562	2.5	37.9
	2.4	240/650	2527	2.4	39.6
	3.1	240/610	2527	2.6	39.4
C35	3.0	230/630	2500	2.0	43.3
	2.3	230/600	2498	2.6	40.2
	3.1	240/580	2530	1.6	43.5
	3.0	230/620	2529	3.2	42.8
	3.1	220/620	2524	2.5	44.2
	2.6	250/650	2498	2.5	45.4
	2.3	240/600	2495	1.8	43.1
	3.2	220/610	2539	2.4	43.8
	2.6	250/640	2500	2.7	45.6
	3.0	220/600	2519	2.4	44.8
C40	2.9	240/610	2500	2.0	50.8
	3.2	230/620	2521	2.9	51.3
	2.8	230/600	2536	3.4	51.5
	3.0	250/630	2563	1.9	51.9
	3.0	240/630	2506	2.1	51.9
	2.6	240/630	2500	2.4	54.5
	2.8	230/620	2510	2.8	55.1
	2.5	245/610	2551	1.7	53.5
	2.8	230/620	2510	2.8	55.1
	3.0	220/600	2519	2.4	53.4
C45	2.5	230/610	2507	1.6	56.8
	3	220/620	2521	1.9	55.2
	2.8	240/620	2530	3.0	56.5
	2.7	250/610	2557	2.3	57.4
	2.7	230/630	2518	2.5	56.6
	3.2	230/610	2527	1.4	59.0
	2	230/620	2500	2.6	57.3
	2.4	230/630	2501	2.7	59.5
	2.9	220/610	2541	2.4	60.9
	3.8	240/600	2507	1.0	58
C50	2.5	230/610	2507	1.6	56.8
	3	220/620	2521	1.9	55.2
	2.8	240/620	2530	3.0	56.5
	2.7	250/610	2557	2.3	57.4
	2.7	230/630	2518	2.5	56.6
	3.2	230/610	2527	1.4	59.0
	2	230/620	2500	2.6	57.3
	2.4	230/630	2501	2.7	59.5
	2.9	220/610	2541	2.4	60.9
	3.8	240/600	2507	1.0	58
C55	2.8	240/640	2542	2.0	67.8
	3.3	220/620	2530	2.0	67.4
	2.5	230/600	2545	2.1	68.6
	3.1	230/600	2589	1.3	68
	4	240/590	2504	2.1	69.1
	2.9	250/650	2506	3.4	69.4
	3.1	230/620	2500	3.2	67.5
	3.0	220/600	2523	2.0	68.5
	3.1	230/600	2539	2.3	69.9
	3.2	210/600	2530	2.9	71.2
C60	2.5	240/640	2506	1.9	71.7
	2.8	230/620	2527	2.4	72
	2.6	220/620	2537	2.6	73
	2.4	240/630	2542	3.0	72.4
	2.6	240/650	2509	3.0	72.4
	4.2	230/610	2525	2.5	74.8
	2.6	250/600	2518	2	74.1
	3.1	250/620	2598	1.6	73.5
	2.7	230/650	2545	2.4	75
	3.2	230/620	2551	3.0	74.9

3 Test result analysis

(1) The influence of stone ratio and mortar on the strength of different grades concretes

Compression results show, the 56d strength of C15-C30 powder concrete of seven modules installed was higher than six molds', C35-C60 powder concrete of six modules installed was higher than seven molds'. For example, the 56d average strength of concrete which loaded by seven molds was 18.6, 24.3, 31.9,39.5MPa, loaded by six molds was 17.6, 23.5, 31.5, 37.5MPa for C15-C30, other which loaded by seven molds was 44.2, 53.8, 57.7, 64.1, 68.7, 73.4MPa, loaded by six molds was 44.7, 54, 59.5, 64.7, 69.8, 74.5MPa for C35-C60.

Some cement was replaced by coal ash in low grade concrete. Strength of coal ash increased slower. So when it was subjected to external load, the weakest part of the mortar layer and the interface transition zone was easy to reach the limit load and cracked, with few gravel crushing. There was more mortar in concrete loaded by seven molds. So its mortar got higher compressive ability. A small cement was replaced by coal ash in high grade concrete. Its strength increased quicker. So when it was subjected to external load, the mortar layer and the interface transition zone was more difficult to crack,with large gravel being crushed. There was more gravel in concrete loaded by six molds. So its gravel got higher compressive ability[6-10].

When concrete composition was designed, we increased the proportion of sand in low grade concrete to improve the strength of concrete. For high grade concrete, we increased the proportion of gravel to increase its the strength.

(2) Effect of powder on concrete performance index: strength, gas content, rewinding time

On the whole, the strength of the concrete was slightly reduced after mixing with the powder, but the mixing capacity was improved with the increase of the gas content, the decrease of the consistency, the increase of the rewinding time and the slump.

1) Effect of powder on concrete strength

The strength of 56d concrete mixed with micro-powder was slightly lower than that of unconfined concrete, but its long time strength growth rate was faster.

2) Effect of powder on concrete gas content

The air content of concrete mixed with powder was higher than that of unmixed concrete overall.For example,the content of C30 unmixed concrete was 1.9% and the content of C30 mixed concrete was 2.6%. The same phenomenon also existed in other grades concrete.

3) Effect of powder on concrete rewinding time

The experimental result showed that the rewinding time of concrete decreased with the addition of powder. For example,the rewinding time of C25 unmixed powder concrete was 3.0s and the mixed powder concrete was 2.4s.

(3) The relationship between air content, wet bulk density and strength of concrete

1) The relationship between air content and strength of concrete

When the gas content was high, the other strength may be high or low. The same phenomenon also existed when the gas content was low. The effect of the content powder concrete on the strength was not very obvious when it was under 3%. For example, the 56th day strength of C20 powder concrete filled with seven molds was 28.1MPa,26.5MPa when the gas content was 1.2%,1.7%. For another one, the 56th day strength of C20 powder concrete filled with six molds was 28.1MPa,26.5MPa when the gas content was 1.2%,1.7%.

2) The relationship between wet bulk density and strength of concrete

The effect of the wet bulk density concrete on the strength was not very obvious when it was between 2500~2560kg/m3. When the wet bulk density was high, the other strength may be high or low. The same phenomenon also existed when the wet bulk density was low. For example, the 56th day strength of C20 powder concrete filled with seven molds was 28.1MPa,26.5MPa when the wet bulk density was 1.2%,1.7%. For another one, the 56th day strength of C20 powder concrete filled with six molds was 28.1MPa,26.5MPa when the wet bulk density was 1.2%,1.7%.

It proved that there was no interrelation between performance index of air content, bulk density and concrete strength. We cannot predict the strength of concrete through the evaluation of these indicators.

4 Conclusion

To increase concrete's strength, we can increase sand ratio in low grade concrete and increase gravel ratio in high grade concrete in designing concrete composition.

After mixing powder into concrete, mixing performance of concrete got improved notably. So powder is good admixture for concrete.

The gas content and strength of concrete had little interrelation with each other in the range of 3%. The wet bulk density had no relevance to strength of concrete when the wet bulk density was between 2500~2560kg/m^3.

As the test was purely completed in the laboratory artificial test, there was some error in the weighing, the precision of the instrument, environmental factors and manual operation and so on. To reduce the elimination of error to get more accurate data, more research need to be conducted in-depth.

Fund Project: Natural Science Foundation of Guizhou Province GZ Word [2015]3013.

Reference:

[1] Li Xingang, xu Libin,etc 2015 Effect of superfine powder doped with phosphorus slag on properties of high strength mechanism sand concrete. *Rare Metal Materials And Engingeering*(44):365-368.
[2] Neville A M, Liu S S, Leng F G and Chen X 2011 Performance of Concrete.China Architecture & Industry Press,Beijing.
[3] Yuan R Z 2012 Cementitious Materials Subcject. Wuhan University of Technology Press, Wuhan.
[4] Liu S H, Leng F G and Li l h 2010 Supplementary Cementing Materials Used in Concrete. China Building Materials & Industry Press, Beijing.
[5] Leng F G and Feng N Q 2000 Research on Permeability and Density of PHC and its Method of Evaluation. *Low Temperature Architecture Technology* 4:14-16.
[6] Zhao T J 1997 Research on permeability of PHC. Dissertation, Tsinghua University
[7] Wang Y J, Yang C H, Ye J X, et al 2010 Research Progress of Application Study of Phosphorous Slag in Concrete. *Concrete*(11):95-97.
[8] Li Q M and Cao J X 2007 The way of green concrete production--the use of phosphorous slag.Guizhou Chemical Industry10(5): 5.
[9] Qi G H and Peng X Q 2011 Analysis on the pozzolanic effects of phosphorus slag power in concrete. *Key Engineering Materials*(477):112-117.
[10] Qi G H and Peng X Q 2011 Analysis on the pozzolanic effects of phosphorus slag power in concrete. *Key Engineering Materials*(477):112-117.

IOP Conf. Series: Materials Science and Engineering 170(2017) 012024 doi:10.1088/1757-899X/170/1/01202

Hydrometallurgical Treatment for Mixed Waste Battery Materials

L W Ma, X L Xi, Z Z Zhang, Z Q Huang, J P Chen

College of Material Science and Engineering, Beijing University of Technology, Beijing 100124, China

E-mail:mlw_56@163.com

Abstract. Hydrometallurgical experiments are generally required to assess the appropriate treatment process before the establishment of the industrial recovery process for waste battery materials. The effects of acid systems and oxidants in metal leaching were studied. The comprehensive leaching effects of the citric acid were superior to the sulfuric acid. The potassium permanganate inhibits the dissolution of metals. Thermodynamic calculations showed that metals precipitate more easily in sulfuric acid system than in citric acid system. The Fe precipitation efficiency in sulfuric acid system was 90% at pH 3.5, but with considerable losses of Co (30%) and Ni (40%). The proper pH and organic/aqueous (O/A) ratio for Fe and Zn removal with Di-(2-ethylhexyl) phosphoric acid extraction were 2 and 0.5, respectively; while for the removal of Cu and Mn, the best pH and O/A ratio were 3 and 0.75, respectively. Crude manganese carbonate and a cobalt-nickel enriched liquid were obtained by selective precipitation in raffinate using an ammonium bicarbonate solution. In citric acid systems, the precipitation efficiency of Co, Ni, Mn, Fe, Cu and Zn were less than 20% at pH 7. The proper pH and O/A ratio for the separation of the metals in two groups (Ni/Co/Cu and Mn/Fe/Zn) were 1.5 and 2. The cobalt-nickel-copper enriched liquid was finally obtained.

1. Introduction

Recovery of metals such as cobalt, nickel and manganese from secondary resources has been an important area of "urban minerals" exploitation. Cobalt and nickel secondary resources usually include spent lithium-ion batteries, nickel-metal hydride battery waste, metal scraps, complex mineral wastes and smelting slags [1]. Hydrometallurgical experiments for waste materials are often required to roughly acquire the treatment process before the establishment of any industrial recovery process. The recovery process often contains these steps: acidic leaching, metal separation and product synthesis. Due to the complex and diverse composition of secondary resources, the leaching methods are different. Sulfuric acid is commonly used in the leaching of spent batteries, and the extraction ratio of Co can achieve at least 70% [2]. In addition, some organic acids such as citric acid ($C_6H_8O_7$), malic acid ($C_4H_5O_6$), glycine, EDTA and DTPA, also were used to leach or extract metals of spent batteries or waste catalysts [3,4]. The leaching efficiency of major metals for these organic acids can also reach more than 70%. Apart from leaching agent, the oxidant is a key factor in promoting the dissolution of valuable metals during leach process. The use of hydrogen peroxide as an oxidizing agent can promote leaching of nickel, cobalt and rare earth from waste capacitors and batteries [5]. The use of potassium permanganate or hydrogen peroxide oxidation during acid leaching of indium sulfide minerals can significantly reduce the apparent activation energy and reaction order, increasing the reaction rate [6]. Potassium permanganate can also play the role of Mn remover [7]. Different leaching systems will affect the subsequent separation and recovery of metals. Theoretical studies have shown that the metal separation behaviors are different in different solution systems of Me-OH^-, Me-OH^--CO_3^{2-}, Me-OH^--S^{2-}, Me-OH^--NH_3 and Me-OH^--NH_3-CO_3^{2-} [8]. Applying different solution system different metals can be selectively precipitated. According to thermodynamic calculations, the Me-OH^--NH_3-CO_3^{2-} solution system can remove Fe efficiently with a low cobalt and nickel loss, and the iron precipitate showed a good filtration performance [9]. Mo and Al can be effectively separated from waste Ni-Mo catalysts in EDTA leaching system using a selective precipitation method, and $Ni_3(PO_4)_2$ was recovered by ion salt replacement method with a purity over 96% [10]. According to these results, theoretical and experimental studies on metal separation are effective and feasible for the comprehensive recovery of metals from secondary resources.

In this paper, waste battery material was used as raw materials. Sulfuric and citric acids with an oxidizing agent were used to leach the waste battery material to obtain the metal solution. Thermodynamic calculations were performed for the two leaching systems to study the metal separation behavior. Solvent extraction was then used to further separate the metals, and the metal pregnant liquor or metal compound was finally obtained.

2. Experimental

1.1. Leaching

Waste battery material obtained from a recycling company (GEM High-Tech Co., Ltd., China) was crushed and screened. Powders with a particle size of 200 mesh were leached with agitation of 800 rpm at 85℃ for 2 hours, at a liquid to solid ratio (L/S) of 40:1, in four leaching systems: (1) 2.5 mol/L sulfuric acid + 5 vol% hydrogen peroxide (H_2O_2), (2) 2.5 mol/L citric acid + 5 vol% H_2O_2, (3) 2.5 mol/L sulfuric acid +5 vol% H_2O_2 + 20 wt% potassium permanganate ($KMnO_4$), and (4) 2.5 mol/L citric acid + 5 vol% H_2O_2 + 20 wt% $KMnO_4$. After leaching, the metal leaching efficiency (L) was calculated according to formula (1):

$$L = \frac{M \times C_0 \times V_0}{m \times w\%} \times 100\% \tag{1}$$

where M and C_0 are the molecular mass and concentrations of metal ions in the leach liquor, respectively; V_0 is the leach liquor volume; m is the mass of cobalt-bearing waste and w% is the metal mass fraction.

2.2. Precipitation

The metals were precipitated from the leach solutions and the solutions were filtered at different pH values (3-7). The metal precipitation efficiency (P) was calculated according to formula (2):

$$P = (\frac{C_0V_0 - C_1V_1}{C_0V_0}) \times 100\% \tag{2}$$

where C_0 and C_1 are the metal ion concentrations in the solution before and after precipitation; V_0 and V_1 are the solution volumes before and after precipitation.

2.3. Solvent Extraction

Di-(2-ethylhexyl) phosphoric acid (D2EHPA or P204) was used to further extract metals from the sulfuric and citric acid systems at 25 °C, with a P204 concentration of 30% and a saponification ratio of 65%. Organic/aqueous (O/A) phase volume ratios of 0.25-2.5 and pH values of 1.0-3.0 were investigated. The metal ion extraction efficiency (*E*) was calculated according to formula (3):

$$E = \frac{C_1V_1 - C_2V_2}{C_1V_1} \times 100\% \tag{3}$$

where C_1 and C_2 are the metal concentrations in the aqueous phase before and after the extraction; V_1 and V_2 are the aqueous phase volumes before and after the extraction.

2.4. Analysis

The waste battery material composition was analyzed by x-ray fluorescence (XRF-1800, Shimadzu, Japan). The metal concentration of the leachate, the filtrate and the affinate were determined by inductively coupled plasma atomic emission spectroscopy (ICP-AES, Optima 7000DV, PerkinElmer, U.S.A.). The structure of the raw material and the product were investigated by x-ray diffraction (XRD, 7000 S/L, Rigaku, Japan).

3. Results and discussion

3.1. Material Characterization

The composition and the structure of the waste battery material were studied by XRF and XRD (shown in Fig. 1). The main components of the waste material are Co (0.24 wt.%), Ni (1.01 wt.%), Mn (35.22 wt.%), Zn (21.86 wt.%), Fe (3.64 wt.%) and Cu (0.12 wt.%). The main phases of the waste battery material are (Co,Mn)(Co,Mn)$_2$O$_4$, $Co_{1.29}Ni_{1.71}O_4$, ZnO, $Cu_{0.95}Co_{2.09}O_4$ and $FeMnO_3$, suggesting that this waste is a complex mixture of cobalt-nickel batteries and zinc-manganese batteries.

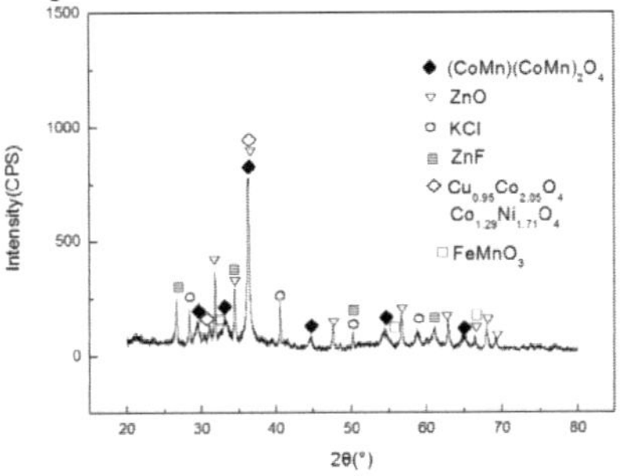

Fig. 1 Mixed waste battery material XRD analysis

3.2. Leaching

The leaching efficiency of metals in sulfuric acid-hydrogen peroxide system is shown in Fig. 2(a). Most of the metals can be leached thoroughly. The leaching efficiency of Ni and Mn are almost 100%, the Co leaching efficiency is 82.9%, and the Fe, Cu and Zn leaching efficiencies are over 70%, indicating that the waste metals can be easily dissolved in a strong acidic environment with hydrogen peroxide as oxidant.

The leaching efficiency of metals in citric acid-hydrogen peroxide system is shown in Fig. 2(b). The metals exhibit different leaching behaviors. The Ni leaching efficiency is almost 100%, the Co leaching efficiency is 85.1%, and the Fe, Cu and Zn leaching efficiencies are over 50%. The Mn leaching efficiency displays the lowest value (28.3%). The organic acidic environment with hydrogen peroxide as oxidant favors valuable metals dissolution (Co and Ni) but does not promote the dissolution of Mn, Fe, Zn and the other metals. The citric acid system is beneficial for the selective dissolution of metals in the mixed battery waste as compared to the sulfuric acid system.

As the Mn content in the mixed waste battery material increases, the addition of H_2O_2 and $KMnO_4$ for Mn removal during the leaching process can reduce subsequent purification difficulty. The H_2SO_4-H_2O_2-$KMnO_4$ leaching results are shown in Fig. 2(c). The leaching efficiency of Mn, Fe, Zn, etc. are all below 42%, and the leaching efficiency of Mn is only 10.3%. Meanwhile, the leaching efficiency of both Co and Ni decrease to 32.9% and 69.8%, respectively, as compared to the systems without H_2O_2-$KMnO_4$. The metal dissolution behavior will be changed by adjusting the oxidant apart from changing the leaching agent.

The leaching efficiency of metals in citric acid-H_2O_2-$KMnO_4$ system are shown in Fig. 2(d). The metal dissolution behaviors are similar to that in H_2SO_4-H_2O_2-$KMnO_4$ system. Since the organic acid and the $KMnO_4$ do not encourage the dissolution of the related metals, the leaching efficiency of the metals is further reduced as compared with the H_2SO_4-H_2O_2-$KMnO_4$ system. The leaching efficiency of Mn, Fe and other metal is dropped to less than 20%, but the leaching efficiency of the valuable metals, Ni and Co, are also very low, only 23.1% and 48.2%, respectively, which is unfavorable for the recovery of Ni and Co.

Based on the above results, the citric acid is better than the sulfuric acid because the first favors the leaching of Co and Ni but is not conducive to the dissolution of the other metals. Using $KMnO_4$ as an oxidant, the Mn leaching efficiency can be significantly reduced but the dissolution of the other metals also decreases. Therefore, the citric acid-H_2O_2 system is the best leaching system for the mixed waste battery material used in this study.

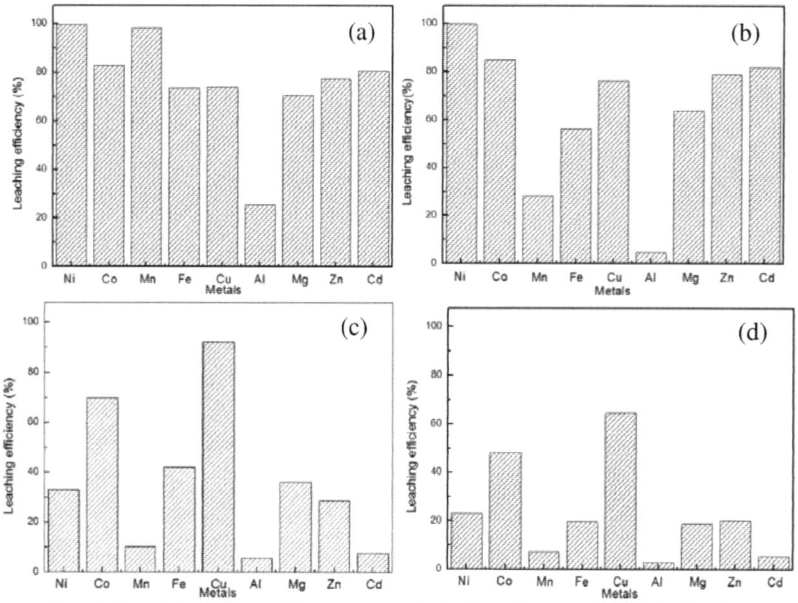

Fig. 2 Metal leaching efficiency in (a)H_2SO_4-H_2O_2; (b) H_3Cit-H_2O_2; (c) H_2SO_4-H_2O_2-$KMnO_4$; (d) H_3Cit-H_2O_2-$KMnO_4$

3.3. Precipitate

3.3.1. Theoretical Analysis According to the thermodynamic data from Ref. [11], the metal precipitation trends can be predicted by thermodynamic calculation. The relationships between the logarithm of metal ion concentration ([Me]) and pH in the sulfuric acid system and the citric acid system are shown in Fig.3. As can be seen from Fig. 3(a), the metal concentration decreases with an increase of pH up to the value of 9.5; i.e., theoretically, all of the metals achieve the maximum precipitation at pH 9.5. In sulfuric acid system, Fe precipitates completely at pH 3 and can be separated from the other metals; Cu precipitates and can be separated from the other metals at pH 5.5-6.5, accompanied by some loss of cobalt; most metals precipitate completely at pH>7. As can be seen from Fig. 3(b), the concentration of each metal also decreases with an increase of pH in the citric acid system, but the decline is not as pronounced as compared with the sulfuric acid system. Theoretically, in the citric acid system, Fe precipitates completely at pH>7, while the concentrations of Co and Ni remain stable, and do not precipitate completely at pH 6-12, indicating that these metals are much easier to precipitate in the sulfuric acid system than in the citric acid system.

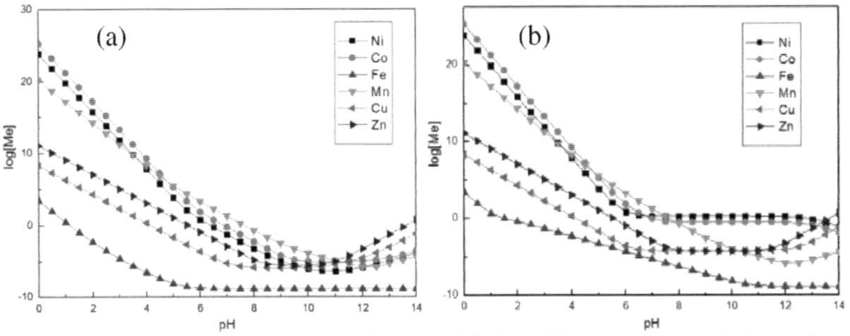

Fig. 3 Theoretical log[Me]-pH curve in (a) sulfuric acid system; (b) citric acid system

3.3.2. Experimental Study Fig. 4(a) shows the experimental metal precipitation efficiencies in sulfuric acid leaching solution at different pH values. The precipitation efficiency of each metal increases with pH, which is consistent with the theoretical prediction. The Fe precipitation efficiency is 90% at pH 3.5. Since the content of Co and Ni are relatively low in the mixed waste battery material, the low precipitation efficiency of 30% for Co and 40% for Ni will result in a great loss of these valuable metals; therefore, a lower pH is recommended for Fe pre-removal, followed by the removal of most impurities by extraction. Other studies have shown that Fe^{3+} and Cu^{2+} can be pre-precipitated with a pH adjustment to 5.5 [12], but for this material with low Co and Ni contents it is not suitable. As Mn, Cu and Fe can be removed by P204 extraction in the next stage, a extraction after leaching is also feasible. Fig. 4(b) shows the experimental metal precipitation efficiencies in citric acid leaching solution under different pH conditions. The precipitation efficiency of all the metals increase gradually with pH. At pH 5, the precipitation efficiency of Ni, Co, Fe, Mn, Cu and Zn is 11.5%, 12.7%, 16.3%, 12.5%, 15.4% and 14.4%, respectively, and at pH 7 their precipitation efficiencies are still less than 20%. As can be seen, metals separation is difficult in the citric acid system, which is consistent with the theoretical prediction; therefore, a metal separation using direct extraction should be considered.

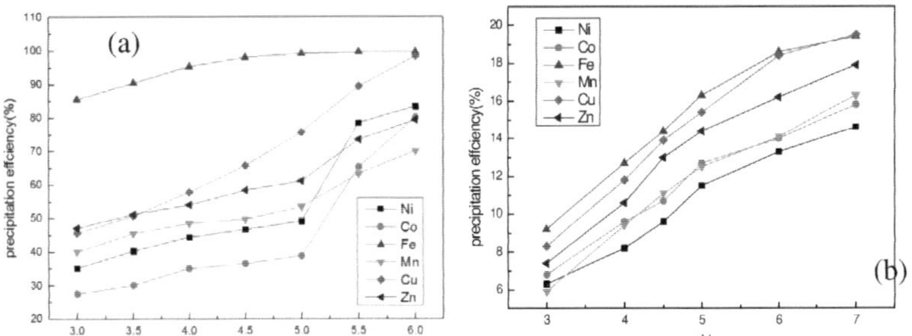

Fig. 4 Metal precipitation efficiencies at different pH (a) in sulfuric acid system; (b) in citric acid system

3.4. Extraction

Fig. 5(a) shows the effect of pH on the extraction efficiency of different metal ions in sulfuric acid system. The metal extraction efficiency increases with pH. The extraction order according to pH values from low to high is Fe, Zn, Mn, Cu, Co and Ni, but the extraction efficiency of Fe always remains over 80% while the Zn extraction efficiency increases to 93% at pH 2. The extraction efficiency of the other metals are lower than 16%. For the removal of Fe and Zn, the suitable pH is 2, but for the removal of Cu and Mn the best pH is 3. Fig. 5(b) shows the effect of O/A ratios on the extraction efficiency of different metal ions in sulfuric acid system. The metal extraction efficiency increases with O/A ratio. Lower O/A ratios generally mean a fewer organic phase compared to a certain amount of aqueous phase and a limited extraction capacity. When the O/A ratio is low, metal ions such as Mn, Fe, Cu, etc., with a strong affinity to P204, can enter into the P204 organic phase preferentially and be separated from Co and Ni. When the O/A ratio reaches a certain level, the metals with a weak affinity to P204 also entered into the organic phase, and there is no separation effect among the metals. According to Fig. 11, the extraction efficiency of Fe and Zn are high while the extraction efficiency of the other metals are still low at an O/A ratio of 0.5, so 0.5 is the suitable O/A ratio for Fe and Zn separation, while 0.75 is the proper O/A ratio for Cu and Mn separation.

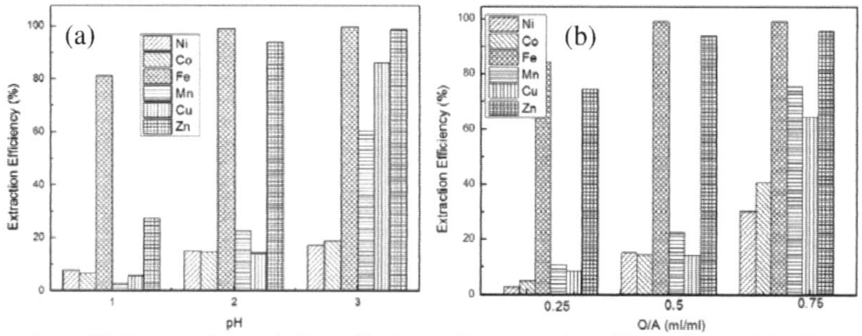

Fig. 5 Extraction efficiency of metals in sulfuric acid system (a) effect of pH; (b) effect of O/A ratio

Fig. 6(a) shows the effect of pH on the extraction efficiency of different metals in citric acid system. In this system, the extraction efficiency of all the metals increased slowly as the pH increased. Since the extraction ability of P204 is very sensitive to pH and metal ion valence, and the citric acid is a ternary weak acid, which will release H^+ step by step to stabilize the pH of the solution, it is not beneficial for the metal extraction reaction. It is apparent from Fig. 12 a segregation effect between the Ni/Co/Cu group and the Mn/Fe/Zn group. The proper pH is 1.5 where the extraction efficiency of Ni, Co, Cu, Mn, Fe, Zn is 15.8%, 16.3%, 22.4%, 80.2%, 64.8% and 76.8%, respectively; however, the separation effect among Ni, Co, Cu, or Mn, Fe, Zn are very poor. Fig. 6(b) shows the O/A ratio effect on the extraction efficiency of different metal ions in citric acid system. Similar to the sulfuric acid system, the extraction efficiency of each metal increases with O/A ratio, but with a smaller degree as compared with the sulfuric acid system, indicating that more organic phase is needed in the extraction process. This could be due to the formation of metal complexes in the citric acid system, which is difficult to extract. In addition, the excess citric acid can combine with the extractant, requiring more P204. The extraction efficiency of Mn, Fe and Zn are relatively high and the extraction efficiency of Ni, Co and Cu are relatively low at O/A ratio of 2.0, indicating a relatively good separation effect.

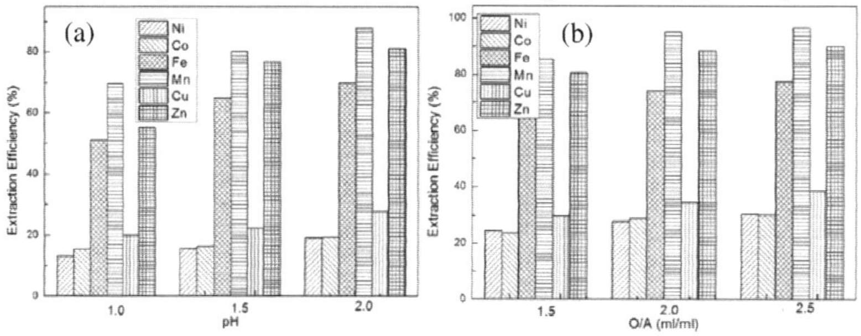

Fig. 6 Extraction efficiency of metals in citric acid system (a) effect of pH; (b) effect of O/A ratio

3.5. Recovery

Based on the theoretical and experimental leaching results obtained for the mixed waste battery material in sulfuric and citric acid systems, it is believed that the metal separation in the citric acid system is not as good as in the sulfuric acid system due to complexing effect. Finally, only Co-Ni-Cu-enriched liquid was obtained after solvent extraction in the citric acid system, and Co-Ni-enriched liquid and crude manganese carbonate (Fig. 14) were obtained after selective precipitation by ammonium carbonate solution.

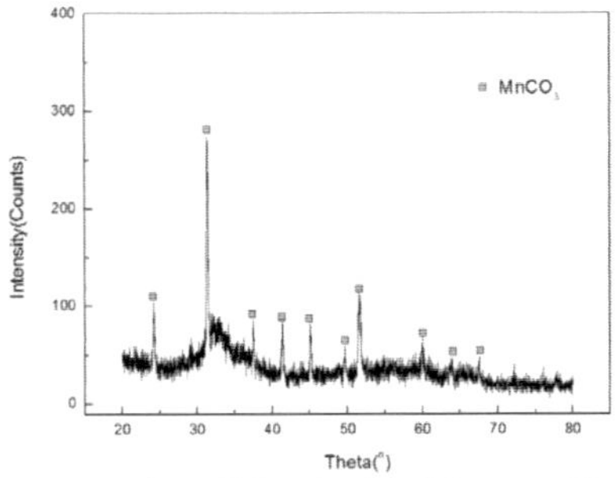

Fig. 14 XRD pattern of $MnCO_3$ obtained from sulfuric acid system

The overall metal separation process is shown in Fig. 15. Since the waste material used in this study is a complex mixture of cobalt-nickel and zinc-manganese batteries, the Mn content is very high and the content of Co and Ni are relatively low. This situation leads to a difficult separation and recovery of Co and Ni, but an easy recovery of Mn. This study provides technical guidance for industrial recovery processes.

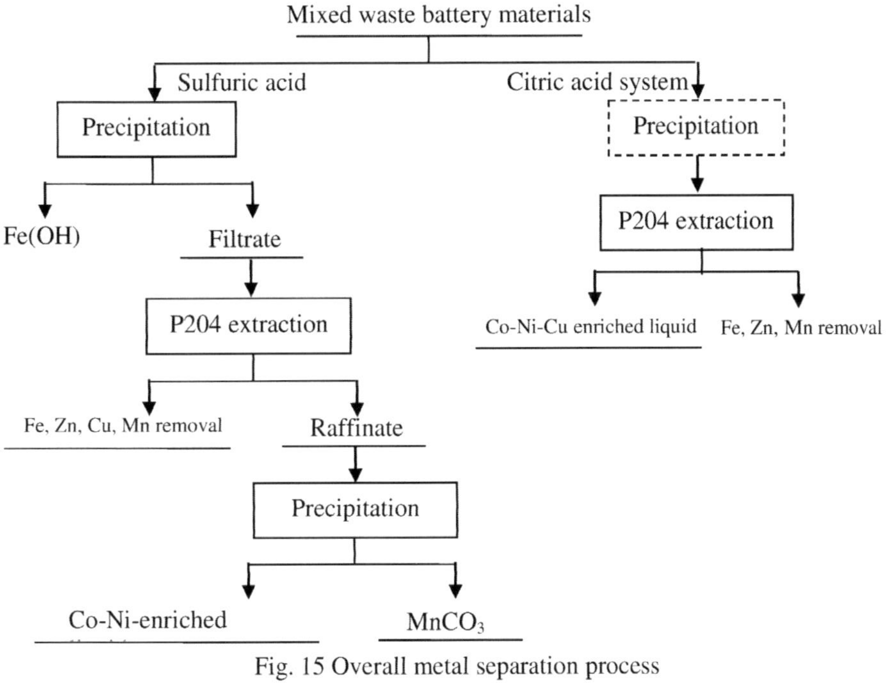

Fig. 15 Overall metal separation process

4. Conclusions

(1) Sulfuric and citric acids with hydrogen peroxide and potassium permanganate were used as medium to leach mixed waste battery material. Citric acid favors the dissolution of cobalt and nickel but is not conducive for the dissolution of the other metals. Potassium permanganate does not promote the dissolution of metals as an oxidant. The suitable leaching media is the citric acid-hydrogen peroxide system.

(2) According to the theoretical analysis, the metals do not precipitate in the citric acid system as easy as in the sulfuric acid system. Experimental studies have shown that the highest precipitation efficiency of Fe is 90% and occurred at pH 3.5. Since the content of Co and Ni are low in the raw

material, these metals precipitate 30% and 40% in the sulfuric acid system, respectively. In the citric acid system, the precipitation efficiency of Co, Ni, Mn, Fe, Cu and Zn are all less than 20% at pH 7, indicating that metal separation is difficult by precipitation.
(3) P204 was used in the sulfuric acid system and the citric acid system. The extraction order with increasing pH was Fe, Zn, Mn, Cu, Co and Ni in the sulfuric acid system. For Fe and Zn removal, the proper pH is 2 and the O/A ratio is 0.5, while for the removal of Cu and Mn the proper pH and O/A ratio is 3 and 0.75, respectively. The raffinate was selectively precipitated by ammonium carbonate solution to obtain crude $MnCO_3$ and Co-Ni-enriched liquid. At a pH of 1.5 and an O/A ratio of 2.0, the citric acid system favors metal separation in two groups; namely, Ni/Co/Cu and Mn/Fe/Zn, and a Co-Ni-Cu enriched liquid was finally obtained.

Acknowledgements
This work was supported by National Natural Science Foundation of China (51304010, 51422401) and Technology Project (Z151100003415008).

References

[1] Kang J, Sohn J and Chang H 2010 Preparation of cobalt oxide from concentrated cathode material of spent lithium ion batteries by hydrometallurgical method, *Adv. Powder Technol.* **21** 175-179.
[2] Ferreira D A, Prados L M Z and Majuste D 2009 Hydrometallurgical separation of aluminium, cobalt, copper and lithium from spent Li-ion batteries, *J. Power Sources* **187** 238-246.
[3] Goel S and Gautam A 2010 Effect of chelating agents on mobilization of metal from waste catalyst, *Hydrometallurgy* **101** 120-125.
[4] Li L, Ge J and Chen R J 2010 Environmental friendly leaching reagent for cobalt and lithium recovery from spent lithium-ion batteries, *Waste Manage.* **30** 2615-2621.
[5] Meshra P, Pandey B D and Mankhand T R 2015 Recovery of valuable metals from cathodic active material of spent lithium ion batteries: Leaching and kinetic aspects, *Waste Manage.* **45** 306-313.
[6] Wei Y, Li X and Ma C 2014 Kinetics Comparison of Oxidizing Acid Leaching and Conventional Acid Leaching of Indium Sulfide, Metal Mine **453** 165-170 (in Chinese)
[7] Chen X P, Xu B, Zhou T, Liu D P, Hu H and Fan S Y 2015 Separation and recovery of metal values from leaching liquor of mixed-type of spent lithium-ion batteries, Sep. Purif. Technol. **144** 197-205.
[8] Ma L, Nie Z, Xi X and Han X 2013 Thermodynamic equilibrium in Co-Ni-Fe-Mn complexation-precipitation system, Chin. *J.Nonferrous Met.* **23** 516-527. (in Chinese)
[9] Han X, Xi X, Ma L, Nie Z and Zhou Z 2013 Studies of iron–removal from cobalt and nickel secondary resources with "complexation-precipitation" system, Chin. *J.Nonferrous Met.* **23** 3218-3224. (in Chinese)
[10] Pinto L S S, Sadeghi S M and Soares H M V M 2015 Separation and recovery of nickel, as a salt, from an EDTA leachate of spent hydrodesulphurization catalyst using precipitation methods, *Chem. Eng. Sci.* **122** 130-137.
[11] Dean J A, Translator: Wei J, et al 2003 Lange's Handbook of Chemistry, second ed. Science Press, Beijing 8.80-8.104.(in Chinese)
[12] Chen L, Tang X and Zhang Y 2011 Process for the recovery of cobalt oxalate from spent lithium-ion batteries, *Hydrometallurgy* **108** 80-86.

Modeling analysis of temperatures at points in oxide film of grinding wheels

H L Zhang[1,a], J C Kuai[2,b]

[1]School of Mechanical Engineering, Nantong University, Nantong, China
[2]School of Mechanical and Power Engineering, Henan Polytechnic University, Jiaozuo, Henan, China

E-mail: hitgjc@163.com

Abstract. In this work, a temperature at points model of the abrasive grains in ELID grinding is presented for the analysis of the temperature of single cutting grain between the workpiece and the oxide film on the ELID wheel surface. The corresponding calculation results show that temperature of single grinding point is significantly higher than the average temperature through the grinding zone.

1. Introduction

Most temperature models of grinding are based on the assumption that grinding process is conducted by zonal heat sources. The average temperature of the entire grinding area then can be obtained by integral operation [1]. But these methods are limited only to local area and are often not suitable to illustrate changes in temperature for single grinding grain and its surrounding area.

The formation and transition of oxide films in grinding process is dependent upon grinding temperatures, especially upon temperature around a single abrasive grain. More iron oxides convert into α-Fe2O3 in oxide films under the effect of the high temperature around the abrasive grains, which comes up to or excess to transition temperature [2].

To better assess the formation and transition of α-Fe2O3 around abrasive grains, we need data that reflect the temperature distribution around the grains measured in an accurate way. However, speed of grinding wheels is relatively high for plain grinder. Thermometry, such as thermocouples and IR temperature measurement technology, has been reported [3], but to the best of our knowledge, their sub-millisecond response time cannot fully satisfy the measurement require for temperature at point on the single grain. Here we present a theoretical study on temperature-at-point model in grinding process, for obtaining reliable prediction on temperature distribution around abrasive grains.

2. Flash Temperature Modeling

Taking the temperature as the performance, the grinding power per unit time is given by:

$$P=F_t V_s \tag{1}$$

where P is the total grinding power, F_t is the tangential grinding force, and V_s is the grinding velocity. According to the grinding theory, the total grinding power is contributed by the workpiece, chips, and the abrasive wheels. Then the other arguments are given by

$$P=P_{work} + P_{wh} + P_{chip} \tag{2}$$

where P_{work} is the grinding power contributed to the workpiece,P_{wh} is the grinding power contributed to the grinding wheels, and P_{chip} is the grinding power contributed to the chips. Assumed V is the volume of chips removed per unit time, and the corresponding grinding power contributed to the chips is given by

$$P_{chip} = V\rho c\theta_{chip} \quad (3)$$

where ρ is density of the chip, c is specific heat of the chip, and θ_{chip} is the temperature rise of the chip.

The grinding power is consequently contributed to the grinding wheels and the workpiece, except the partially to the chips. Assuming that the number of actual contacting grains and cutting grains in the grinding area is N, and the individual heat sources between the abrasive grains and the workpiece, is considered as a circular disc as shown in Figure. 1. The grinding power for each cutting grain is given by

$$P_{contact} = (P - P_{chip})/N \quad (4)$$

where $P_{contact}$ is the grinding power in contact area between the abrasive grains and the workpiece.

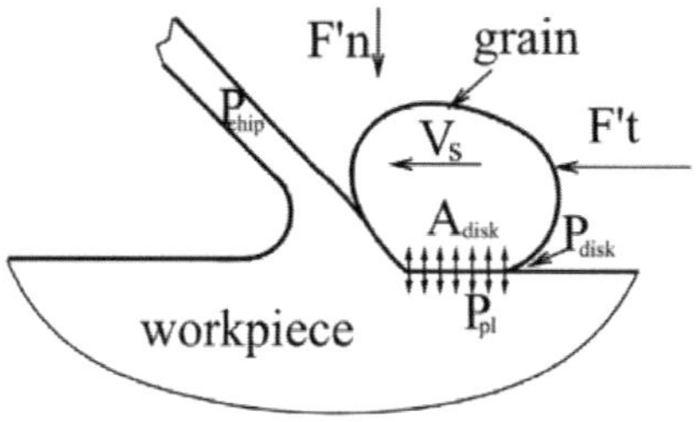

Figure. 1 Schematic of a circular disc heat sources in grinding

The theoretical number of contacting grains in the grinding area N0 is given by

$$N_0 = 6V_g / \pi d_g^2 \quad (5)$$

where V_g is the volume ratio of the grains to the wheel surface, and d_g is the grain radius. Thus, the effective number of the contacting grains per unit area, N_{cont} is given by

$$N_{cont}/N_0 = (d_g - d_b) / d_g \quad (6)$$

where d_b is the average diameter of peeling-off grains.

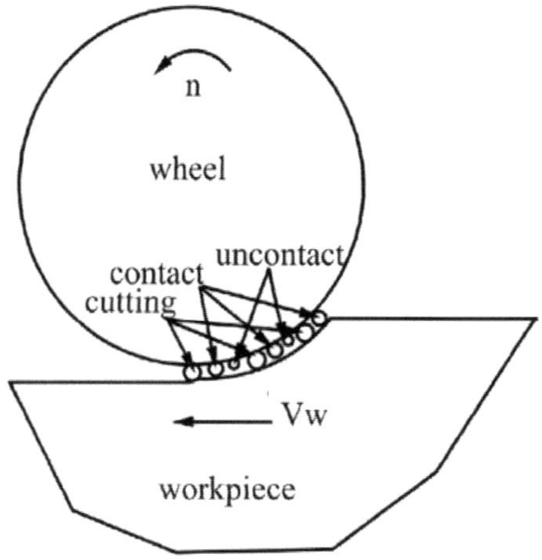

Figure.2 Schematic of the actual contacting grains in the grinding zone

The range of diameters of the actual contacting grains is represented as dg - db, covering from the maximum sizes to the minimum size of the grains for ELID grinding wheels, as shown in Figure. 2.

The cutting time of the grains, t1 , is given by

$$t_1 = l / V_s \tag{7}$$

where l is arc length of contact area. The time of contact of any point at the contacting interface on the workside, t_2 is given by

$$t_2 = 2r_0 / V_s \tag{8}$$

where r_0 is the radius of the circular area between the grains and the workpiece.

We obtain the energy density of the contact circular as

$$p_{pl} = P_{\text{circular disc}} / A_{\text{circular disc}} \tag{9}$$

where $A_{\text{circular disc}}$ is the area of the contact circular.

Referring to the heat sources model [4],

$$\theta = \frac{P_{pl} \times z}{k\sqrt{\pi}} \left[\frac{\sqrt{4\alpha t}}{z} e^{-z^2/4\alpha t} - \sqrt{\pi} erf\, c\left(\frac{z}{\sqrt{4\alpha t}} \right) \right] \tag{10}$$

where θ is the grinding temperature, k is the heat conductivity, and α is the thermal diffusion coefficient. Substituting z=0 in Eq. (10), we obtain,

$$\theta_{z=0} = \frac{P_{pl}}{k\sqrt{\pi}} \sqrt{4\alpha t} \tag{11}$$

Eq. (11) represents the grinding temperature between the grains and the workpiece. The energy of the circular transfers to both the workpiece and the abrasive grains. Similarly, to obtain the grinding temperatures on the grain side, we need to know the heat partition fractions for the workpiece and the abrasive grain. For the grains:

$$\theta_1 = \frac{B_1 P_{pl}}{k_1\sqrt{\pi}} \sqrt{4\alpha_1 t_1} \tag{12}$$

For the workpiece:

$$\theta_2 = \frac{B_2 P_{pl}}{k_2\sqrt{\pi}} \sqrt{4\alpha_2 t_2} \tag{13}$$

where θ_1,θ_2 are temperatures of the grains and the workpiece, respectively. B_1, B_2 are the heat distribution coefficient from heat sources to the grains and to the workpiece, respectively. a is the corresponding heat diffusion coefficient, k is the corresponding heat conductivity, and t is the corresponding contact time for the grains and the workpiece.

Since θ_1=θ_2 at the interface of contact, we obtain

$$\frac{B_1 P_{pl}}{k_1\sqrt{\pi}} \sqrt{4\alpha_1 t_1} = \frac{B_2 P_{pl}}{k_2\sqrt{\pi}} \sqrt{4\alpha_2 t_2} \tag{14}$$

We can simplify Eq. (14) by B_2=1-B_1, thus,

$$B_1 = \frac{\sqrt{4\alpha_2 t_2}}{k_2} \left[\left(\sqrt{4\alpha_1 t_1}/k_1 \right) + \sqrt{4\alpha_2 t_2}/k_2 \right]^{-1} \tag{15}$$

Using Eq. (15), the heat partition fractions for the abrasive grains and the workpiece can be calculated. Substituting Eq. (15) to Eqs. (12)-(13), temperatures for the abrasive grains and the work can be obtained, and are referred as instantaneous contact temperatures, as well as flash temperature or point temperature in grinding. And the Eqs (12)-(13) are referred as the flash temperatures model of the contacting grains.

3. Results And Discussion

Parameters in the flash temperature model are shown in Table 1 and Table 2. And the results are depicted in Figure. 3.

Table 1 Values of the grinding parameters in the model

η	d_g-d_b	F_t	V_s	ε	ρ''	a
	μm	N	m/s		g/cm^3	mm
6.25-37.5	40-28	10-100	15.7	0.1	3.12	0.005-0.02

Table 2 Heat conductivity and heat diffusion coefficients of some materials

	ρ	c	α	K
	g/mm^3	J/g● ℃	mm^2/s	W/(mm● ℃)
Si3N4	3.44e-3	0.71	6.9	16.7*10^{-3}
8Y-YZP	6.04e-3	0.45	10	27*10^{-3}
Soda-lime glass	2.5e-3	1.1	0.345	0.95*10^{-3}

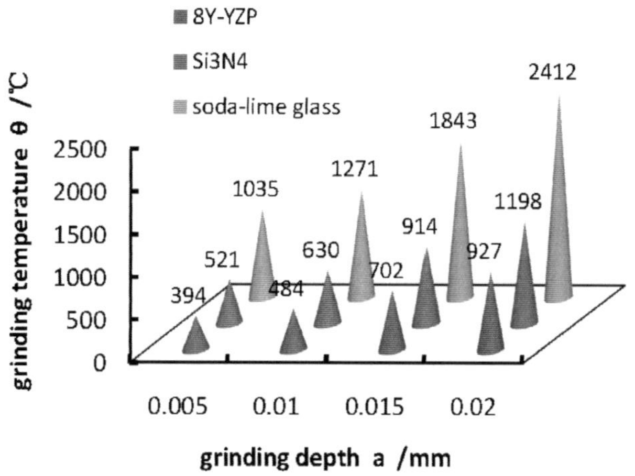

Figure.3 Flash temperatures vs. grinding depth of specific materials

Temperatures at grinding points as a result of calculation are determined to be higher than the average temperature of grinding zone [5]. The oxide films of ELID grinding wheels are in high temperature surrounding, making them more easily dehydrated and having tendency to convert their chemical composition. Heat energy of lattice water and adsorption water, which absorbed in the oxide films coated around the abrasive grains, are 10^{-6}J and 10^{-5}J, respectively, far less than cutting heat. The cooling effect on single abrasive grain is negligible, which is coincident with plain grinding, and can be ignored. It demonstrates that cutting fluid is hard loaded into the cutting area, and hence, temperatures around the abrasive grains cannot be cooled down efficiently. However, only a small amount of the abrasive grains for the whole grinding zone participate in the grinding [6]. The majorities of the abrasive grains may not actually contact with the workpiece. Some may only rub or plough. Consequently, other area except the actual cutting grains covered with the oxide films. And the oxide film is actually contacting with the workpiece at these area, with adsorbed water and lattice water's effective cooling on the grinding zone. Hence, the adsorbed water and the lattice water within the oxide film are estimated to distinctly absorb the grinding heat and lower the temperature through the grinding zone.

4. Conclusions

A temperature at points model of ELID grinding is proposed in this paper based on grinding heat distribution and the grinding temperatures around the abrasive grains. The corresponding results are performed below:

(1) This thermal model considers the grinding heat partition and temperatures of the abrasive grains. Data demonstrated that temperatures of the abrasive grains were significantly higher than the average temperature at the grinding zone, which enables to theoretically explain the composition transition of the oxide films in ELID wheels.

(2) The dynamic cooling effect of the oxide films in the contact temperatures was also presented. It predicted that the grinding heat was barely reliable to cool down, while the cooling effect of the oxide films on the abrasive grains can be neglectful. However, the average temperature at the whole grinding zone can be chill-down with the effort of the oxide films coated on the grinding wheels.

5. Acknowledgment

This work was supported by the National Natural Science Foundation of China (Grant 51475147).

References

[1] Tahvilian A M, Champliaud H, Liu Z and Hazel B 2013 *Procedia CIRP* **12** 205
[2] Kuai J C 2013 *Advanced Materials Research.* **669** 91
[3] Guo L, Li B, Sheng X and Huang H 2007 *China Mechanical Engineering* **18** 2388
[4] Carslaw H S and Jaeger J C 1992 *Conduction of Heat in Solids* (New York: Oxford University Press) p 103
[5] Malkin S 1989 *Grinding Technology Theory and Application of Machining with Abrasives* (New York: Prentice Hall) p 114
[6] Hou Z B and Komanduri R 2004 *International Journal of Machine Tools & Manufacture* **44** 247

Development of Secondary Recrystallization in high permeability grain oriented Silicon steel produced by Thin Slab Casting and Rolling process

L F Fan[1,2], S M Liu[1], F Guo[1], J Z He[2], B Lu[2], R F Dong[1]

1.College of Materials Science and Engineering, Inner Mongolia University of Technology, Hohhot 010051, China;
2.Technology Center of Baotou Steel Group Company, Baotou 014010, China;

E-mail:Fanlifeng829@163.com

Abstract. Texture evolution during high temperature annealing process in high permeability grain oriented(Hi-B)silicon steel produced by Thin Slab Casting and Rolling(TSCR) process was investigated using macro-/microtexture analysis. Goss texture appears in ODF at 800℃, the intensity of Goss texture is very weak until 950℃;Goss grains grow abnormally during 980℃~1000℃, grain size reached centimeter level at 1000℃, Goss grains instead of Primary grains fully at 1020℃， Secondary Recrystallization completed at 1040℃; The deviation angle of Goss grain orientation decreases gradually with temperature increasing, and it reached the minimum at 1040℃.

1. Introduction

Grain-oriented silicon steel is an important magnetic material that is widely used as core material in transformers,it is the only product manufactured in the steel industry that applies the secondary recrystallization phenomenon[1]. The most successful texture control has been achieved in the industrial scale,and People have made a large number of research on the mechanism of secondary recrystallization[2-8].

Thin Slab Casting and Rolling (TSCR)process is a new manufacturing technique,some scholars focus on the research on the grain oriented silicon steel with TSCR Technology at present[9-11]. The purpose of this study is to investigate the evolution of Goss Texture at high temperature annealing process in Hi-B silicon steel produced by TSCR Process.

2. Experimental details

Table 1 shows the chemical composition of the specimen. The steel was melt by vacuum induction furnace, and poured into water-cooled copper mould with dimension of 50mm×100mm×400mm, ensuring that the temperature was at least 950℃ before removal from the mould. In the subsequent process, the slab was kept at 1150℃ for 0.5h in a holding furnace, and then hot rolled to 2.3mm-thick strip. This was then rolled to 0.3mm by single-stage cold rolling after two-step normalizing. Subsequently, the specimen were then subjected to decarburization annealing in wet hydrogen at 875℃ and high temperature annealing at 1200℃. The final Iron loss is 1.231W/kg, Magnetic is 1.908T.

Fig.1 shows the high temperature annealing process. In order to simulate the on-site production， as soon as possible to reduce the heating rate. the heating rate is 15℃/h， and atmosphere is 75%H_2+25%N_2， purposive sampling from 700℃ to 1100℃ at regular temperatures. The evolution of Goss texture in the high temperature annealing process is analyzed by ZEISS SUPRA 55VP SEM with EBSD system.

Table 1. Chemical composition of specimen wt %.

Steel grade	C	Si	Mn	Cu	S	Als	N
Hi-B	0.04	3.00	0.10	0.15	0.0045	0.030	0.011

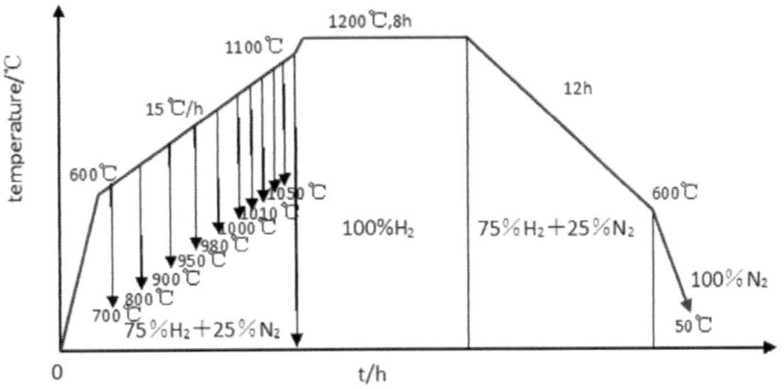

Figure 1. High temperature annealing process

3. Results and discussion

Fig.2 shows the macro-/ microstructure at different temperatures. We can see that it has not the abnormal growth grains until 950℃,but some grains reached to 300μm at 980℃,and some grains reached to cm level at 1000℃,Goss grains instead of Primary grains fully at 1020℃, the texture don't change above 1040℃.

Fig.3 shows the textures at different temperatures. It keeps the texture features of Primary recrystallization until 950℃,it is mainly the γ texture;Goss texture appears in ODF of sample annealed at 800℃, but its intensity is very weak; Goss grains grow abnormally during 980℃~1000℃,and the intensity becomes higher and higher;when Goss grains instead of Primary grains fully at 1020℃,the intensity of Goss texture reached to 118.113;With temperature increasing,the orientation of Goss grains further adjustment,the intensity of Goss texture reached maximum at 1040℃,and the intensity was almost unchanged above 1040℃.

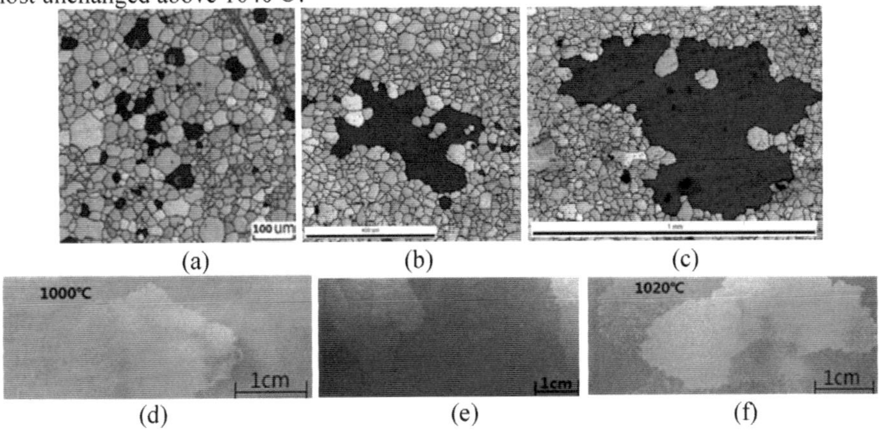

(g) (h) (i)

Figure 2. Macro-/ Microstructure at different extraction temperature (a):950℃;(b):980℃; (c):1000℃; (d):1000℃;(e):1010℃; (f):1020℃; (g):1030℃;(h):1040℃; (i):1050℃.

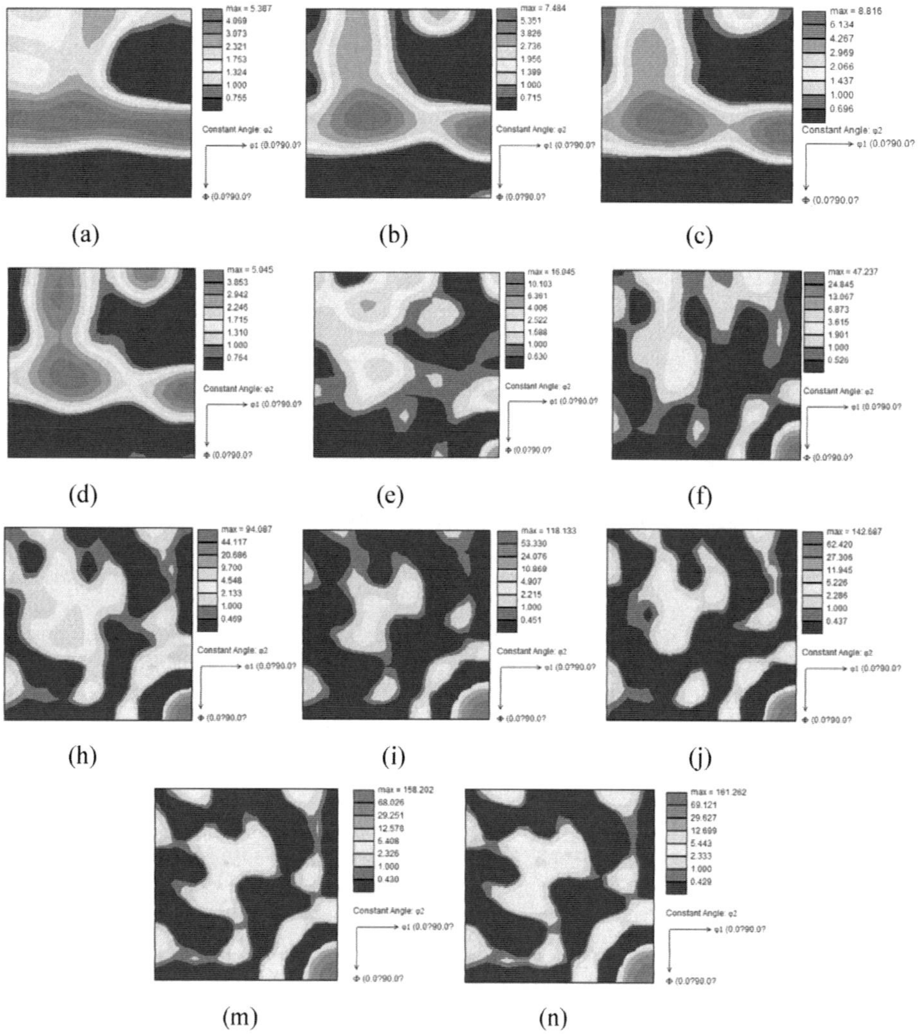

Figure 3. φ2=45°section of ODF at different temperatures (a):700℃(b):800℃(c):900℃ (d):950℃(e):980℃(f):1000℃(h):1010℃(i):1020℃(j):1030℃(m):1040℃(n):1050℃.

Fig.4 shows the relationship between Extraction temperature and B8(all data were the average of 5specimens),it indicates the secondary recrystallization behavior. it can be recognized that the finish temperature of secondary recrystallization is 1040℃ in this test. Even if the Goss grains instead of Primary grains fully at 1020℃,the B8 is only 1.58T. The B8 increased with the temperature increasing until 1040℃,and then come to a constant value.

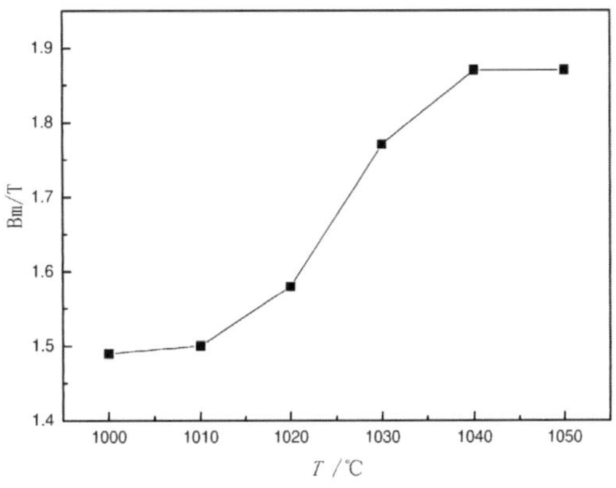

Figure 4. Extraction temperature and B8

The deviation angle of Goss Grain Orientation is the only influencing factor of B8[1].The deviation angle of Goss Grain Orientation has been researched from 700℃ to 950℃,it can't find the Goss grains if the deviation angle of Goss Grain Orientation is limited to 10°;some Goss grains appear if deviation angle is limited to 20°,the deviation angle of Goss grains changed from 10.10° to 19.33°,and the average deviation angle is 15.38°. Making statistical analysis for the deviation angle of abnormal growth Goss grains at 1020℃, the average deviation angle is 9.7°;the average deviation angle is 7.39°at 1030℃,and the average deviation angle is 3.7° at 1040℃,the deviation angle don't change at 1050℃. So with temperature increasing, the variation of deviation angle takes the manner as 15.38°→9.7°→7.39°→3.7°. The results show that the deviation angle is more than 10° at first,and then decreases gradually with temperature increasing,it reached the minimum at 1040℃. The deviation angle is relative to the B8,although the Goss grains instead of Primary grains fully at 1020℃,but the value of deviation angle is rather large,it lead to the low B8.

4. Conclusions

In summary,by suitably sampling at high temperature annealing process,the definite temperature of secondary recrystallization and the Evolution Law of deviation angle were obtained. The results provide an efficient way to optimize the production technology in practice.

(1)Goss texture appears in ODF at 800℃, Goss grains grow abnormally during 980℃~1000℃, Secondary Recrystallization completed at 1040℃

(2)The deviation angle of Goss grain orientation decreases gradually with temperature increasing, and it reached the minimum(3.7°) at 1040℃.

Acknowledgements

This research is supported by the Chinese National Natural Science Foundation(51661026), China Postdoctoral Science Foundation (2016M602914XB), IMAR (Inner Mongolia Autonomous Region) Natural Science Foundation(2015BS0508).

References

[1] HE Z Z, ZHAO Y and LUO H W 1960 Electrical Steel. Beijing: Metallurgical Industry Press,2012
[2] Walter J L, and Dunn C G. *Acta mtall,***8(8)**:497-503
[3] Walter J L, and Dunn C G 1960 *AIME Trans.***218**,1033-38
[4] Walter J L, and Dunn C G 1595 *AIME Trans.***215**:465-71
[5] Dunn C G, and Lionetti F 1949 *AIME Trans.***185**:125-132
[6] Dunn C G, Daniels F W, and Bolton M J 1950 *AIME Trans.***188**:368-77.
[7] Hayakawa Y, Szpunar J A and Palumbo G 1996 *Journal of Magnetism and Magnetic Materials.***160**: 143-48

[8] Flowers J W 1979 IEEE Transactions on Magnetics ,**15(6)** : 1601-03.
[9] Xiang L, QIU S T, ZHAO P 2012.*Word Metals,* **09**.17
[10] QIU S T, XIANG L, YUE E B and ZHAO P 2008 *Iron and Steel,***43(9)**: 1-7.
[11] YU Y M, LI C S and WANG G D 2007 *Iron and Steel*,**42(11)**: 45-47

IOP Conf. Series: Materials Science and Engineering (2017) 012027 doi:10.1088/1757-899X/170/1/012C

Influence of Fused $CaZrO_3$ Addition on Properties of Chrome-free Castables for RH Degassers

X W Zhu[1], Y H Liang[2], J H Nie[3], P Yue[4] and R Q Cui[5]
[1,2,3,4,5]The Hubei Province Key Laboratory of High Temperature Ceramic and Refractory of Wuhan University of Science and Technology. Wu Han, 430081,CHINA

Abstract. This thesis considered specific operating conditions of RH degassers, using corundum and spinel as the main raw material, adding proper micropowder and the efficient compound water reducing agent, discussed the influence of fused CaZrO3 addition on RH degassers used chrome- free castable's properties, which includes the castable's strength, permanent line rate, wear resistance and performance of the slag resistance. The results showed: (1) Adding proper fused CaZrO3 can improve the strength of castable and wear resistance;(2) With the increase of the content of CaZrO3, volume density of castable declined slightly, apparent porosity increases, the line rate increased obviously;(3) With the increase of the amount of CaZrO3, slag erosion resistance of castable declined;(4) From the perspective of the comprehensive performance of castable, fused CaZrO3 suitable addition amount of 2 ~ 4%.

1. Introduction

RH degassing vacuum cycle method is more popular among steel plants with its advantages as good degassing performance, taking off less heat from molten steel, wide application scope, flexible and reliable operation compared with a variety of other refining outside the furnace method[1].Therefore, RH vacuum refining is widely applied by big steel plants. However, the demanding environment of RH refining forces its lining to use the direct binding or combine magnesite chrome brick, but using magnesite chrome based refractory will produce poisonous hexavalent chromium. Thus, chrome-free refractory materials are widely researched in present domestic universities, enterprises and research institutes. The general research is in corundum-spinel system since this system of materials is of high strength, good abrasion resistance, high temperature volume stability , excellent thermal shock resistance and good resistance to slag erosion, etc[2]..But the class and conditions of iron and steel smelting are complex in RH degassers. Materials of corundum-spinel system is no longer the ideal material facing the great thermal stress change and high basicity (R quartile 2.5) of the molten slag erosion. Therefore, it becomes the focus of attention as how to make corundum-spinel chrome-free material that adapts to RH refining environment, especially with higher resistance to erosion of high basicity slag. The characteristics of fused calcium zirconate are microstructure uniformity, high strength, good resistance to chemical and mechanical attack. Pollution free and environmental friendly, it becomes ideal alternative raw materials to magnesium chromium, magnesium and aluminate spinel[3].This paper based on corundum-spinel system, using fused $CaZrO_3$ and as the research object, purposed to improve its abrasion resistance, spalling resistance and resistance to high basicity slag erosion, thereby to improve the service life of chrome-free refractory in RH degassers.

2. Experimental

2.1 Raw materials in experiment

Using densely corundum (grain size of 5~3 mm), 3~1 mm, 1~0 mm, ≤0.074 mm fused white corundum, 1~0 mm, 325 mesh fused spinel, CA-80 cement, active alpha Al_2O_3 micro powder as the main raw materials, added 325BS fused $CaZrO_3$, the physical and chemical indexes are shown in table 1.

Table 1. Chemical compositions of main raw materials(ωt %)

Raw material	Al_2O_3	SiO_2	CaO	MgO	Fe_2O_3	ZrO_2
Densely corundum	98.31	0.21	0.03	0.02	0.15	-
Fused white corundum	99.51	0.22	0.03	0.02	0.30	-
Fused spinel	77.56	-	-	20.35	-	-
Activated alumina	98.90	0.23	-	-	0.08	-
CA-80 cement	77.89	-	19.12	-	-	
$CaZrO_3$	-	-	26.90	-	-	70.84

Table 2. Slag compositions (ωt %)

compositions	SiO_2	Al_2O_3	Fe_2O_3	FeO	CaO	MgO	TiO_2	MFe	P	S
Content（ωt %)	10.4	23.69	5.46	4.69	34.64	14.6	0.36	1.36	0.13	0.029

2.2.Recipe of experiment

Table 3. Recipe (ωt%)

Raw material	1#	2#	3#	4#	5#	6#
Fused corundum（5~0mm）	76	74	72	70	68	66
Fused spinel（325BS~1mm）	10	10	10	10	10	10
Active alumina	10	10	10	10	10	10
CA-80 cement	4	4	4	4	4	4
$CaZrO_3$（325BS）	0	2	4	6	8	10
additive	2.2	2.2	2.2	2.2	2.2	2.2

2.3 The experiment process and performance testing

2.3.1 Experiment process

Weighs all sorts of aggregates, powder, binding agent and water reducing agent in proportion according to the recipe. then pour the materials into mortar blender. Dry mixing for 1~2 minutes until well distributed and then slowly add right amount of water to wet mix for 2~3 minutes to reach the standard of molding. Put mixed wet material quickly into the sample mould（160×40×40mm） which was set in advance. Charging during the vibration process until there is little small bubbles and the pulp appears. Keep the castable in mould for 24h and then demould to measure its body size. Maintain the demould sample in room temperature for 24h. Put the sample in electrical drum wind drying oven in 110℃ for 24 hours for drying. Finally after drying of sample perform 1100℃×3h,, 1500℃×3h heat treatment.

2.3.2 Performance testing

1. Adapting the three point bending method, according to GB/T 3001-2000, measure the normal temperature flexural strength; According to GB/T 3997. 2-1998, measure the samples' cold modulus of rupture and cold crushing strength.
2. Adapting the Archimedes drainage method, according to GB/T 2997-2000, measure the samples' apparent porosity and volume density.
3. Measure the samples' linear expansivity rate after heat treatment according to GB/T 5203-1993.
4. Measure the samples' wear resistance according to GB/T18301-2001 test methods.
5. Slag resistance ability test is performed using static crucible method, put proper amount refining ladles slag into crucible, analyze the eroded crucible by dissecting its fracture surface after heat treatment in 1550℃ for 3 hours.

3. Results and discussion

3.1 Effect of $CaZrO_3$ on strength of castables

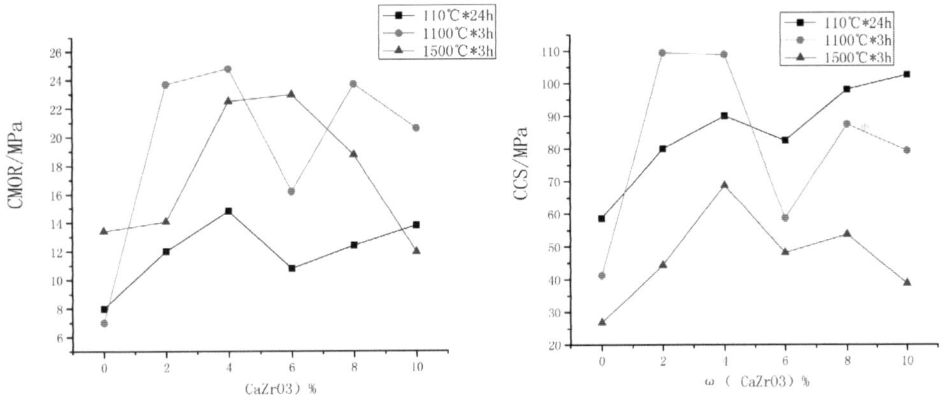

Figure 1. Influence of $CaZrO_3$ on strength of castables

Figure 1 shows, when $CaZrO_3$ addition amount of 4%, the sample flexural strength and compressive strength achieves maximum, than the strength of sample without $CaZrO_3$ has improved significantly.But, when the addition of calcium zirconate quantity within 4 ~ 10%, with the increase of MZ addition amount, the strength of sample has declined, this is mainly because: (1) the castable in shape, with the increase of $CaZrO_3$, water content increasing, the liquidity is reduced, cause 110 ℃ x 24 drying the samples after the porosity increases, the intensity is reduced, but just because the reason of forming the strength reducing trend of small;(2) 1100 ℃ for 3 h sample, because the sample did not happen a lot of chemical reaction, so it was; and the strength of the 110 ℃ trend;(3) samples at

1500 ℃ sintering, a small amount of CaO and Al_2O_3 reaction to generate CA6 in matrix, the compactness of material increase, the microstructure was optimized.So when $CaZrO_3$ addition amount was 4%, the strength of the specimens significantly higher than that of the original, but $CaZrO_3$ in 4-10% range, along with the increase of the content of $CaZrO_3$,CaO content is increased, so the generated CA6 volume also increased significantly, leading to sample, loose structure, expansion, resulting in significantly reduced intensity[4]. From the perspective of the intensity of the sample, addition amount of 4% of $CaZrO_3$ achieve more ideal effect.

3.2 Influence of $CaZrO_3$ on castables' permanent linear change

As seen in figure 2, under 110℃ and 1100℃, with the increase of the content of $CaZrO_3$, permanent line rate of castables is not obvious, but the sample under 1500 ℃, with the increase of the content of $CaZrO_3$, permanent linear change of castables is obvious increase,the main reason is that under the condition of high temperature, along with the increase of the content of $CaZrO_3$, CaO content increased, so the generated CA_6 volume also increased significantly, causing sample quantity increase.From the permanent line rate of castables, when the addition amount of 2% of $CaZrO_3$, sample volume is more stable.

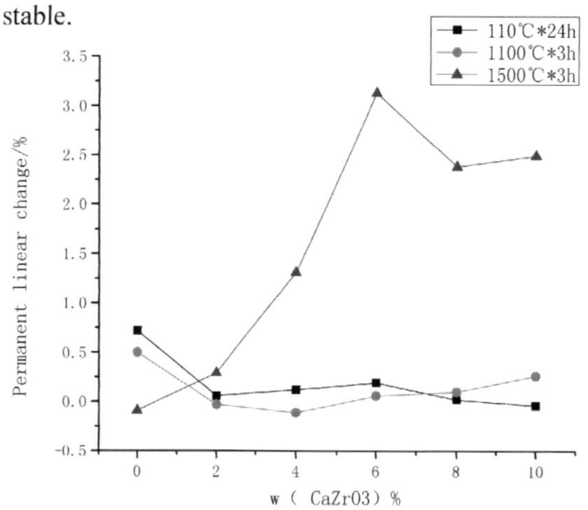

Figure 2. Effect of $CaZrO_3$ on permanent linear change of castables

3.3 Influence of $CaZrO_3$ on castables' wear-resisting

According to the national standard, set the 100×100×30 mm standard sample's test surface vertically to sand blasting tube, blow the wear medium (black silicon carbide) through the gush arenaceous pipe on to sample using compressed air and then measure each sample's wearing volume. Through calculation, the wear-resisting results is shown in figure 3.

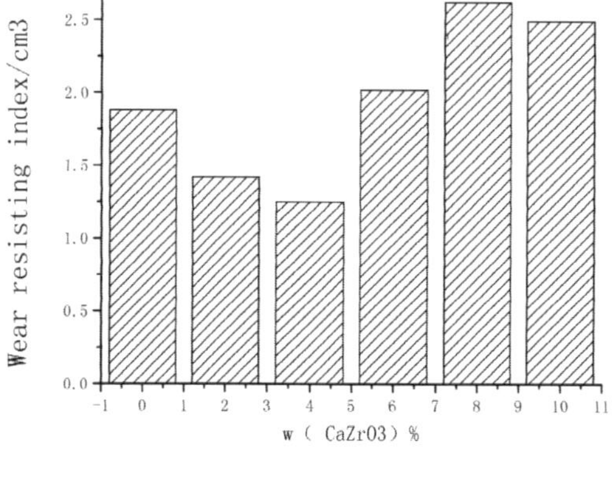

Figure 3. Influence of $CaZrO_3$ on castables' wear resistance

As seen from the figure 3, for the amount of $CaZrO_3$ from 2% to 4% range, its wear resistance is better than the same $CaZrO_3$ (0%);When the content of $CaZrO_3$ was 4%, the wear resistance of sample is best, its wear resistance increases by about 33.5% than before.Main reason is the system of a small amount of CaO at high temperatures and matrix of Al_2O_3 reaction CA_6, its microstructure was optimized, thus improving strength and wear resistance increased. But the content of $CaZrO_3$ in 6%~10% range, along with the increase of the content of $CaZrO_3$, CaO content is increased, so the generated CA_6 volume also increased significantly, leading to sample, loose structure, expansion, resulting in wear resistance is decreased obviously.From the point of abrasion resistance index of the sample, $CaZrO_3$ addition amount of 2%~4%.

3.4 Influence of $CaZrO_3$ on castables' slag resistance

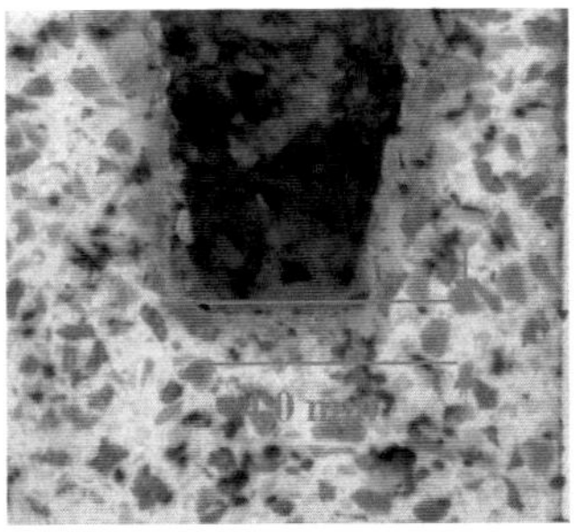

A) 0% (permeation depth 4.0 mm)

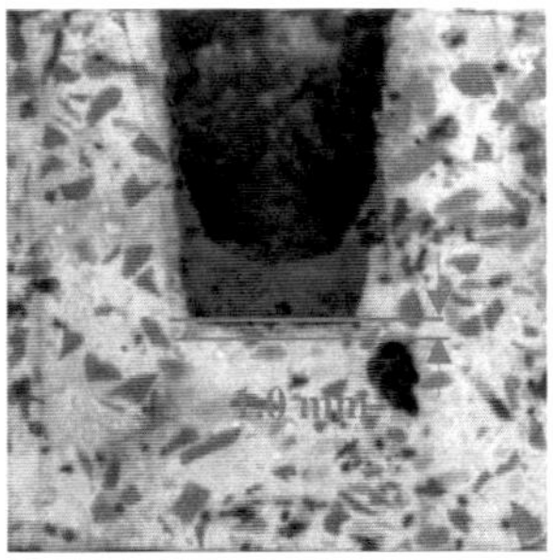

B) 2% (permeation depth 1.0 mm)

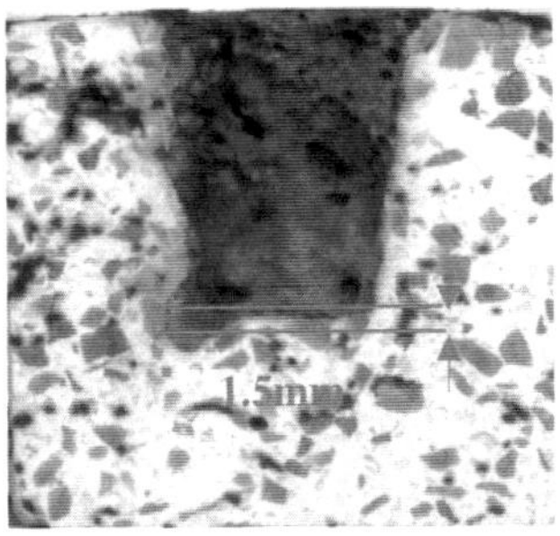

C) 4% (permeation depth 1.5 mm)

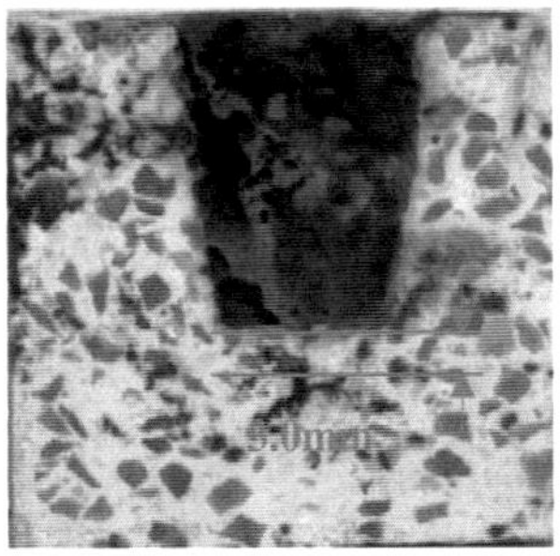

D)6 % (permeation depth 3.0mm)

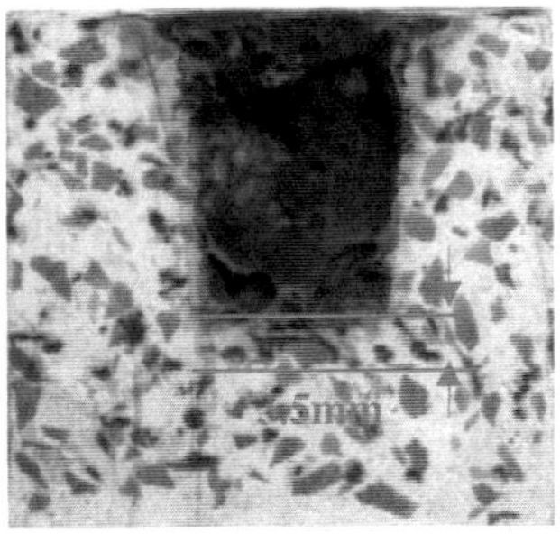

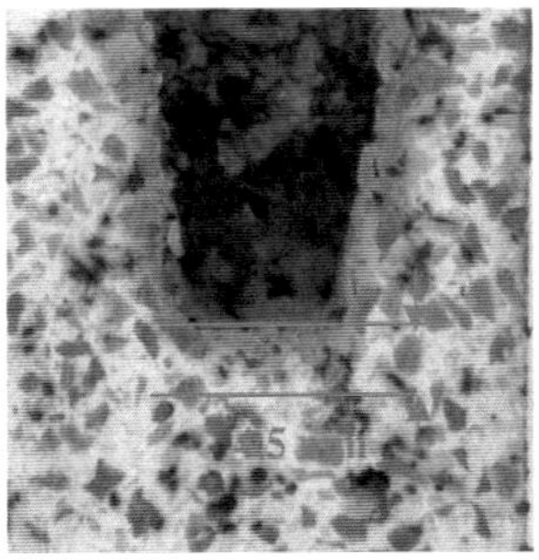

E) 8% (permeation depth 3.5mm) F) 10% (permeation depth 4.5mm)

Figure 4. Effect of MZ content on castables' slag resistance

It can be seen clearly from the figure 4, when adding 2%$CaZrO_3$, its best slag resistance, erosion depth of only 1.0 mm, the main reason has two: (1) Small amount of CaO and Al_2O_3 reaction to generate CA_6 in matrix, the microstructure of the material was optimized, so as to improve the slag resistance of sample;(2) ZrO_2 and CaO itself high temperature resistance, chemical stability, good resistance to alkali slag erosion performance[5].However, can be seen from the figure 4, when the amount of $CaZrO_3$ from 6% to 10% range, with the improvement of $CaZrO_3$ content, sample volume expansion, internal structure is loose, porosity, bulk density is reduced, resulting in resistance to slag erosion resistance showed a trend of decrease.But in general, when adding 2%$CaZrO_3$, the sample has good resistance to slag erosion resistance.

4. Conclusions

(1) Adding 2%~4% $CaZrO_3$ can improve the strength of castables and wear resistance performance;

(2) With the increase of the content of $CaZrO_3$ sample volume density is declined slightly, apparent porosity is increased, the permanent linear rate is increased obviously;

(3) With the addition of more $CaZrO_3$, slag corrosion resistance of castables is declined;

(4) From the perspective of the comprehensive performance of castable, fused $CaZrO_3$ suitable addition amount is 2 ~ 4%.

References

[1] Xu G Q 2006 *Steelmaking* **Vol 37** 12-15
[2] Ren J H , Ren Y and Zhang J Z 2009 *Silicate Bulletin,* **Vol 28** 90-93
[3] Zhang Y, Chen J J and Dou S J 1999 *Refractory Materials,* **Vol 6** 310-312, 319
[4] Jia Q L , Ye F B and Zhong X C 2008 *Silicate Bulletin,* **Vol 27** 1104-1108
[5] He Z Y , Hong Y R and Li L 2004 *Refractory Materials,* **Vol 38** 73-75

Determination of gas length for gas-assisted extrusion forming of polymer melt

Z Ren[1,2], X Y Huang[2]

[1]Key Laboratory of Optic-Electronic and Communication, Jiangxi Science and Technology Normal University, Nanchang, 330038, China
[2]School of Mechanical and Electrical Engineering, Nanchang University, Nanchang, 330031, China

E-mail: renzhong0921@163.com

Abstract. To determine the optimal gas length for the gas-assisted etrusion forming of melt, numerical investigations about the gas length on the extrudate swell of melt were performed by using the finite element method. Meanwhile, the geometric model of gas-assistd extrusion forming was established. The full slip boundary condition was used as the gas-assisted condition. Numerical results show that the gas length should be shortened with increasing of the inlet volumme flow rate of melt. In addition, under the given inlet volume flow rate of melt, the extrudate swell ratio, X velocity and shear stress of melt greatly decreases with increasing the gas length. Accroding to the numerical results and experiences reported past time, under the inlet volume flow rate of $0.5cm^3/s$, the optimal gas lenth of gas-assisted extrusion forming is about 10mm.

1. Introduction

In the process of the plastic extrusion forming, the products of plastic easily occur some problems, such as extrudate swell [1], melt fracture [2], and extrudate distortion [3] due to the high elasticity and high shear rate. To eliminate these extrudate problems, there are some methods have been studied and applied, such as, the polymer addictive [4], mechanical vibration extrusion [5], ultrasonic-assisted extrusion [6], and gas-assisted extrusion. The gas-assisted extrusion is a most-promising and well-established extrusion method due to the advantages of convenient, environmental protection, and high efficiency. The gas-assisted extrusion method has been validated by many experimental and numerical studies. Brzoskowski [7] firstly proposed the gas-assisted extrusion technique. Liang [8] firstly applied the gas-assisted extrusion technique into the polymer melt extrusion. Laterly, Huang [9] successfully achieved the gas-assisted extrusion of the rob plastics. Huang[10] successfully studied the gas-assisted coextrusion of the polymer melt. Deng [11] successfully performed the studies of the gas-assisted coextrusion of the L cross-typed profile plastics. Liu et al. [12] numerically studied the gas-assisted extrusion of the melts by using the finite element method. These studied all validated that the gas-assisted extrusion method is an efficient way of eliminating the extrudate swell, melt fracture, and extrudate distortion. However, it was also demonstrated that the reasonable controlling of the gas parameters and gas length are important to form the stable gas-assisted layer. In this paper, the determination of the optimal gas length for the gas-assisted extrusion of melt was studied by using the numerical method.

2. Numerical simulations

2.1 Numerical model

Figure 1(a) is the geometric model of gas assisted extrusion for the melt. In Figure 1(a), the ABCD-EFGH is the part of no gas-assisted die, its length is 10mm. EFGH-IJKL is the part of gas-assisted die, its length is 30mm. IJKL-MNOP is the part of free surface outside of die, its length is 10mm. Figure 1(b) is the finite element mesh of the geometric model (Figure 1(a)). The meshes were appropriately reinforced near the inlet and outlet surface of die, and the exit surface of melt. The mesh number was 3200.

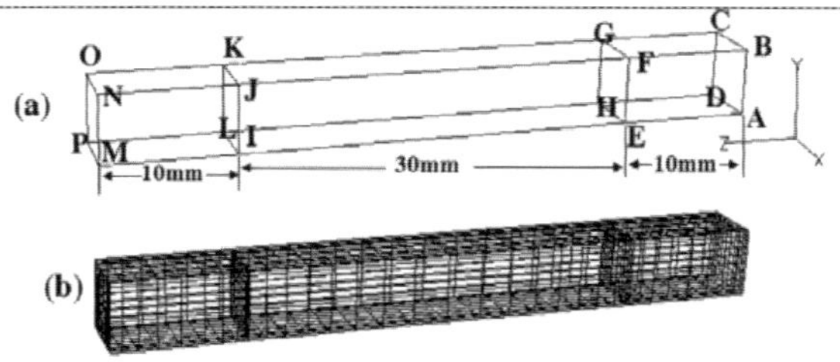

Figure 1. Geometric model and finite element mesh. (a) Geometric model; (b) Finite element meh.

2.2 Governing equations

In the numerical simulations, the melt was regarded as non-compressible, iso-thermal, steady, laminar and visco-elastic Non-Newtonian fluid. The inertia, gravity force, viscous dissipation, and surface tension effects were neglected.

The mass and momentum conservation equations are given as follows:

$$\nabla v = 0 \tag{1}$$

$$-\nabla p + \nabla \tau = 0 \tag{2}$$

where ∇ is Hamilton operator, v is velocity of melt, p is pressure drop, τ is extra stress tension.

In simulations, Phan-Thien-Tanner (PTT) constitutive model [13] was used, which is given as follows:

$$\exp\left[\frac{\varepsilon\lambda}{(1-\eta_r)\eta} tr(\tau_1)\right]\tau_1 + \lambda\left[\left(1-\frac{\xi}{2}\right)\overset{\nabla}{\tau_1}+\frac{\xi}{2}\overset{\Delta}{\tau_1}\right] \tag{3}$$
$$= 2(1-\eta_r)\eta D$$

where $\eta_r = \eta_2 / \eta$ is viscosity ratio, η is the total viscosity of melt, η_2 is Newtonian viscosity component of melt, λ is the relaxation time, ε and ξ are the parameters of melt correlated with material tensile and shear characteristics, respectively. $\overset{\nabla}{\tau_1}$ is the upper convected derivative of the extra stress tensor τ_1, η_1 is the Non-Newtonian component viscosity of melt, D is the strain-rate of tensor.

2.3 Material parameters

In the simulations, the material parameters are given in Table 1.

Table 1 The parameters of PTT construction equation

Parameters	η(Pa.s)	λ(s)	ε	ξ	η_r
Values	2700	0.2	0.18	0.23	0.12

2.4 Boundary conditions

(1) inlet boundary: ABCD is the inlet surface of melt. Supposed that the flow of melt is full-developed, steady and laminar flow, the dynamic and kinematics conditions are given as follows:

$$\partial v_z / \partial z = 0, v_x = v_y = 0 \tag{4}$$

(2) wall boundary: ABFE, and BCGF are the wall boundary, i.e., the no gas-assisted boundary. The following relationship should be satisfied, i.e.,

$$v_n = v_s = 0 \tag{5}$$

(3) gas-assisted boundary: EFJI, and FGKJ are the gas-assisted boundary. The following relationship should be satisfied, i.e.,

$$v_n = 0, f_s = 0 \tag{6}$$

(4) Free boundary: IJNM, and JKON are the free boundary. The following relationship should be satisfied, i.e.,

$$v \cdot n = 0, f_n = 0 \text{, and } f_s = 0 \tag{7}$$

(5) Exit boundary: Supposing that no trance force and entangle velocity are imposed. $f_n = f_s = 0$

3. Numerical results and discussion

3.1 Effect of inlet volume flow rate on the pressure of melt

To determine the gas length in the gas-assisted extrusion of melt, in the range of the inlet volume flow rates from 0 to 1cm^3/s,four different inlet volume flow rates(0.1, 0.2, 0.5, and 1 cm^3/s) were imposed on the inlet surface of the die. The pressure distribution of melt under four different inlet volume flow rates were shown in Figure 2.

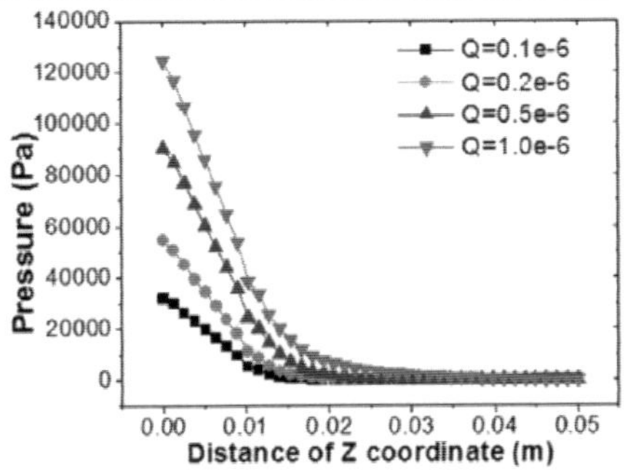

Figure 2. Pressure distributions of melt under four different inlet volume flow rates

From Figure 2, it can be seen that, for the inlet volume flow rate of 0.1 cm^3/s, the pressure decreases to zero when the distance of Z coordinate is about 15mm, i.e., the gas length is about 5mm. When the inlet volume flow rate of melt is increased, it can be seen that the pressure of melt increase, and the gas length when the pressure of melt decreased to zero is longer. When the inlet volume flow rate of melt increases to 1cm^3/s, the gas length is about 30mm, which demonstrates that the gas length should be lengthened in the gas-assisted extrusion forming.

3.2 Effect of gas-length on the extrudate swell

To study the effect of the gas length on the extrudate forming of the melt, three different gas lengths were used, i.e., 5mm, 10mm, and 15mm. the same inlet volume flow rate and same boundary conditions were imposed on the corresponding boundaries. In this simulation, the inlet volume flow rate was 0.5 cm^3/s. The extrudate swell ratio change for the gas-assisted melt extrusion under three different gas lengths are shown in Figure 3.

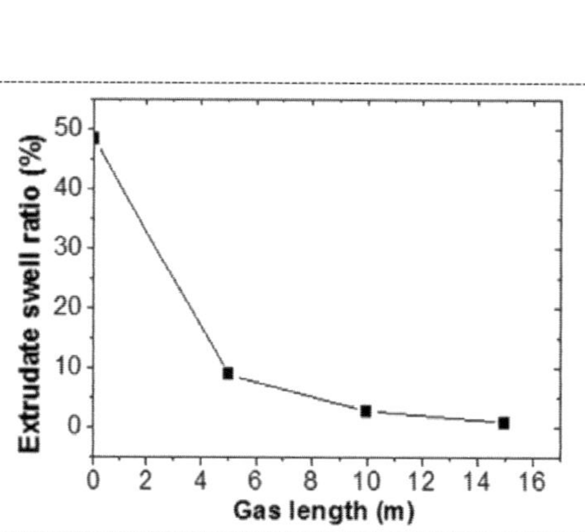

Figure 3. Extrudate swell ratios of melt under different gas lengths

From Figure 3, it can be seen that the extrudate swell ratio of melt decreases with increasing the gas length. Moreover, when the gas length is about 10mm, the extrudate swell ratio is nearly equal to zero.

3.3 Effect of gas-length on the flow velocity

Figure 4 is the X flow velocity of melt under four different gas lengths.

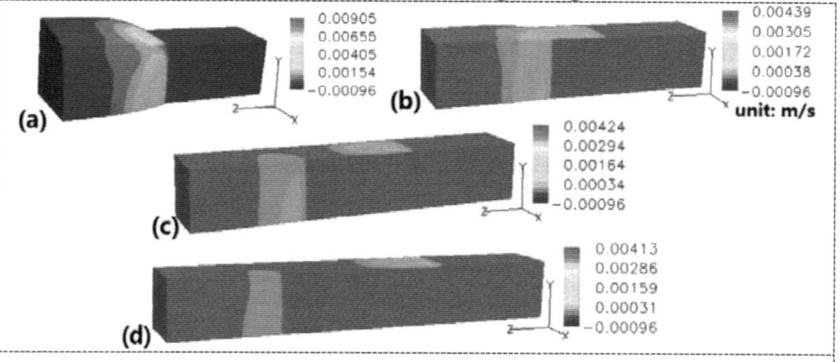

Figure 4. X velocity distributions of melt under four different gas lengths. (a) gas length=0mm; (b) gas length=5mm; (c) gas length=10mm; (d) gas length=15mm.

From Figure 4, it can be seen that the X velocity is largest for the no gas-assisted extrusion mode, which results in the largest extrudate swell phenomenon. With the increasing of the gas length, the X velocity distribution of melt at the outlet of the die was greatly decreased. When the gas length is increased to about 10mm, the X velocity of melt at the outlet of die nearly equals to zero. However, when the gas length is continuous to increase, the X velocity of melt at the outlet of die nearly does not change.

3.4 Effect of gas-length on the shear stress

Figure 5 is the shear stress contours of melt under four different gas lengths.

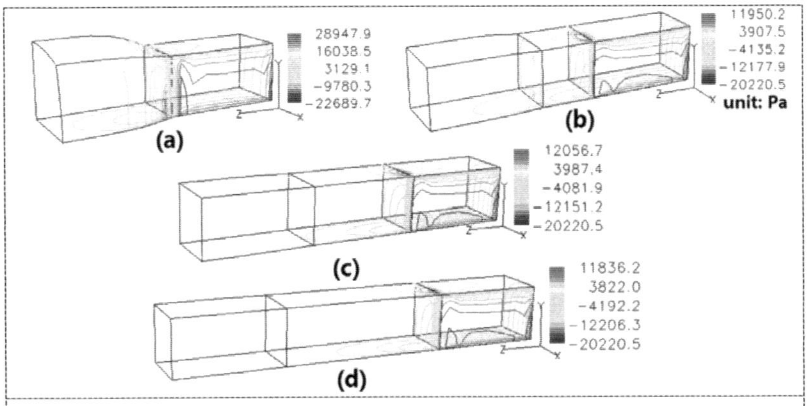

Figure 5. Shear stress contours of melt under four different gas lengths. (a) gas length=0mm; (b) gas length=5mm; (c) gas length=10mm; (d) gas length=15mm.

From Figure 5, it can be seen that the shear stress of melt at the outlet of the die is largest for the no gas-assisted extrusion mode. With the increasing of the gas length, the shear stess of melt was greatly decreased. When the gas length is increased to about 10mm, the shear stress at the outlet of die nearly equals to zero. Then, when the gas length is continuous to increase, the shear stress of melt at the outlet of die does not change, i.e., nearly equals to zero, which demostrates that the extrudate swell effect of melt nearly removed when the gas length is about 10mm. According to the results in the past reports [9-12], under the condition of removed extrudate problems, e.g., extrudate swell, the gas length should be short because the gas length is longer the gas flow is more unstable. Therefore, the reasonable gas length should be choosen about 10mm for the inlet volume flow rate of $0.5cm^3/s$.

4. Conclusions

In this paper, the gas-assisted extrusion method was used into the extrusion forming of melt to remove the extrudate swell phenomenon. Among the process parameters of gas-assisted extrusion forming, the gas length is one of most important factors of impacting the effect of gas-assisted extrusion forming. To determine the optimal gas lengh of the gas-assisted extrusion forming, the square cross-typed model of gas-assisted extrusion forming was established, and the numerical simulations about the effect of gas length on the extrusion forming were performed by using the finite element method. Numerical results showed that the extrudate swell decreased with increasing gas length, which also verified from the viewpoints of X flow velocity and shear stress of melt at the outlet of the die. The optimal gas length was determined based on the numerical simulation results and the practical experiences.

Acknowledgements

This work is supported by the Natural Science Foundation of Jiangxi Province (20151BAB202011), the Science and Technology Pillar Program of Jiangxi Province (20132BBG70103), Top-notch talent foundation project of JXSTNU (2014QNBJRC004), the Scientific Research Foundation of Jiangxi Provincial Education Bureau (No. GJJ150814), and 2011 Cooperative Innovation Center Project of Jiangxi Province (2014XTCX005).

References

[1] Goublomme A, Draily B and Crochet M J 1992 *J. Non-Newton. Fluid Mech.* **44** 171
[2] Othman N, Jazrawi B, Mehrkhodavandi P, et al. 2012 *Rheol. Acta* **51** 357
[3] Kalika D S and Denn M M 1987 *J. Rheol.* **31** 815
[4] Arda R D and Mackley M R 2005 *J. Non-Newton. Fluid Mech.* **126** 47

[5] Liu Y J 2006 *Polym. Bull.* **56** 599
[6] Hung J C and Huang C C 2012 *Int. J. Precis. Eng. Man.* **13** 2103
[7] Brzoskowski R, White J L, Szydlowski W, Weissert F C, Nakajima N and Min K 1987 *Rubber Chem. Technol.* **60** 945
[8] Liang R F and Mackley M R 2001 *J. Rheol.* **45** 211
[9] Huang X Y 2006 Theoretical and experimental study of polymer gas-assisted extrusion from dies (Nanchang: Nanchang University) (in Chinese)
[10] Huang Y B 2011 Theoretical and experimental study on polymer gas-assisted co-extrusion (Nanchang: Nanchang University) (in Chinese)
[11] Deng X Z 2014 Experimental and theoretical study on gas-assisted co-extrusion of plastic profile with an irregular cross-section (Nanchang: Nanchang University) (in Chinese)
[12] Liu H S, He J T, Huang X Y, Deng X Z and Wan Q F 2014 *Polym.Mater. Sci. Eng.* **30** 113 (in Chinese)
[13] Phan T and Tanner R 1977 *J. Non-Newton. Fluid Mech.* **2** 353

Microwave Dielectric Properties of $Na_{1/2}Sm_{1/2}Ti_{1-x}(Cr_{1/2}Nb_{1/2})_xO_3$ Ceramics (x=0~0.025)

Z X Fang[1,2] **B Tang**[1,2] **and S R Zhang**[1,2]

[1]State Key Laboratory of Electronic Thin Films and Integrated Devices,

[2]National Engineering center of Electromagnetic Radiation Control Materials, University of electronic Science and Technology of China, Chengdu 610054, China

E-mail: ZXFANG2015@163.com

Abstract: The compound Na1/2Sm1/2Ti1-x(Cr1/2Nb1/2)xO3 (x =0.00, 0.005, 0.01, 0.015, 0.02, 0.025) (NSTCNx) ceramics were prepared by the conventional solid-state route. The main phase of all NSTCNx ceramics were confirmed as an orthorhombic perovskite structure. The εr and Q×f were improved because of improvement of densification and homogeneous fine grained microstructure with proper substitutions (x ≤0.01). The τ*f* was able be tuned to a relatively low value of 154.3 ppm/°C because of a continuously decreasing tolerance factors. Typically, the NSTCNx (x =0.01) ceramic fired in air at 1450°C for 2h showed good microwave dielectric characteristics of εr=104.3, Q×f=9179 GHz and τf =171.7 ppm/°C.

1. Introduction

Because of the rapid development of wireless communication industry, microwave dielectric ceramics have been intensively researched and widely used in filters, resonators, EMI (Electro Magnetic Interference) filter, RFID (radio frequency identification) antenna active module and many other microwave devices. Recently, the microwave dielectric with a high dielectric constant (ε_r), high quality factor ($Q\times f$) and lower temperature coefficient of resonant frequency (τ_f) have been a very active research field [1-3].

The significance of miniaturization and integration cannot be overemphasized in any hand-held communication devices, since inversely proportional relationship between the physical length of a dielectric resonator and $\sqrt{\varepsilon_r}$. The high permittivity compound $Na_{0.5}Sm_{0.5}TiO_3$ with perovskites type structure has drawn lots of attentions due to its high permittivity (ε_r=100.5), high quality factor ($Q\times f$=8,993GHz) and relatively lower τ_f (τ_f =+199ppm/°C[4]. Generally, the dielectric ceramics with high permittivity always have a large positive τ_f, which makes them impractical in the electric devices. Two methods are commonly employed to lower the τ_f. One easy and common method is to combine two compounds with opposite τ_f values, forming compositional systems, but it always leads to a serious deterioration of ε_r and $Q\times f$ value. Another effective approach is changing the tilting of the octahedral by substituting at B site in perovskites [5, 6], and the tilting of the octahedral is explained by the tolerance factor (t): $t = \frac{r_A + r_O}{\sqrt{2}(r_B + r_O)}$, where the r_A, r_B and r_O represent the radii of ions at A, B, and C sites.

Great efforts have been made to decrease the τ_f as well as maintaining a high permittivity and quality factor in many microwave dielectric systems. For example, the τ_f and $Q\times f$ were improved by doping Ceria (CeO_2) because the Ce^{4+} entered the A-site of $Na_{1/2}Sm_{1/2}TiO_3$ ceramics and changed the lattice structure of perovskites. What's more, X. Guo and B. Tang claimed that the τ_f was effectively changed to zero value, and the $Q\times f$ value were significantly improved by $Cr_{1/2}Nb_{1/2}{}^{4+}$ substituting for Ti^{4+} owing the continuous decrease of tolerance factor in $Ba_{6-3x}Nd_{8+2x}Ti_{18-y}Cr_{y/2}Nb_{y/2}O_{54}$ [7]. Unfortunately, $Cr_{1/2}Nb_{1/2}{}^{4+}$, as an effective aids, hasn't been researched in the $Na_{1/2}Sm_{1/2}TiO_3$ system.

Therefore, in this study, the method of substitution of $Cr_{1/2}Nb_{1/2}{}^{4+}$ for Ti^{4+} is used to enhance the properties of NST systems, and the effects of $Cr_{1/2}Nb_{1/2}{}^{4+}$ replacement upon microstructure were researched, as well as dielectric characteristics.

2. Experimental procedures

Specimens of $Na_{1/2}Sm_{1/2}Ti_{1-x}Cr_{x/2}Nb_{x/2}O_3$ (x =0.00, 0.005, 0.01, 0.015, 0.02, 0.025) ceramics were synthesized by the conventional solid-state ceramic route. High-purity powders (at least 99.9 %) of Na_2CO_3, Sm_2O_3, Cr_2O_3, Nb_2O_5 and TiO_2 were used as the starting materials. The raw oxide materials were weighed according to the stoichiometry proportions with addition of 6 wt. % Na_2CO_3 powder. The mixture of starting materials was ball-milled in alcohol medium for 5 h in nylon jar using zirconia balls. The mixed slurry was dried, passed through a100-mesh sieve and then calcined in air at 1,150 °C for 3 h. The fine powder was mixed with a 6 wt% of a 10% solution of polyvinyl alcohol (PVA) as a binder. The obtained powders were axially pressed into cylindrical disks with thickness of 0.75 cm and 1.5 cm in a diameter under pressures of 200kg/cm^2. These samples were fired at 1450 °C for 2 h.

The bulk densities of the fired samples were tested using the Archimedes' principle. The phase composition were examined by X-ray diffraction (XRD) using CuKα radiation (DX–1000 CSC, Japan) and the lattice parameters of samples were collected from XRD data using the Rietveld method by using the ''Material Analysis Using Diffraction'' (Maud). To examine surface morphology, scanning electron microscopy (SEM) (FEI Inspect F, United Kingdom) was utilized to study the thermally etched surface morphology of the specimens. The dielectric properties at microwave frequencies were tested by the Hakki–Coleman dielectric resonator method in the $TE_{01\delta}$ mode utilizing a network analyzer (Agilent Technologies E5071C, USA) and a temperature chamber (DELTA 9023, Delta Design, USA). The τ_f were obtained by equation: $\tau_f = (f_{t_2} - f_{t_1})/(f_{t_1} \times (t_2 - t_1))$, where f_{t_1} and f_{t_2} were the resonant frequencies at the measuring temperature t_1 (25 °C) and t_2 (85 °C), respectively.

3. Results and discussions

The XRD patterns of $NSTCN_x$ (x= 0.00~0.025) ceramics fired at 1450°C are presented in Fig. 1. The orthorhombic solid solutions in space group $Pm\bar{3}m$ are obtained and there is no secondary phase is determined for all compositions. The weak superlattice reflections were also detected. And these weak peaks can be indexed as (1, 1/2, 0), (1,1,1/2), (3/2, 1/2, 1/2), (2, 1/2, 0), (3/2, 3/2, 1/2), (2, 1, 1/2) and (2, 1, 2/3), which were in doubled index superlattice reflections (an even-even-odd and odd-odd-odd manner) and assigned as the space group $Pm\bar{3}m$ [4, 8, 9]. The diffraction peaks slightly shift to lower 2θ angles as $Cr_{1/2}Nb_{1/2}{}^{4+}$ content increases, reflecting the increments in the cell volumes on the substitution of Ti^{4+} (0.605 Å) by the larger $Cr_{1/2}Nb_{1/2}{}^{4+}$(r=0.6275 Å[10]). The refinements were performed using Maud based on the XRD of all NSTCNx ceramics.

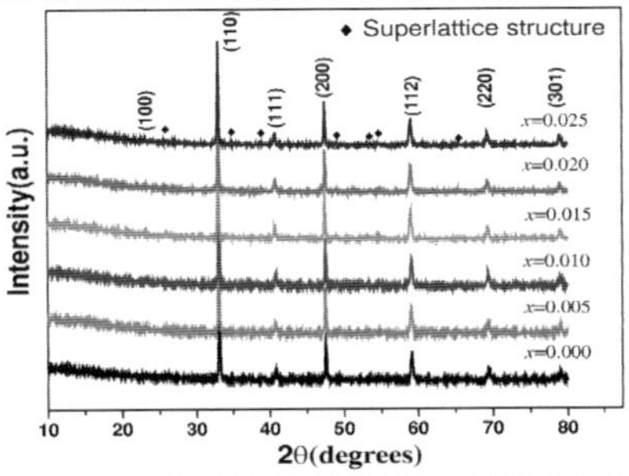

Figure 1. (a) XRD patterns and (b) the refined lattice parameters of $NSTCN_x$ (x= 0.00~0.025) ceramics fired at1450°C for 2 h in air.

Fig. 2(a) shows the results of the observed and calculated XRD patterns for x=0.01. The refined parameters for this orthorhombic perovskite phase in the $Na_{1/2}Sm_{1/2}Ti_{0.99}(Cr_{1/2}Nb_{1/2})_{0.01}O_3$ were a= 3.83406 Å, b= 3.8429 Å and c= 3.8386 Å with a space group $Pm\bar{3}m$. The lattice parameters for all NSTCNx ceramics are also described in Fig. 2(b). With increasing x, the unit cell volumes expanded. These refined results also confirmed that the $(Cr_{1/2}Nb_{1/2})^{4+}$ could partially substitute for Ti^{4+}.

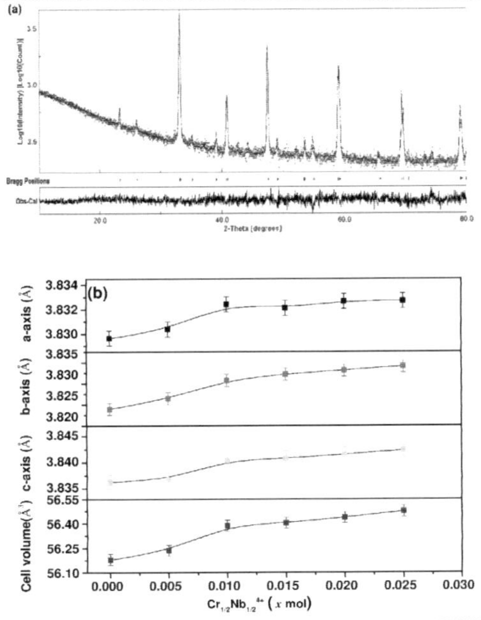

Figure 2. (a) Experimental (Blue dots) and calculated (Black line) X-ray powder diffraction profiles for NSTCNx(x= 0.01) (Rw(%)=4.34098, Rnwb(%)=10.9347, Rb(%)=4.639707 and Rexp(%)=5.3061733). The short vertical lines below the patterns mark the positions of Bragg reflections. The bottom continuous line is the difference between the observed and the calculated intensity; (b)the refined lattice parameters of $NSTCN_x$ (x= 0.00~0.025) ceramics fired at1450°C for 2 h

Figure 3 shows the back scattered electron images for thermally etched surface of the $NSTCN_x$ ceramics fired at 1450 °C for 2 h with (a) x= 0.00 to (f) x = 0.025. Obviously, large amounts of pores around the boundaries were observed with x=0.00, and the pores could be effectively eliminated with addition of $Cr_{1/2}Nb_{1/2}^{4+}$ content. But when the $Cr_{1/2}Nb_{1/2}^{4+}$ content exceeded 0.01, a certain amount of pores was detected again. On the other hand, the grain growth was promoted with the assistance of $(Cr_{1/2}Nb_{1/2})^{4+}$ content, and when $x \leq 0.01$, the microstructure gradually became uniform. It was apparent that the densest and most uniform microstructure was obtained at x=0.01.

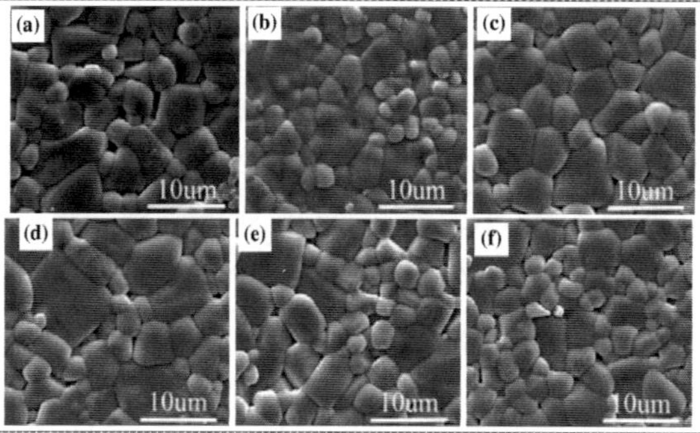

Figure 3. SEM photographs of the $NSTCN_x$ (x= 0.00~0.025) ceramics fired at 1450 °C for 2 h with (a) x= 0.00 to (g) x = 0.025.

Fig. 4 illustrates the microwave dielectric characteristics of $NSTCN_x$ (x= 0.00 to 0.025) fired at 1450°C for 2 h. For the apparent density and permittivity shown in Fig. 4(a) and 4(b), they presented a similar varying trend, which was attributable to the same reason that density and permittivity were much affected by the densification and pores. It was clear that there is a big drop of permittivity from 104.3 to 95.6 as superfluous $Cr_{1/2}Nb_{1/2}^{4+}$ contents (x ≥0.015) were added. The decrease of density was the main reason. On the other hand, the decrease of permittivity should be related to the decrease of microscope polarizability and expansion of molar volume correlated to ionic radius[11]. And in this work, the average ion polarizability of $Cr_{1/2}Nb_{1/2}^{4+}$ (α = 2.73 Å^3) is much weaker than the ion polarizability of Ti^{4+} (α = 2.94 Å^3)[12], and the cell volume increased with the substitution of larger $Cr_{1/2}Nb_{1/2}^{4+}$.

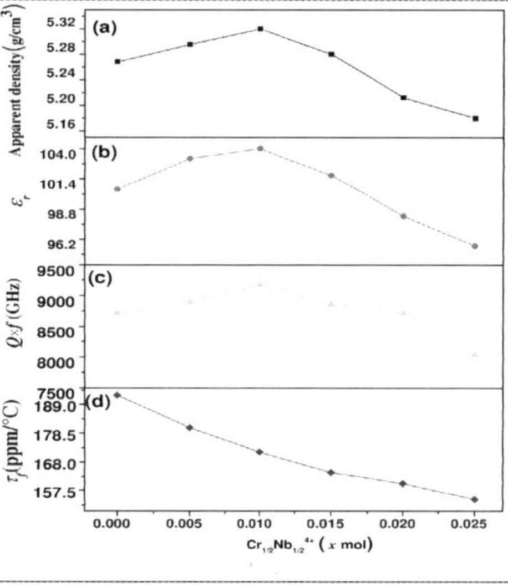

Figure 4. (a) Apparent density, (b) permittivity, (c) Quality factor and (d) Temperature coefficient of resonant frequency of $NSTCN_x$ (x= 0.00~0.025) fired at 1450°C for 2 h.

As illustrated in Fig. 4 (c), the $Q\times f$ value firstly increased to 9179 GHz at x=0.01, but sharply decreased to 8038 GHz at x=0.025 thereafter. The average grain size increased and homogeneous microstructure was obtained with a proper substitutions ($x \leq 0.01$), which enhanced the quality factor[13]. And the substitution of Cr^{3+} for Ti^{4+} restrained Ti^{4+} reduction which was attributable to improvement of $Q\times f$:$Cr_2O_3 \xrightarrow{TiO_2} 2Cr'_{Ti} + V_O^{\cdot\cdot} + 3O_O^x$. However, the inhomogeneous microstructure would lead to a deterioration of $Q\times f$ values when excessive substitutions (x ≥0.015) were carried out.

According to the formula: $Nb_2O_5 \xrightarrow{TiO_2} 2Nb_{Ti}^{\cdot} + \frac{1}{2}O_2 \uparrow + 4O_O^x + 2e'$, the substitution of Nb^{5+} for Ti^{4+} resulted in the accumulation of electron and damaged the $Q\times f$ [11, 14]. Finally, the appearance of pores shown in Fig. 3(e) and 3(f) should also be attributable to the reduction of $Q\times f$ value.

Fig. 3(d) presents the τ_f of NSTCN$_x$ (x= 0.00~0.025) ceramics fired at 1450°C for 2h. Remarkably, the τ_f was changed from 199.1 to 154.3 ppm/°C with substitutions increased. In perovskites type ceramics, the change of τ_f was closely related to the tilting of Ti-O octahedral which could be strongly explained by the variation of t. The changing trend of τ_f was consistent with the t decreasing from 0.9575 to 0.95738 [2, 6, 15], as the substitutions rose sequentially. Therefore, the τ_f was effectively changed to a relatively low 154.3 ppm/°C owing to a continuous decrease in tolerance factors.

4. Conclusions

The microwave dielectric characteristics and microstructures of NSTCNx ceramics (x=0~0.025) have been researched. The XRD patterns associated with the refined data confirmed that all samples were indexed as the orthorhombic perovskites structure in a in space group $Pm\bar{3}m$. The proper amount of B-site substitutions of $Cr_{1/2}Nb_{1/2}{}^{4+}$ for Ti^{4+} adjusted the τ_f for the decreasing tolerance factor, as well as improving permittivty and $Q\times f$. The improvement of densification mainly contributed to increasing permittivity, and $Q\times f$ was enhanced because of the homogeneous fine grained microstructure at x=0.01. Ultimately, good microwave dielectric properties of NSTCNx were achieved at 1450°C for 2 h when x=0.01: ε_r=104.3, $Q\times f$=9179 GHz, and τ_f=171.7 ppm/°C.

Acknowledgement

This work is supported by the Open Foundation of National Engineering Research Center of Electromagnetic Radiation Control Materials (ZYGX2016K003-5).

References

[1] Sebastian M T 2010 *Dielectric materials for wireless communication* Elsevier.
[2] Fang Z, Tang B, Li Y, Si F and Zhang S 2015 *Journal of Elec Materi* **44** 4236-4242
[3] Zhou D, Pang L X and Qi Z M 2014 *Inorganic Chemistry* **53** 9222-9227
[4] Fang Z X, Tang B, Si F and Zhang S R 2015 *J Mater Sci: Mater Electron*, 1-7
[5] Yu S, Zhang S, Tang B, Zhou X and Fang Y 2012 *Ceramics International* **38** 613-618
[6] Liu T, Zhao X Z, Chen W 2006 *Journal of the American Ceramic Society* **89** 1153-1155
[7] Guo X, Tang B, Liu J, Chen H and Zhang S 2015 *J Alloys Compd* **646** 512-516
[8] Glazer A 1972 *Crystallographica Section B* **28** 3384-3392
[9] Sun, Pai-Hsuan, Nakamura T O, Shan Y J, Inaguma Y, Itoh M 1997 *Ferroelectrics* **200** 93-107
[10] Shannon R D and Prewitt C T 1969 *Acta Crystallographica Section B* **25** 925-946
[11] Zhang S L, Ma P P, Liu X Q and Chen X M 2015 *Journal of the American Ceramic Society* **98**, 3185-3191
[12] Shannon R T 1976 *Acta Crystallographica Section A: Crystal Physics, Diffraction, Theoretical and General Crystallography* **32** 751-767
[13] Chazono H and Kishi H 2000 *Journal of the American Ceramic Society* **83**, 101-106
[14] Nakayama Hiroyuki and Hiroshi Katayama Y 2001 *Jap J Appl Phys* **40** L1355
[15] Tseng C F, Wei T C and Lu S C 2014 *Ceramics International* **40**, 7081-7085

Effect of Swelling on Performance of Surface-Imprinted Composite Membranes

N C Bing, Q Chen, L P Zhu, L L Wang and L J Wang

School of Environmental and Materials Engineering, Shanghai Polytechnic University, Shanghai 201209, P. R. China

Email: ncbing@sspu.edu.cn

Abstract. Molecularly imprinted composite membranes (MIMs) are developed by thermally initiated copolymerization on the surface of polyvinylidene fluoride (PVDF) hollow fiber membranes for the selective binding to levofloxacin. The prepared membranes before and after surface polymerization are characterized by SEM and FT-IR. The water flux, swelling changes and selectivity of the membranes are evaluated. The highest specific separation factor of about 1.27 is achieved for MIMs to separate levofloxacin and ofloxacin. The results show that imprinted polymerization has obvious influence on the performance of membranes and swelling results in the decrease of specific recognition.

1.Introduction

Molecularly imprinted membranes (MIMs) have gained increasing attention in the last few years and been developed for some technically challenging separations, including the separation of chiral drugs or biomolecules, owing to the advantages of the highly specific identification to molecules and membrane separation. At present, MIMs are always prepared as free-standing membranes either from conventional molecularly imprinted polymers (MIPs) or from simultaneous formation of imprinted structure and membrane morphology and surface imprinting polymerization on support membranes [1]. Comparing to the simultaneous preparations, surface imprinted polymerization can optimize MIPs recognition and membrane transport properties at the same time. Generally the approaches for synthesis of MIMs are heterogeneous plasma polymerization[2], photografting[3] and thermal initiated copolymerization[4] on the surface of membranes in different solvents. Usually support membranes and MIPs have different degrees of swelling in solvents, which will bring the changes of permeability and selectivity of MIMs.

Levofloxacin, as an important quinolone, is well tolerated and displays impressive pharmacokinetics attributed to excellent bioavailability and widespread tissue penetration. Asymmetrical synthesis to produce levofloxacin is always reported has very low yield and difficult purification [5]. In China, the throughput of ofloxacin occupies 98% of yield of fluoroquinolones together with norfloxacin, ciprofloxacin [6]. Therefore, it has huge potential to separate ofloxacin for levofloxacin in industrial application.

In this study, levofloxacin-imprinted composite membranes by thermal initiated copolymerization are developed on PVDF hollow fiber membranes. We aim to provide an alternative technique for obtaining levofloxacin and study the effect of swelling on selectivity of MIMs.

2.Experimental

2.1Materials and instruments. PVDF hollow-fiber membranes (self-made; porosity=70%,MWCO with BSA67000 as a standard solution=94.7%). Ofloxacin and levofloxacin were from Kunshan Double-crane Pharmaceutical Co. LTD. Trimethylolpropanetria- crylate (TMPTA) was from Shanghai Kewang Chemical Co. LTD. 2, 2'-Azobis-(isobutyronitrile) (AIBN), Methacrylic acid (MAA), Glacial acetic acid, chloroform, methanol and ethanol, solvents of analytical-reagent grade, were all obtained from Shanghai Sinopharm Chemical Reagent Co. Ltd. MAA was further purified by using active carbon.

A UV762 spectrophotometer (Shanghai Precision & Scientific Instruments Co., Ltd), JSM-6360 LV scanning electron microscope (JEOL, Japan) and V70 Fourier transform infrared spectrometer (Bruker, German) were used in this study.

2.2Membrane preparation. Prepolymerization solutions recipe contained 1.0 mmol of levofloxacin, 4.0 mmol of MAA, 15 mmol of TMPTA, and 0.2 mmolof AIBN in a chloroform solution. All the solutions were mixed for 30 min by ultrasonic wave before casting on the support PVDF membranes. Polymerization was performed by heating the membranes in a vacuum oven at 60℃ for 24 h. Then, the membranes were rinsed with acetic/methanol (1:9 v/v) and ethanol to extract the templates and dried at room temperature. For comparison, non-imprinted polymer membranes (NIMs, without template) were also subjected to same treatment.

Polymerization process can be described as scheme1.

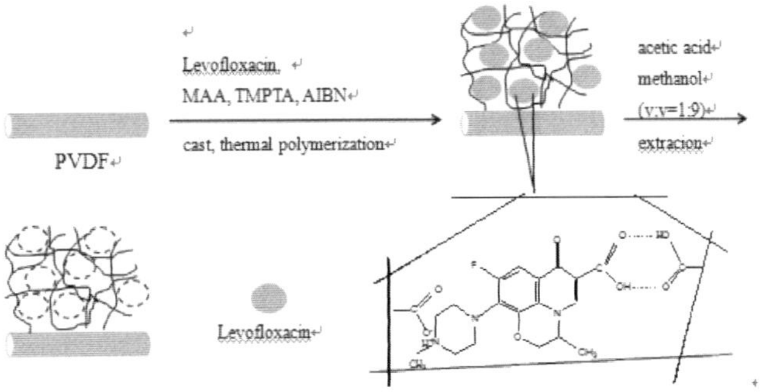

Scheme1 Polymerization process of MIMs

2.3Swelling analysis. Swelling experiments were performed as described in literature [7]. The definite mass of dry membranes (m_0) were soaking in methanol at atmospheric conditions. After 24h equilibration at 25℃, the excess of solvent was removed from the MIMs by applying reduced pressure for 1 min, and the weight of the swollen membrane(m_s) was measured. The swelling ratio (*Sr*) of the membrane was calculated from the following equation:

IOP Conf. Series: Materials Science and Engineering **170** (2017) 012030 doi:10.1088/1757-899X/170/1/012030

$$Sr = \frac{m_s - m_0}{m_0}$$

2.4Performance evaluation of membranses. Evaluation device and separation process of membranes were referred to literature [1]. The water flux was investigated under 25 ℃and 1 bar. The binding capacity of the prepared membranes to levofloxacin and ofloxacin was determined by levofloxacin/ofloxacin concentration of 0.02 mmol/L in methanol under 25 ℃and 1 bar. The concentration of the solution was monitored by a UV spectrophotometer at 298 nm.

3.Results and discussion

3.1Morphology of membranes. Fig.1 shows the SEM images of external surfaces and cross sections of membrane before and after imprinted polymerization. Seen from it, support PVDF membranes (Fig1. A, B) are completely covered by a polymer layer after surface polymerization (Fig1. C, D). The surface of modified membranes appears crosslinked network and the thickness increases, which indicated a polymer layer has anchored the surface of PVDF membranes successfully.

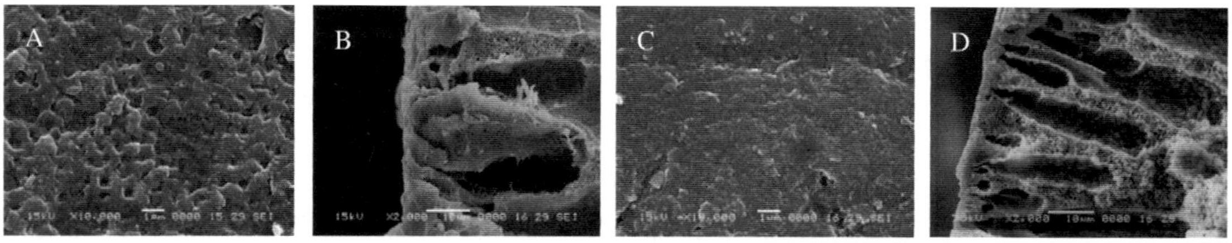

Fig.1 SEM images of PVDF membrane (A, B) and MIMs (C, D)

3.2ATR-FTIR spectra of membranes. Fig.2 shows the ATR-FTIR spectra of support membrane and extracted MIMs. In Fig. 2a, the absorption peaks of 1,408, 1,043, 878 and 835 cm^{-1} are the typical absorption peaks of PVDF. The peak at 2,930 cm^{-1} are corresponding to dissymmetry stretching vibration of $-CH_2-$. Seen from Fig. 2b, the absorption peak 1,722cm^{-1} should correspond to absorption bands of $-C=O$ stretching vibration. It is concluded that the surface of the support PVDF membrane has appeared imprinted polymer layer.

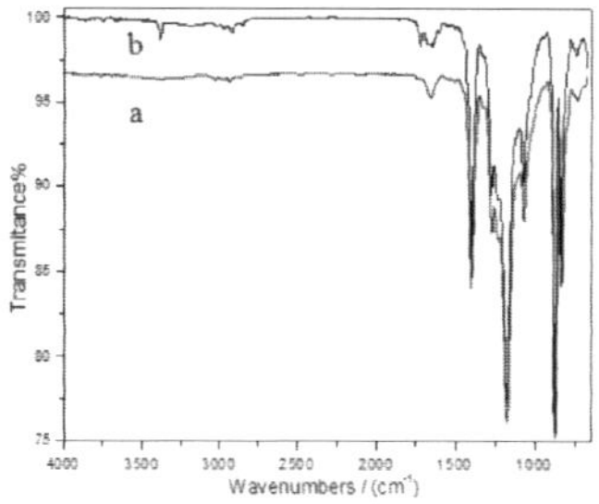

Fig.2 ATR-FTIR spectrograms of PVDF membrane (a) and MIMs (b)

3.3Parameters of membranes. Pure water flux measurements (L_p, at 1bar and 25℃) and swelling changes in methanol of PVDF membranes, MIMs and NIMs are listed in Table 1. The results shows that surface polymerization causes a decreasing water permeability due to the block of some cavities. At the same time, the decreasing *Lp* of NIMs than MIMs can be due to the formation of new cavities after removal of the template. A higher Sr of MIMs and NIMs than PVDF membrane indicated the polymer layers are infinite to methanol. Often unreacted double bonds remains in the polymer will lead to the change of pore volume, surface areas, and swelling properties [7]. It is clear that surface polymerization has a significant effect on the performance of the membranes and there remain many unreacted double bonds under the experiment conditions in this study.

Table1 Changes in character of membrane after and before surface polymerization

Sample	L_p (L/m^2·h· bar)	*Sr* in methanol
PVDF membrane	12.00	0.4556
MIMs	6.14	0.7764
NIMs	3.87	0.6044

3.4Membrane adsorption binding studies. The binding capacity of PVDF membranes, MIMs and NIMs to levofloxacin and ofloxacin was evaluated. Seen from Table2, MIMs and NIMs have a higher binding capacity compared to the support PVDF membranes, which shows both imprinting polymerization and swelling play the roles on recognition. Active polymer layer contained imprinted recognition sites and channel after swelling attributed to the sorption of levofloxacin. Owing to ofloxacin is racemic mixture, MIMs also has the adsorption to ofloxacin, specific separation factor $\alpha_{Levofloxacin/ofloxacin}$is 1.27.

Table2. The results for levofloxacin and ofloxacin binding to the initial and modified membranes

Sample	$C_{Levofloxacin}$/mmol·L^{-1}	$C_{ofloxacin}$/mmol·L^{-1}
PVDF membrane	0.0009	0.0012
MIMs	0.0202	0.0156
NIMs	0.0056	0.0041
Specific adsorption	0.0146	0.0115

4.Conclusions

In this study, levofloxacin-imprinted composite membranes by thermal initiation copolymerization are developed on PVDF hollow fiber membranes. Imprinting polymerization results in the change of morphologies, water flux, swelling ratio and binding capacity of the prepared membranes. Both imprinting polymerization and swelling play the roles on recognition. When polymerization recipe contains levofloxacin, MAA, TMPTA (at a molar radio of 1:4:15) and chloroform as solvent, the highest separation factor of about 1.27 was achieved. Optimized experiment process may decrease swelling and increase specific binding capacity. This approach for MIMs preparation and swelling analysis offers a foundation to levofloxacin and other chiral drug separation.

5.Acknowledgements

The authors gratefully acknowledge the research funding supported by The key subject of Shanghai Polytechnic University (Material Science and Engineering XXKZD1601) and Shanghai Cooperative Centre for WEEE Recycling (No. ZF1224-08).

6.References

[1] Bing N C, Xu Z L, Wang X J et al 2007 Recognition Properties of Poly(vinylidene fluoride) Hollow-Fiber Membranes Modified by Levofloxacin-Imprinted Polymers [J]. *J Appl. Polym. Sci.*

106: 71-76

[2] Cowieson D, Piletska E, Moczko E et al 2013 Grafting of molecularly imprinted polymer to porous polyethylene filtration membranes by plasma polymerization [J]. *Anal. Bioanal. Chem.* **405(20)**: 6489-6496

[3] He D. M., Tailored M Ulbricht 2009 Grafting-From Functionalization of Microfiltration Membrane Surface Photo-Initiated by Immobilized Iniferter [J]. *Macromol. Chem. Phys.* **210(13-14):** 1149-1158

[4] Zhang Y, Qian L L, Yin W et al 2016 A Dual Read-out Molecularly Imprinted Composite Membrane Sensor Based on Zinc Porphyrin for the Detection of Dimethyl Methylphosphonate [J]. *Chem. Res. Chinese U.* **32(5):** 725-730

[5] Kang S B, Ahn E J, Kim Y et al 1996 A facile synthesis of (S)-(-)-7, 8-difluoro-3, 4- dihydro-3-methyl-2H-1, 4-benzoxazine by zinc chloride assisted mitsunobu cyclization reaction [J]. *Tetrahedron Lett.* **37:**9317-9320

[6] Lv Y M 2004 Production and process of fluoroquinolones medicine[J].*Chinese Pharm. Inform.* **10(5)**:28-31

[7] El-Toufaili F, Visnjevski A and Brüggemann O 2004 Screening combinatorial libraries of molecularly imprinted polymer films cast on membranes in single-use membrane modules [J]. *J Chromatog. B* **804:**135–139

Effects of La_2O_3-B_2O_3-ZnO additions on the low temperature sintering and microwave dielectric properties of $(Ca_{0.61}La_{0.26})TiO_3$ ceramics

Y W Chen, E Z Li, N Niu, M Y Zou, S X Duan, S R Zhang

State Key Laboratory of Electronic Thin Films and Integrated Devices, University of Electronic Science and Technology of China, Chengdu, 610054, China

E-mail:lienzhu@uestc.edu.cn

Abstract. The influence of La_2O_3-B_2O_3-ZnO (LBZ) additions on the sintering behavior, microstructure, phase composition, and the microwave dielectric properties of $(Ca_{0.61}La_{0.26})TiO_3$ (CLT) ceramics have been investigated. The results indicate that the LBZ additions could efficiently lower the sintering temperature of the CLT ceramics from 1400°C to 950°C, and excellent microwave properties remain. Small amount of LBZ glass promotes the densification of the CLT ceramics and enhances the microwave dielectric properties. However, excess amount of LBZ glass deteriorates the dielectric properties because of the increasing glass phase. The CLT ceramic with 3 wt. % LBZ additions, sintered at 950°C, exhibit excellent properties: ε_r= 103.12, Q× f = 8826 GHz(f=3.312 GHz) and τ_f=299.52 ppm/°C.

1. Introduction

Low temperature co-fired ceramics (LTCC) technology plays a important role in the integration of the electronic device, which reduces the fabrication expenditure by using cost-effective electrodes Ag or Cu relative to the traditional Pt electrode and simplifies the manufacture process of electronic components [1,2]. In normal conditions, the firing temperature of the LTCC materials is limited below 950°C[3]. High relative permittivities are necessary since the size of the dielectric device is inversely proportional to the square root of the dielectric constant ε_r. While the result should not be achieved with the decrease of the quality factor ($Q \times f$), where $Q \sim 1/\tan\delta$ ($\tan\delta$ is the dielectric loss tangent and f is the resonant frequency)[4–8].

The dielectric properties of the $CaTiO_3$ ceramics were first reported by Kell et al. [9] in1970s. They reported that $CaTiO_3$ had a high permittivity, ε_r=170, and a temperate dielectric losses, Qxf =3500 GHz, which provides a valuable reference for microwave components [9]. Then many studies about the advance of the microwave dielectric properties of $CaTiO_3$-based ceramics have been conducted for the last 20 years [10–13]. Moreover, the substitution of Ca^{2+} by other trivalent lanthanide ions, such as La^{3+}, Sm^{3+} and Nd^{3+}, has also been investigated by several research groups. Kim et al. [14] reported that the partial substitution of Ca^{2+} by La^{3+} led to forming of the phase $Ca_{1-x}La_{2x/3}TiO_3$; And when x = 0.4, the microwave dielectric properties (ε_r = 109, Q × f = 17,600 GHz and τ_f = +213 ppm/°C) of $Ca_{0.6}La_{0.267}TiO_3$ ceramics were reached at 1400°C[15]. As mentioned above, the $CaTiO_3$-based ceramics are generally fired above 1100°C, which is too high to apply in actual application. So the important thing is to lower the sintering temperatures of ceramics without the decrease of dielectric properties. And the study about the low temperature firing of the CLT ceramics with high ε_r and high Q × f value has not been reported so far. Usually, there are two common methods to reduce the firing temperature: low melting temperature sintering aids, the use of ultrafine powders prepared by chemical routes as the starting material of ceramics. And the addition of low melting temperature glasses is the most inexpensive and effective method to get dense ceramics at a low sintering temperature. [16-18].

La_2O_3-B_2O_3-ZnO (LBZ) glass has a low glass transition point[19], and in the previous study[20], the CNT ceramics with 3 wt% LBZ could be fired at 975℃ and exhibited good properties: $\varepsilon_r = 87.8$, $\tau_f = +244$ ppm/℃, Q × f = 8132 GHz. Hence, in this work, the LBZ glass was also selected as the sintering addition agent of the CLT ceramic. And the low temperature firing behavior with the firing addition agents and the dielectric properties were systematically discussed.

2. Experimental procedure

The CLT ceramics were prepared by the solid-state reaction method. $CaCO_3$, La_2O_3, TiO_2, H_3BO_3 and ZnO with purity higher than 99% were adoped as raw chemicals. Firstly, the powders were weighed as the ratio of $(Ca_{0.61}La_{0.26})TiO_3$ and then milled with zirconium balls in the de-ionized water for 8 h. Then the mixtures were calcined at 1100°C for 5 h after drying. To synthesize the LBZ glass, the chemical formula was designed to be $1La_2O_3$-$2B_2O_3$-0.5ZnO, and the above materials were ball milled for 2 h and then melted in a platinum crucible at 1400°C for 2h. Then the melting glass was poured into the copper plate to quench, triturated in mill for about 4 h. After that, the calcined CLT powder and LBZ glass were weighed as CLT+ x wt. % LBZ where x=1, 3, 5,7 and re-milled together for 7 h. Subsequently, the powders were dried and pressed into cylindrical samples of 15 mm in diameter and 8 mm in thickness under a pressure of 20 MPa. After heating from the ambient temperature with a rate of 5°C /min, all samples were sintered from875°C to 975°C, for 4h.

The bulk densities of the samples were measured by the Archimedes method. The shrinkage during heat treatment was measured by the dilatometer with alumina rams and boats (Model DIL402C, Netzsch Instruments, Germany). Melting behavior of LBZ glass with CLT was investigated by placing LBZ on the surface of CLT substrate, then heated from 21°C to 950°C with a rate of 5°C /min, at the same time, changes of the glass of LBZ and CLT were recorded by sintering point testing facility. Identification of crystalline phases was identified by X-ray diffractometry (XRD, Rigaku Industrial Corporation, Japan). The permittivity ε_r and the quality factor (Q × f) were determined by Hakki-Coleman dielectric resonator method with the network analyzer (HP83752A, the United States). The microstructure was determined by scanning electron microscope (SEM, JSM-6490LV, Japan). And the temperature coefficient of the resonant frequency τ_f value could be measured using the formula:

$$\tau_f = \frac{f_{85} - f_{25}}{f_{25} \times 60} \times 10^6 \text{ (ppm /}^\circ\text{C)}$$

where f_{25} and f_{85} are the resonant frequencies at the temperature of 25°C and 85°C, respectively. The temperature measurement was using the temperature chamber (HLT6003P, China).

3. Results and discussion

The linear shrinkage variations of the pure CLT and CLT +3wt. % LBZ during heating process were studied in Figure 1. The shrink of the pure CLT occurs at around 1140°C, while a great shrink takes place at around 850°C with 3 wt. % LBZ glass added, it means that the LBZ glass efficiently promotes the firing process of the CLT ceramics.

Fig. 2 reveals the melting behavior of CLT ceramic with LBZ glass at different temperatures. It can be found that no obvious melting is observed at temperatures below 640°C, while the LBZ shows a significant shrinkage at around 745°C, which agrees with the previous research result of DSC [20] that the endothermic peak at 742°C shows the melting temperature point of the LBZ glass. Then the volume expansion of LBZ appears and stays the same with the temperature increasing. It shows that a certain amount of LBZ can melt and penetrate the CLT substrate, while the thermal expansion of the remaining LBZ occurs during the further heating process. These results indicate that there is a good chemical compatibility between LBZ glass and CLT at around 745°C which the low temperature densification probably owes to the formation of the liquid phase.

Fig. 3 shows the XRD patterns of CLT ceramics with 3 wt. % LBZ fired at different temperatures from 900 °C to 975°C.The XRD results illustrate that all samples contain three phases: the main phase $CaTiO_3$ (JCPDS No.82-0231), a small amount of $LaBO_3$ (JCPDS No.12-0762) and some excess TiO_2(JCPDS No.87-0920) phase. The $LaBO_3$ ceramics sintered at 1300°C exhibit excellent microwave properties with a relative permittivity of ε_r~11.8, a quality factor of Qxf value~76,869 GHz (at~15 GHz), and a negative τ_f~−52 ppm/°C[21]. These results explain that liquid phases which caused by La-B-Zn mixture were formed during the heating procedure. As the liquid enter into the space between the crystal grains, then fills the pores and pulls the particles together via the capillary attraction, accelerates the grow of the grains. And the mixture exists in amorphous state and the $LaBO_3$ phase around the grains in the chilling process.

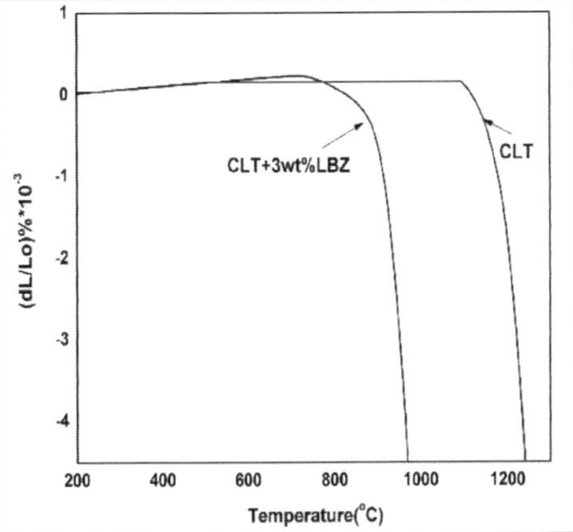

Figure 1. Shrinkage curve of CLT and CLT+ 3 wt. % LBZ.

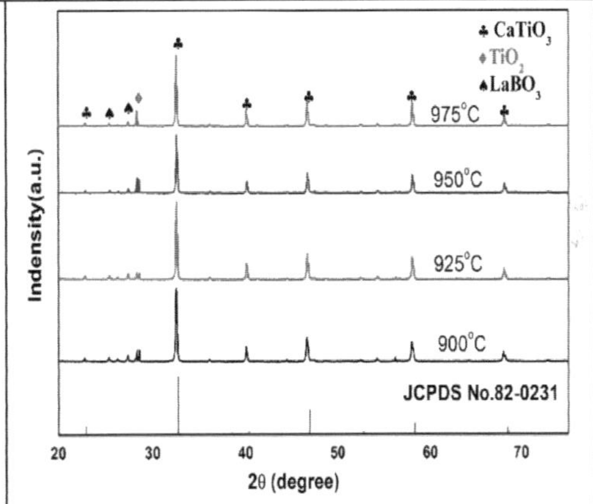

Figure 3. XRD patterns of CLT samples with 3 wt. % LBZ and sintered at different temperatures for 4h.

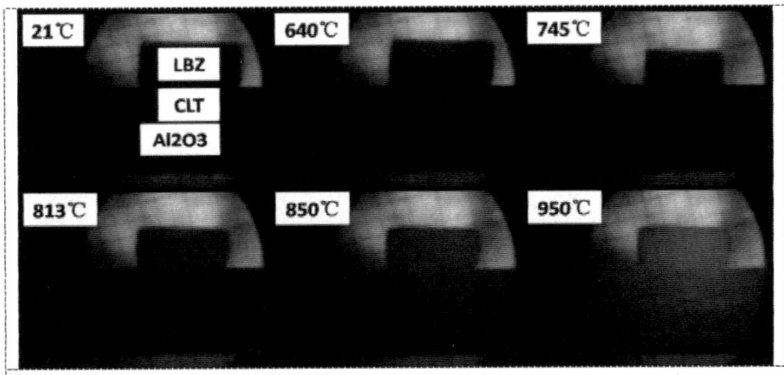

Figure 2. The melting behavior of LBZ glass/ CLT at different temperatures.

The SEM photographs of the CLT specimens with 3 wt. % LBZ glass fired at different temperatures (875°C to 1000°C) are shown in Figure 4. With the firing temperature increasing, a dense microstructure with hardly any pores is observed and the grains of the ceramic samples gradually grow more completely from Fig. 4(d)–(f), owing to the fluid phase which created by LBZ melting (with a fusion point of 750°C) fills the holes then attracts the grains gather. This liquid-assisted interaction facilitates the arrangement of particles and expedites the mass transportation. Thus, the glass of LBZ improve the sintering process and accelerate the shrinkage of the ceramics, which is accordance with the results of Fig. 1. When further raise the temperature, Fig. 4(e) and (f) shows that higher temperature results in the more glass phase separate out which damages the dielectric properties.

Fig. 5 (a) manifests the volume of CLT with different contents (1wt. %-7 wt.%) of LBZ as the change of sintering temperature (875°C -975°C). For the pure CLT ceramic, the density maintains very low. But the densities of the samples with LBZ added elevate significantly and reach a peak value at 950°C, then decrease slightly as further increasing firing temperatures. At this point, the liquid phase formed by LBZ promotes effectively the firing process of CLT ceramics which is consistent with the previous conclusion. As the liquid draws the particles together in the sintering process, therefore, the density increase.

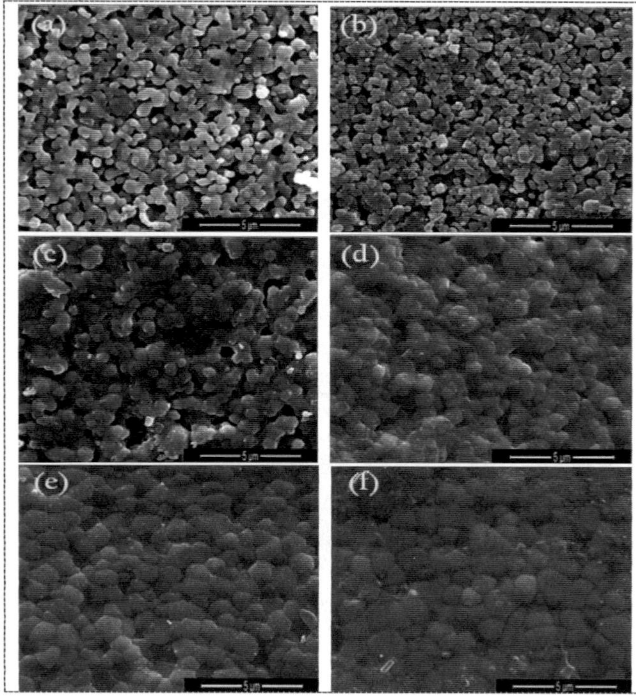

Figure 4. SEM of the CLT samples with 3 wt. % LBZ sintered at different temperatures for 4h: (a) 875 ℃ (b) 900℃ (c) 925℃ (d) 950℃ (e) 975℃ (f) 1000℃.

Figure 5(b) and (c) indicate the ε_r and the Q×f of the CLT+ x wt. % LBZ(x=0 ~7) ceramics at different fired temperature. It shows obviously that the LBZ glass effectively improves ε_r and Q × f value respectively. With 1 wt. % LBZ added, the ε_r of the samples reach the greatest values at 950°C. With more LBZ added (3wt. %, 5 wt. %, 7wt.%), the ε_r enhance follow the increase of firing temperatures (Fig. 5(b)), which is accordance with the trends of density. While the Q × f of the CLT+ 1 wt% LBZ sample changes smoothly at a relative low. When more LBZ is added (x= 3, 5), the Q × f value increases a lot, however, further increasing LBZ content (7 wt. %) results in the obvious diminution of the value (Fig. 5(c)). As we know, the dielectric loss contains intrinsic loss dominated by lattice vibrational and extrinsic loss affected by second phase, densification, crystallite dimension, the oxygen vacancy, etc. [22]. Consequently, microwave dielectric properties of CLT ceramic enhanced a lot at a low temperature since the glass of LBZ promoted the firing process and improved the microstructure. The dielectric properties of the CLT+ x wt. % LBZ samples are generalized in Tab1. Compared with the papers [14, 15], the addition of suitable LBZ efficiently decreases the firing temperature of the CLT ceramic without damaging the dielectric properties. And the enhance of the ε_r and Qxf values due to the densification of the samples induced by LBZ firing agent, but excess glass phase would damage the properties.

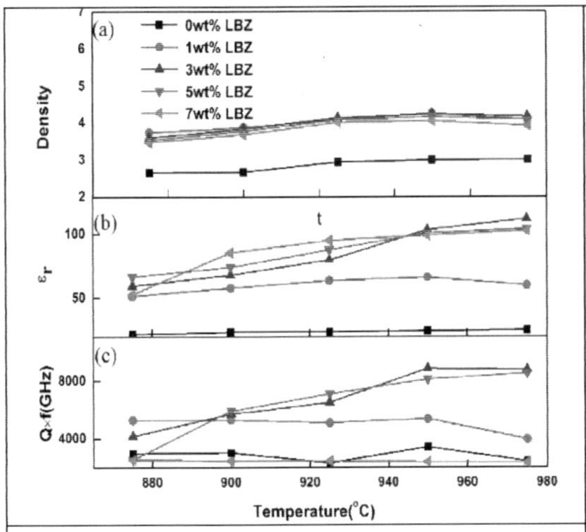

Figure 5. Density and dielectric property of CLT+ x wt.% LBZ(x=0,1,3,5,7) ceramics at different sintering temperatures for 4h (a) density (b) ε_r (c) Qxf.

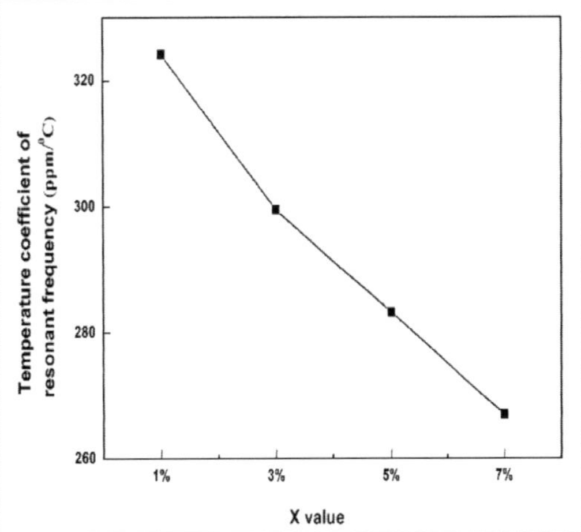

Figure 6. Temperature coefficient of resonant frequency of CLT+ x wt.% LBZ (x=1,3,5,7) ceramics sintered at 950°C for 4h.

Table 1. The properties of CLT + x wt. %(x =0, 1, 3, 5, 7) LBZ samples.

Composition	Sintering Temperature (°C)	ε_r	Qxf (GHz)
CLT[18]	1400	109	17600
CLT+1wt%LBZ	950	112.26	2298
CLT+3wt%LBZ	950	103.12	8826
CLT+5wt%LBZ	950	100.66	8098
CLT+7wt%LBZ	950	98.99	2391

Figure 6 shows the temperature coefficient of resonator frequent of CLT+ x wt. % LBZ (x=1, 3, 5, 7) samples sintered at 950°C. Since of the τ_f value of LBZ glass is negative (~ -40 ppm/°C)[1], the τ_f value of CLT samples decreases with the increase of LBZ glass, from324.19ppm /°C to 267.1ppm /°C.

4. Conclusions

The functions of LBZ glass on the dielectric properties and microscopic structure of the CLT ceramics were studied. LBZ glass is a good sintering additive, which can promote the firing process of the CLT ceramics and effectively reduce the temperature to 950℃. The appropriate amount of LBZ glass can form the liquid phase during sintering, leading to a densification microstructure, therefore enhances the dielectric properties. Typically, 3 wt.% LBZ-doped CLT ceramics sintered at 950°C present the best dielectric properties of ε_r = 103.12, Q× f = 8826 GHz, τ_f=299.52ppm/°C , which are hopeful for the application of LTCC modules.

References

[1] Valant M, Suvorov D, Pullar R C, Sarma K and Alford N M 2006 A mechanism for low-temperature sintering *J. Eur. Ceram. Soc.* **26** 2777-2783.

[2] Anjana P S and Sebastian M T 2009 Microwave Dielectric Properties and Low-Temperature Sintering of Cerium Oxide for LTCC Applications *J. Am. Ceram. Soc.* **92** 96-104.

[3] Li E, Duan S, Sun S, Li H, Mi Y, Zhou X and Zhang S 2013 Influence of Li-B-Si Additions on the Sintering and Microwave Dielectric Properties of Ba-Nd-Ti Ceramics *J. Electron. Mater.* **42** 3519-3523.

[4] Wilhelm S P, Kay R W, Mohammed M I, Lacrotte Y and Desmulliez M P 2013 Lamination based embossing technique for LTCC *Microsyst. Technol.* **19** 801-807.

[5] Zhou D, Wang H, Pang L X, Yao X and Wu X G 2009 Low temperature firing of $BiSbO_4$ microwave dielectric ceramic with B_2O_3–CuO addition *J. Eur. Ceram. Soc.* **29** 1543-1546.

[6] Zhou J 2012 Towards rational design of low-temperature co-fired ceramic (LTCC) materials *J. Adv. Ceram.* **1** 89-99.

[7] Zou D, Zhang Q, Yang H and Li S 2008 Low temperature sintering and microwave dielectric properties of $Ba_2Ti_3Nb_4O_{18}$ ceramics for LTCC applications *J. Eur. Ceram. Soc.* **28** 2777-2782.

[8] Zhou H, Wang H, Zhang M and Yao X 2009 Microwave dielectric properties of the $5.5Li_2O$–Nb_2O_5–$7TiO_2$ ceramics *Appl. Phys. A.* **95** 513-516.

[9] Kell R C, Greenham A C and Olds G C E 1973 High-Permittivity Temperature-Stable Ceramic Dielectrics with Low Microwave Loss *J. Am. Ceram. Soc.* **56** 352–354.

[10] Jancar B, Suvorov D, Valant M and Drazic G 2003 Characterization of $CaTiO_3$-$NdAlO_3$ dielectric ceramics *J. Eur. Ceram. Soc.* **23** 1391–1400.

[11] Ezaki K, Baba Y, Takahashi H, Shibata K and Nakano S 1993 Microwave dielectric properties of CaO-Li2O-LN2O3-TiO2 ceramics *Jpn. J. Appl. Phys.* **32** 4319–4322.

[12] Kim E S and Yoon K H 2003 Microwave dielectric properties of $(1-x)CaTiO_3$–$xLi_{1/2}Sm_{1/2}TiO_3$ ceramics *J. Eur. Ceram. Soc.* **23** 2397–2401.

[13] Liu T and Zhao X Z 2006 A/B Site Modified $CaTiO_3$ Dielectric Ceramics for Microwave Application *J. Am. Ceram. Soc.* **89** 1153–1155.

[14] Kim I S , Jung W H, Inaguma Y, Nakamura T and Itoh M 1995 Dielectric properties of a-site deficient perovskite-type lanthanum-calcium-titanium oxide solid solution system [$(1-x)La_{2/3}TiO_3 - xCaTiO_3 \quad (0.1 \leq x \leq 0.96)$] *Mater. Res. Bull.* **30** 307–316.

[15] Huang C L, Tsai JT and Chen Y B 2001 Dielectric properties of $(1-y)Ca_{1-x}La_{2x/3}TiO_3$-$y(Li,Nd)_{1/2}TiO_3$ceramic system at microwave frequency *Mater. Res. Bull.* **36** 547–556.

[16] Feng G, Jiaji L, Rongzi H, Zhen L and Changsheng T 2009 Microstructure and dielectric properties of low temperature sintered $ZnNb_2O_6$ microwave ceramics *Ceram. Int.* **35** 2687-2692.

[17] Hu M, Xiong J, Gu H, Chen Y and Wang Y 2012 Low temperature cofirable $Ca[(Li_{1/3}Nb_{2/3})_{0.95}Zr_{0.15}]O_{3+\delta}$ microwave dielectric ceramic with ZnO–B_2O_3–SiO_2 frit *Ceram. Int.* **38** 3175–3183.

[18] Kim M H, Lim J B, Kim J C, Nahm S, Paik J H, Kim J H and Park K S 2006 Synthesis of $BaCu(B_2O_5)$ Ceramics and their Effect on the Sintering Temperature and Microwave Dielectric Properties of $Ba(Zn_{1/3}Nb_{2/3})O_3$ Ceramics *J. Am. Ceram. Soc.* **89** 3124–3128.

[19] Takada T and Kageyama K 2005 Synthesis and Microwave Dielectric Properties of La_2O_3–xB_2O_3-Based Melt Mixtures for Low-Temperature Cofired Ceramics *Jpn. J. Appl. Phys.* **44** 6629–6635.

[20] Li E Z, Niu N, Duan S X, Yuan Y and Tang B 2016 Influence of La–B–Zn glass on the sintering and microwave dielectric properties of Ca–Nd–Ti ceramics *J. Mater. Sci: Mater. Electron,* **27** 3164-3169.

[21] Chen X Y, Bai S X, Li M and Zhang W J 2013 Synthesis, characterization, and dielectric properties of low loss $LaBO_3$ ceramics *J. Eur. Ceram. Soc,* **33** 3001–3006.

[22] Takada T, Kageyama K 2005 Synthesis and Microwave Dielectric Properties of La_2O_3–xB_2O_3-Based Melt Mixtures for Low-Temperature Cofired Ceramics *Jpn. J. Appl. Phys.* **44** 6629-6635.

IOP Conf. Series: Materials Science and Engineering **170** (2017) 012032 doi:10.1088/1757-899X/170/1/012032

Experiments of eliminating the destructive effects of excessive Fe inclusions for Al secondary products

D Q Sun[1], G H Dai[2], F Geng[1,2] and K Yang[1]
[1] School of Chemical & Materials Engineering, Changshu Institute of Technology, 215500,China
[2] Jiangsu Key Laboratory of Recycling and Reuse Technology for Mechanical and Electronic Products, 215500,China)

E-mail:sundeqin@cslg.cn

Abstract: Excessive Fe content in Al alloys caused the serious decline of mechanical properties, such as the ductility and impact toughness. Carried out the experiments of eliminating the destructive effects of excessive Fe content by flux-adding technology, which including removing a part of Fe content from Al scrap melt and modifying the morphology of Fe rich precipitates. The experimental results showed that, the ratio of removing Fe element was above 20%, and the morphology of Fe rich precipitates changed from Lamellar to bulk or lath precipitations under the process parameters: the fluxing-agents composed of borax, and $MnCl_2$ (mixed by mass ratio of 1:1) ,and the adding amount of fluxing-agents was about 1.5%; thrown the fluxing-agents into the Al scrap melt by powder injection process and kept for 30 min.

1. Introduction

Promoting the technological progress and improving production efficiency of Al scrap recycling industry were complied with the developing requirements of green, ecological and circular economy. And manufacturing the high performance Al alloy productions by Al scrap recycling process was the chasing target of the scrap recycling industries all over the world. According to industry experts predicted, by 2030, the amount of recycled Al production globally will reach 22 to 24 million tons, and 15% ~ 20% of these products would be applied in the field of high technology [1].

But the critical problem of Al scrap recycling process was that, too much inclusion elements were brought into from the Al scrap raw materials. Because the wide various types of alloy scrap materials were mixed together in the course of purchasing and acquisition process, inevitably some metals and non-metallic inclusions were involved, and resulted that the contents of impurity elements in Al secondary products mostly exceeded the composition allowable range of national standard, especially Fe and Si. At present, developed large amount of research works about the technologies of removing Fe element from Al scrap melt or improving Fe-rich precipitates morphology, mainly including: (1) dilution methods[2], which adding a certain amount of pure aluminum ingots in the Al scrap melt to reduce the content of Fe elements; (2) element neutralization technology[3], which modified the precipitation morphology of primary Fe phase by adding Mn, Cr and other elements; (3) electromagnetic separation process[4], which removing the Fe rich precipitates from the Al melting by utilizing the difference of their magnetic properties; (4) Fe removing technology by chemical reaction between borides and Fe element in Al melting[5]; (5) purification and modification mechanisms of RE elements[6]. Although some technologies were widely applied for Al recycling process, large amount of inclusion elements were still brought into Al scrap melt. Therefore, in order to eliminate destructive effects of Fe inclusions and ensure the Al secondary products with excellent quality, some technologies about Fe removal from Al scrap melt and improving the morphology and distribution of Fe-rich precipitates, should be paid enough attentions [6].

2. Materials and experimental procedures

Excessive Fe content had always been considered as one of the main harmful impurities within Al alloys (especially in wrought Al alloys) [7]. When the Fe content in Al alloys exceeded a certain value, the brittle Fe-rich precipitates, such as α-Fe_2SiAl_8 or β-$FeSiAl_3$ and others, which possessed very high hardness and emerged in the shape of elongated and flaky appearance, would be produced in Al alloy organizations. The existence of brittle precipitation phases resulted in the fragmentation effects on Al matrix, and reduced the strength, plasticity, toughness, machining performances, and so on.[8]. In this paper, carried out feasible experiments of removing Fe element and modifying the precipitates morphology by the technology of adding fluxing-agents into the Al melt.

Materials used in experiments included: (1) Al scrap materials derived from Ye-Chiu Metal Recycling (China) Ltd. (2) The flux components, including borax, manganese chloride($MnCl_2$), were AR chemical reagents. Its physical characteristics displayed as shown in Tab.1. Instruments for performance measurements and microstructure analysis included: used the 5kg induction furnace to carry out the smelting process, SEA1000A X-ray fluorescence analyzer to measure the chemical composition, D/max-2200/PC X-ray powder diffraction to perform the slag phase analysis, optical microscopy or scanning electron microscopy (SEM) to analysis the microstructure, and TWV-1S type micro-accelerometers to measure the surface hardness and interfacial hardness distribution, WDW-200 material tensile testing machine to measure the strength and elongation percentage.

Experimental procedures as follows: (1) filled Al scrap materials into the furnace, heated and melted into Al liquids; took samples for measuring Fe content, microstructure analysis, and tensile tests; (2) thrown the flux agents into Al melting and took measures such as stirred methods, to make the reaction between the fluxing-agents and Al scrap melts completely; (3) carried out the refining process, and promoted the reaction molten slag to float up to the surface; (4) removed the slags, took samples for measuring Fe content, microstructure analysis, and tensile tests; (5) took the slag samples and analyzed the slag composites.

Table 1 Composition and physical properties of the flux.

Flux components	Chemical formula	Melting point, ℃	Density, g/cm^3	specifications	remarks
borax	$Na_2B_4O_7$	742	1.73	AR	
manganese chloride	$MnCl_2$	650	2.977	AR	

3. Results and discussion

3.1 Effects of borax added amount

The chief functions of borax embodied in removing Fe elements from Al melts, and the basic principle was [9]:

$$Na_2B_4O_7(s)+4Al(l)+8Fe(l)=2Al_2O_3(s)+4Fe_2B(s)+Na_2O(s) \quad (1)$$

Individually added borax into Al scrap molten, whether by immersion process or by powder injecting method, the iron-removing efficiency was limited, the relative rate of Fe removal δ was about 12%, the absolute amount of Fe removal Δ was only about 0.08%, as shown in Fig.1 and Fig.2. As the density of borax (2 g/cm^3) was smaller than Al melt (2.7 g/cm^3), borax powder was easily floated up from Al melting, and had no adequate time to carry out the chemical reaction completely. Because the Fe content in Al scrap products generally exceeded 0.6%, some even more than 1%, obviously, adding borax individually would not meet the requirements of manufacturing high performance productions by Al scrap recycling.

$$\delta = \frac{c_0 - c_1}{c_0} \quad (2)$$

$$\Delta = c_0 - c_1 \quad (3)$$

Where δ was relative rate of Fe removal; Δ was absolute amount of Fe removal; c_0 was the original Fe content; c_1 was the Fe content after flux treatments;

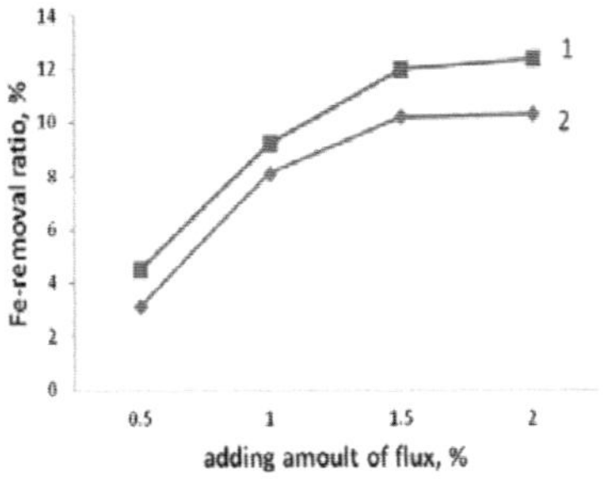

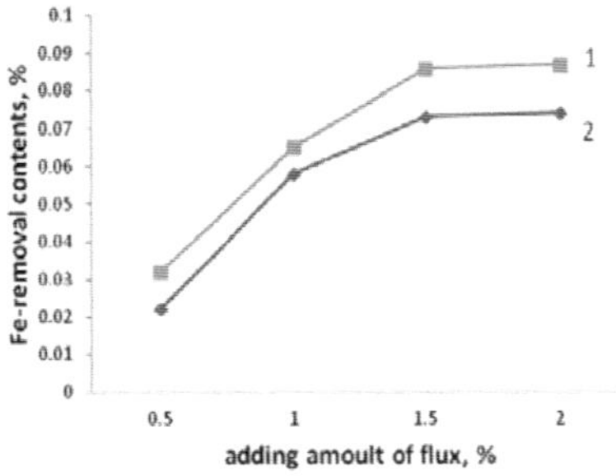

Fig.1 Fe removal ratio in different process **Fig.2** Fe removal amount in different process

1- by powder injecting method; 2- by immersion process

3.2 Effect of adding $MnCl_2$ powder individually

The chief functions of $MnCl_2$ added into the flux were that, the chemical reaction occurred between the manganese chloride and Al melts, as following [10]:

$$3MnCl_2 + 2Al = 3Mn + 2AlCl_3 \uparrow \quad (4)$$

The reaction products included $AlCl_3$ and Mn. The boiling point of $AlCl_3$ was 183℃, and in Al melt, $AlCl_3$ was vaporized and formed small air bubbles which can bring out the precipitates containing Fe phase or other reaction products successfully.

In addition, Mn, as another reaction product, improved the acicular and lamellar Fe precipitates to be transformed into the regular containing Mn compounds in the form of squares, fishbone and petal shape, and reduced the harmful effects.

When $MnCl_2$ powder was added into Al scrap molten individually, the effect on iron-removing was also limited, as shown in Fig.3. The principal reason was that, $MnCl_2$ powder can't occur chemical reaction with Fe element existed in Al slag molten, thus only played a certain role in Fe-removing by the floating up of $AlCl_3$ bubbles and taking some inclusions containing Fe out of Al scrap molten. But the effects of improving the Fe-rich precipitates morphology were noticeable, as illustrated in Fig.4.

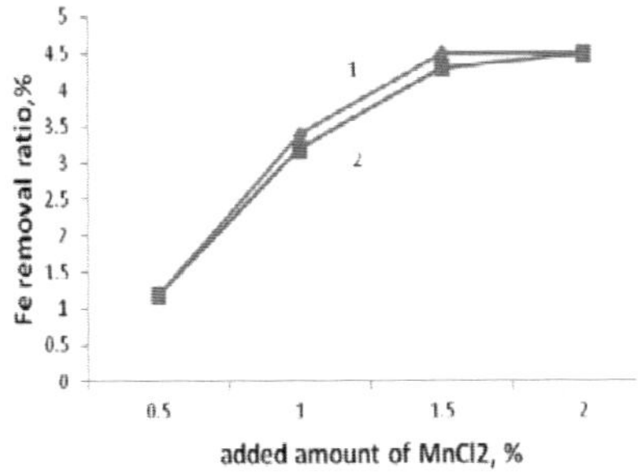

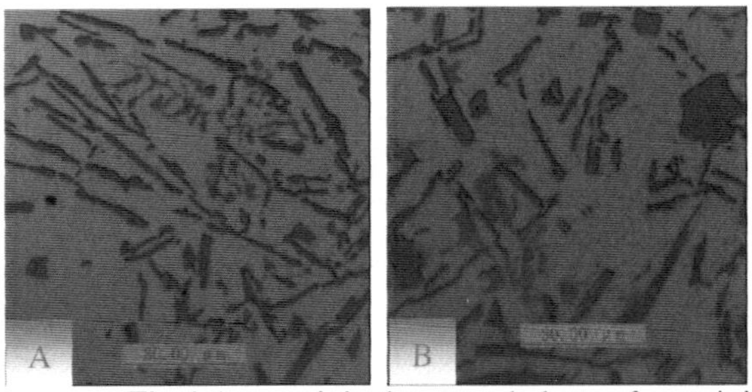

Fig.3 Fe removal ratio of added $MnCl_2$ **Fig.4** the precipitation morphology of containing Fe phases

1- by powder injecting; 2- by immersion process A- original microstructure; B- after $MnCl_2$ flux treatment

3.3 Effect of flux composites

The density of composite fluxes composed of borax and $MnCl_2$ powder (mass ratio of 1:1) was significantly increased, reached to 2.35 g / cm^3, closer to the melting point of Al scrap melt. Therefore, when added into the melt, the flux mixture would stay in the melt for a longer time, and promote the chemical reaction between the flux components and Fe inclusions in Al scrap melt to carry out more completely. Thus, adding the composite fluxes into Al scrap melt increased the Fe-removal ratio compared with individually adding borax. When the adding amount of composite fluxes was about 1.5% of Al melting mass, achieved good Fe removal efficiency; the absolute amount of Fe removal reached 0.15%, the relative rate of Fe removal was about 20%, as shown in Fig.5and Fig.6. The other reason was that, the birth of $AlCl_3$ by the reaction between $MnCl_2$ and Al melt could easily bring the reaction products, such as FeB compounds, out of Al melt, therefore adding $MnCl_2$ into the flux was benefit for removing Fe inclusions. In addition, Mn element was produced by the reaction of composite flux and Al melt, could improve the precipitates morphology, as shown in Fig.7. Thus, adding the composite fluxes composed of borax and $MnCl_2$ into Al scrap melt could have dual functions of Fe removal and modifying the precipitates morphology.

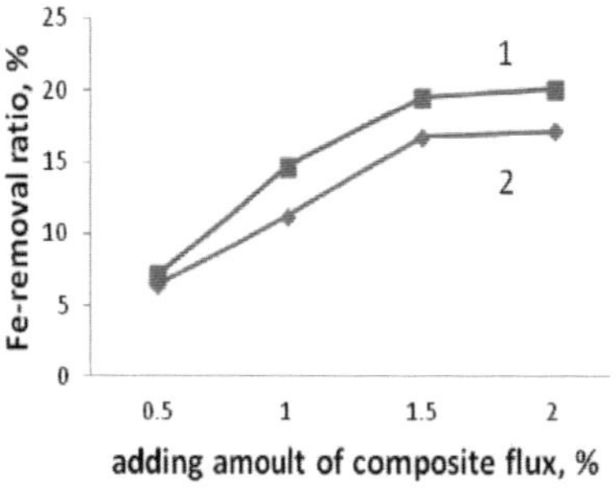

Fig.5 Fe removal ratio in different process

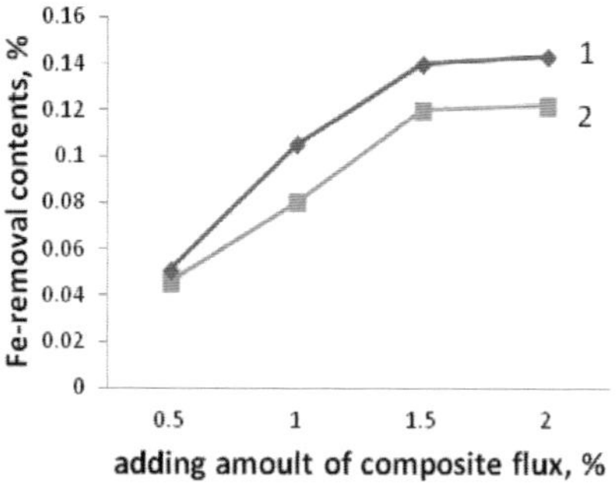

Fig.6 Fe removal amount in different process

1- by powder injecting; 2- by immersion process

But with the increase of the flux addition amount, the Fe removal ratio was not improved obviously, the main reasons reflected in two aspects, (1) with the development of the Fe-removal process, the contents of Fe in Al scrap melt decreased, the ability of Fe-removal reaction by composite fluxes would be dropped down; (2) An excessive dose of composite flux easily was concentrated together, and floated up out of Al melt.

Although the generation of Mn element could reduce the destructive effects of excessive Fe content by producing (Fe,Mn)Al_6 et al. The Mn contents was increased with the increase of adding composite flux, which could make the contents of Mn element beyond the scope of the national standards; and excessive Mn element played limited roles in improving the precipitation morphology, therefore, the adding amount of composite flux should not be too much.

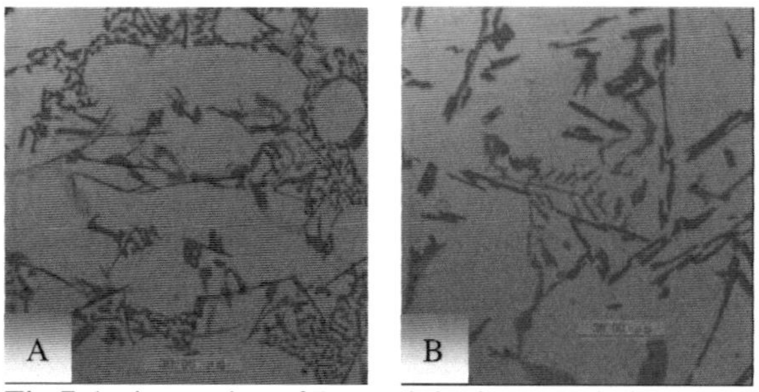

Fig.7 the improving of Fe precipitation morphology
A- original microstructure; B- after $MnCl_2$ flux treatment

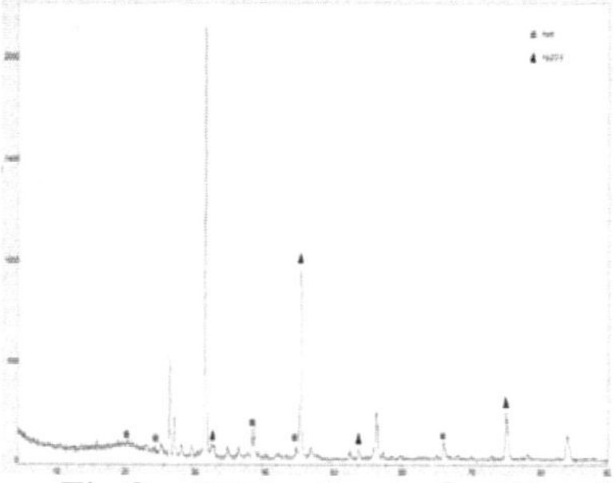

Fig.8 XRD patterns of molten slag

By XRD analysis of the slag sampling, as schematic in Fig. 8, found the reaction product of FeB in the slag samples, and proved that the borax played main role in the Fe-removal reaction.

3.4 effect of standing time after adding composite flux

The standing time was the delayed duration after the composite Flux was thrown into the Al scrap melt. With the extension of the standing time within 30 min, the Fe-removal efficiency was constantly raised up; But over 30 minutes after, made little impact on the Fe-removal ratio, as shown in Fig.9. The composite fluxes was thrown into the Al scrap melt, although the addition of $MnCl_2$ increased the density, the flux was still light-weighted compared with the Al melt, the flux would gradually float up out of the melt; therefore, the residence time of the composite flux in Al scrap melt was limited, and the Fe-removal reaction stopped slowly.

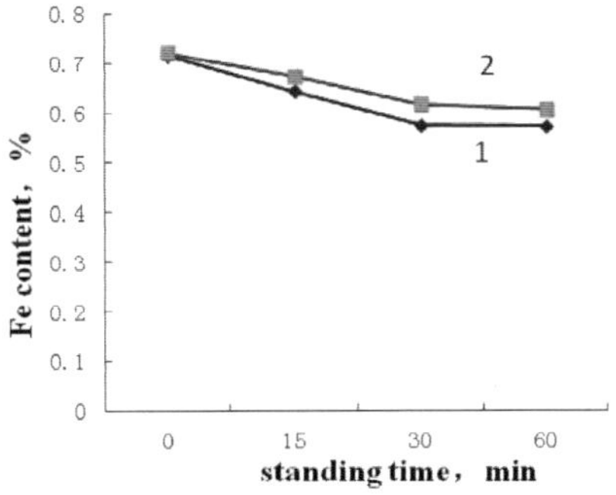

Fig.9 effect of standing time
1- by powder injecting; 2- by immersion process

Effects of the standing time on improving the Fe-rich precipitates morphology were not significant. At first, the reaction between $MnCl_2$ and Al melt occurred violently, Mn element was produced rapidly; secondly, with the prolongation of time, the Fe-rich precipitates, which composed of Fe, Mn and other elements, would occur the aggregation and settlement.

3.5 The quality of Al secondary ingots

After dealing with composite flux to the Al scrap melts, the quality of Al secondary ingots had been significantly improved, as shown in Fig.10 and Fig.11. The tensile strength increased from 107 MPa to 145 MPa, and the elongation increased from almost 0 to 8% or so.

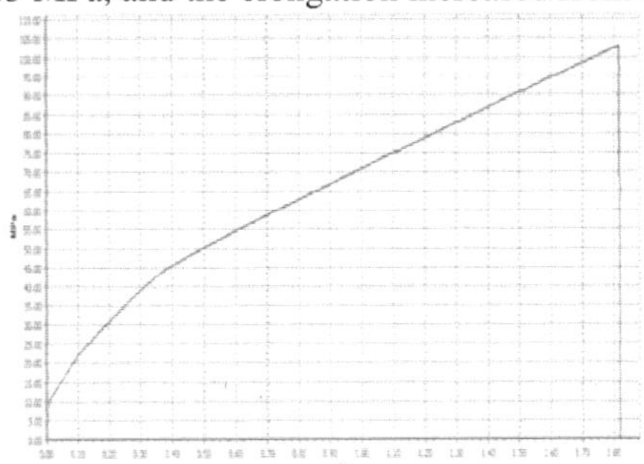

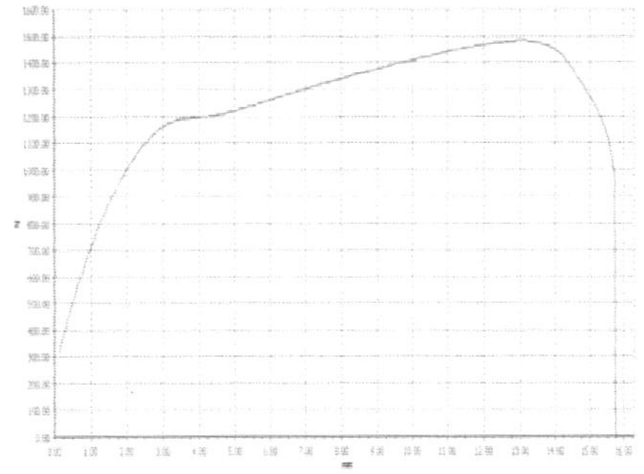

Fig 10 stress-strain curve of original Al scrap ingot **Fig.11** stress-strain curve of ingot by fluxing process

4 Conclusion

Carried out feasible experimental research on the technology of decreasing the destructive impact of excessive Fe contents on Al secondary productions by adding composite fluxes into Al scrap melt, and drawn the following conclusions:

1. The best technological parameters was that: (1)the composite fluxes composed of borax and $MnCl_2$ mixed with the mass ratio of 1:1; (2)added the composite fluxes in the Al scrap melt at 730-750 ℃, by powder injection method; (3) standing for 30 min can obtain excellent effects;

2 Achieved the Fe-removal ratio of 20%, basically eliminated the lamellar or coarse Fe-rich precipitates, and changed into bulky or other round morphology;

3. Increased mechanical properties of the Al secondary ingots, the tensile strength raised up to 145Mpa, and the elongation reached 8%, can greatly improve the work performances.

Acknowledgements

The authors are grateful to the financial support of National Natural Science Foundation of China (Grant Nos. 11405014) and Jiangsu Science and Technology Department of China (BY2015043-02).

References

[1] Verran G O, Kurzawa U. 2008,*Resources Conservation Recycling*, **(52)**: 731-36

[2] Kvithyld Anne. 2011, *Aluminium International Today*, **23** (**4**): 26-29

[3] Sun Y M. 2009, *Foundry technology*, **30(4)**:520-22

[4] Mesina M B, T P R de Jong, Dalmijn W L. 2007, *International Journal of Mineral Processing*, **82(4)**:222-32.

[5] Sun D Q, Xu Y, Dai G H, et al. 2016, *Foundry*, **65(1)**:80-83

[6] Wang X X, Mao Y L. 2010,*Foundry technology*, **31(12)**: 1617-20

[7] Zhang L F, Wang S Q, Dong A P,et al. 2014,*Metallurgical and Materials Transactions B*, **45 (6)**: 2153-85

[8] Xu Z M, Li T X, Zhou Y H. 2003, *Journal of Materials Science*, **38 (22)**:4557-65

[9] Wang Y B, Yi G B, Yang X J. 2014, *Foundry*, **63(10)**:1024-27

[10] Tan X P, Zheng K H, Qi W J, et al. 2013, *Foundry technology*, **34(11)**:1446-48

The micromechanics model analysis of the viscosity regulation of ultra-high strength concrete with low viscosity

M Zhu[1,a], F G Wang[2,b], F Z Wang[1,c], Y P Liu[1,d]

[1]State Key Laboratory of Silicate Materials for Architectures, Wuhan University of Technology, Wuhan, Hubei 430070, China;

[2].Jiangsu Subote New Material Co., Ltd., Nanjing, Jiangsu 211013, China

E-mail:[a]mingz99@163.com, [b]337698651@qq.com,[c] fazhwang@whut.edu.cn, [d]420426327@qq.com

Abstract: The plastic viscosity of mortar and concrete with different binder content, sand ratio, water-binder ratio, microbead dosage and different class and dosage of fly ash were tested and calculated according tomicromechanics model proposed by A.Ghanbari and B.L.Karihaloo, The correlations between these parameters and fresh concrete workability were also investigated, which showed i. high consistence with the objective reality. When binder content, microbead dosage, fly ash dosage or the water-binder ratio was increased or sand ratio was reduced, the fresh concrete viscosity would decrease correspondingly. However their effects were not that same. The relationships between T50 a, V-funnel and inverted slump time with fresh concrete viscosity were established, respectively.

1. Introduction

Ultra-high strength concrete (UHSC), with the advantages of remarkably decreasing the construction weight and effectively improving the concrete durability, can reduce the energy consumption and construction cost. However, due to its low water consumption and low water-binder ratio, the fresh concrete viscosity is relatively high, which . frequently results in pumping accidents.

Fresh concrete can be regarded as a particle suspending system, comprised of liquid phase and solid phase (particle phase) with particle dimension from submicron size to millimeter size. Therefore, the rheological properties of fresh concrete are affected by several bonds or forces from different constituents [1]. Generally, in mortar system the cement grout is regarded as liquid phase and the fine aggregate is regarded as particle phase, while in concrete system the mortar is regarded as liquid phase and the coarse aggregate is regarded as particle phase. In fact, this classification of phase is simplified. On sub-microcosmic level, cement grout can be regarded as a particle suspending fluid and its rheology is dependent on the suspending fluid with other equivalent size fine particles. When rheological properties of fresh concrete are characterized by Bingham body model ($\tau=\tau0+\eta u$) [2], yield shear stress ($\tau0$) is the maximum stress to prevent plastic deformation. With application of external force, fresh concrete will flow if the internal shear stress (τ) is greater than or equal to the yield shear stress ($\tau0$). Plastic viscosity (η) is a parameter to show the flowing property of fresh concrete. The less the viscosity value is, the faster the fresh concrete flows under the same external force. This shows that yield shear stress ($\tau0$) and plastic viscosity (η) are two main rheological parameters to reflect the workability of fresh concrete. BML Viscometer, invented by Wallevik [4], can be used to test the rheological parameters ($\tau0$ and η) of slurry containing fine powders [3]. However, this viscometer is not suitable to test complicated concrete system, which comprised of liquid phase and varied solid particles with a tremendous size difference, from fine aggregate and coarse aggregate to cement and ultrafine additions. So far there has not been an instrument that can be used to correctly and precisely test the rheological property of fresh concrete. Based on the plastic viscosity values of cement slurry or slurry with mineral fillers, A.Ghanbari and B.L.Karihaloo [5] proposed micromechanics model and estimated the plastic viscosity of steel fiber and non-steel fiber self-compacting concrete (SCC). And the calculated results had relatively good consistence with the objective reality. In this paper, this method was employed to evaluate rheological properties of UHSC with low viscosity. The effect of binder content, sand ratio, water-binder ratio, microbead dosage and different class and dosage of fly ash on concrete rheological performances were studied. the plastic viscosity of mortar and concrete with different. And the correlations between these parameters and concrete workability were also investigated.

2. Experimental programs

2.1 Materials

P·I52.5 Portland cement (C) provided by Huaxin Cement Co,. Ltd., Wuhan, China, a Class I fly ash (FA) by Hubei Guodian Qingshan Co-generation Co.Ltd., Wuhan, China, a Grade S95 class ground granulated blast furnace slag (GGBS) by Wuhan Ganghua Co,. Ltd., Wuhan, China, silica fume (SF) by Guizhou Haitian Ferroalloy Abrasive Co., Ltd., Guizhou, China, microbead (M) by Shenzhen Tongcheng New Material Co. Ltd., Shenzhen, China, and ViscoCrete3301 C100 polycarboxylates superplasticizer with a 24% solid content by Sika were used in this study. Taken from Dongting Lake, Hunan, China, medium sand with an apparent density of 2.62g/cm^3 and a fineness module of 2.8 was used as fine aggregate. The medium sand contained less than 1% material finer than 75 μm in natural sand. Continuously graded crushed basalt stone with minimum particle size of 5mm, maximum particle size of 16mm, apparent density of 2.72g/cm^3 and crush value of 2.7% was used as coarse aggregate. The crushed stone, with less than 5% elongated and flaky particles content by weight, contained less than 0.2% material finer than 75 μm in natural sand. The chemical composition and physical properties of materials are presented in tables 1 and 2, respectively.

2.2 Experimental method

The viscosity and yield stress of binder slurry were tested by BROOKFIELD R/S Plus Rotating Rheometer. Then linear regression equation of rheological curve was derived from linear regression method. According to Chinese Code JGJ/T281-2012, T and CECS203-2006, the workability of fresh concrete were tested, such as slump-flow, time to reach 19.68in.(500mm) diameter (T50), V-funnel, inverted slump time and air content. According to micromechanics model proposed by A.Ghanbari and B.L.Karihaloo, the plastic viscosity of mortar and concrete were calculated, respectively..

Table 1 Physical properties of cement

Fineness (〉80μm)%	Setting time (min)		Soundness	Flexural strength (MPa)		Compressive strength (MPa)		Density (g/cm^3)	Water requirement of normal consistency (g)	Blaine specific area (m^2/kg)
	Initial	Final		7d	28d	7d	28d			
1.7	130	185	Qualified	8.8	9.2	50.1	59.5	3.14	135	350

Table 2 Physical properties of mineral fillers

Mineral fillers names	Flexural strength (MPa)		Compressive strength (MPa)		Index of activity (%)		Blaine specific area (m^2/kg)	Density (g/cm^3)	Ratio of water demand (%)	Fineness (%)
	7d	28d	7d	28d	7d	28d				
FA(I)	5.4	8.3	24.8	37.4	67	78	559	2.25	92	5.5
FA(II)	4.7	7.0	24.0	34.7	61	68	460	2.46	102	27.7
GGBS	6.3	8.9	29.6	53.9	76	106	535	2.87	-	-
SF	9.2	10.2	57.8	75	115	126	20000	2.07	-	-
M	7.8	8.8	49.2	65.5	98	110	3300	2.43	85	-

2.3 Experimental theory

The raw material particles are regarded as rigid spheres in the micromechanics model proposed by A.Ghanbari and B.L.Karihaloo [5-7],. The calculation process is given in figure.1: Firstly, the paste is regarded as liquid-solid phase suspension system, among which the cement grout is the liquid phase and the cement replacement material (CRM) is the solid phase, then its viscosity was tested. Secondly, the mortar slurry and concrete is regarded as liquid-solid phase suspension system, respectively.

Their viscosity could be calculated by Eq. (1) to (4).

Note: CRM stands for cement replacement material, namely mineral fillers, such as GGBS and SF.

The plastic viscosity of liquid-solid suspension phase system is obtained by Eq. (1) to (2).

$$\eta_{Ci+1} = \eta_{Ci} f_{i+1}(\phi_{i+1}) \quad (1)$$

$$\phi_i = \frac{v_i}{v_i + v_0} \quad (2)$$

where (η_{CI}) is the plastic viscosity (Pa.s) of slurry mixed with mineral fillers, ϕ_i is the volume fraction of solid phase, v_i is the volume of solid phase (m^3), v_0 is the continuous matrix volume (m^3), and $f_i(\phi_i)$is a function. Depending on the volume fraction value, the value can be obtained by Eq. (3) or (4).

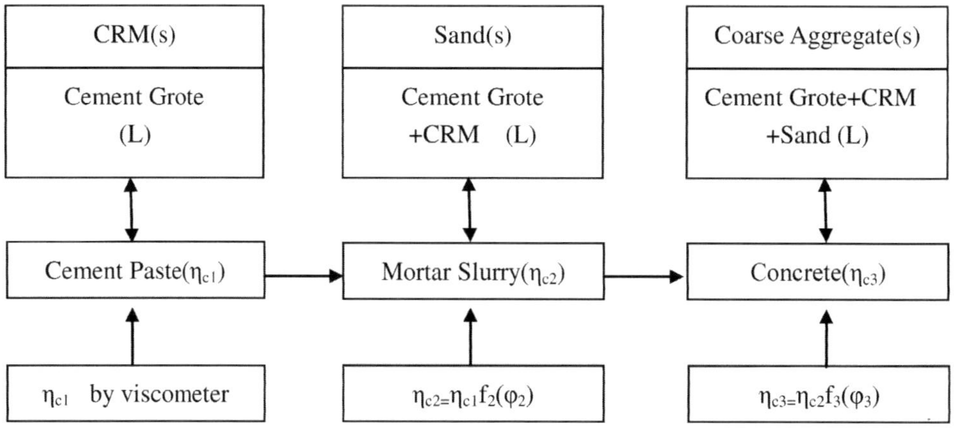

Figure.1. Process of viscosity calculation of fresh concrete

1) When volume fraction of solid phase (ϕ_i)is less than 10%, function $f_i(\phi_i)$ ismerely related to the volume fraction(ϕ_i) and can be obtained by Einstein equation, namely Eq. (3).

$$f_i(\phi_i) = 1 + [\eta]\phi_i \tag{3}$$

where $[\eta]$ is a dimensionless inherent viscosity, a tested value [8] representing the effect of a single particle on viscosity. When the particles are regarded as rigid dense spheres, and the distance among the particles are far more than the mean particle diameter, $[\eta]$ value equals to 2.5.

2) When volume fraction of solid phase (ϕ_i) is more than 10% but less than the probable maximum volume fraction (ϕ_i), function $f_i(\phi_i)$ is not only related to the volume fraction but also to the dispersion degree and morphology of the solid phase in the fluid. And the function $f_i(\phi_i)$ can be obtained by Eq. (4) proposed by Krieger and Dougherty [9].

$$f_i(\phi_i) = (1 - \frac{\phi_i}{\phi_m})^{-[\eta]\phi_m} \tag{4}$$

where ϕ_m is the probable maximum volume fraction, representing the minimal separating distance among particles, namely the conditions with the minimum void ratio and the exceedingly high viscosity.

Both $[\eta]$ and ϕ_m are dependent on the shear rate. With the increase of shear rate, the $[\eta]$ decreases while the ϕ_m increases. When either parameter increases, the other will decrease correspondingly. For rigid dense sphere, $[\eta]\phi_m$ value equals to 2.5.

3. Results and discussion

3.1 Mix proportion design and viscosity calculation of UHSC with low viscosity

With different binder content, sand ratio, water-binder ratio, microbead dosage and different fly ash class and dosage, the corresponding mix proportion design and workability results are presented in table 3.

Table 3 Mix proportions and workability of UHSC with low viscosity

No.	Binder (kg/m^3)	Binder proportions (%)						w/b	Water (kg/m^3)	Superplasticizer (SP/Binder)	Sand ratio (%)	Sand (kg/m^3)	Coarse aggregate (kg/m^3)	Workability (s)		
		C	GGBS	FA (I)	FA(II)	M	SF							T50	V-funnel	Inverted slump time
1	580	63.5	20	10	0	0	6.5	0.21	121.8	0.03	40	685	1027	9	60	9.5
2	630	63.5	20	10	0	0	6.5	0.21	132.3	0.03	40	665	995	8	45	8.5
3	680	63.5	20	10	0	0	6.5	0.21	142.8	0.03	40	645	967	7	35	7
4	730	63.5	20	10	0	0	6.5	0.21	153.3	0.03	40	625	935	6.5	30	7
5	630	63.5	20	10	0	0	6.5	0.21	132.3	0.03	37	615	1047	8.5	46	9
6	630	63.5	20	10	0	0	6.5	0.21	132.3	0.03	43	713	945	8	42	8
7	630	63.5	20	10	0	0	6.5	0.21	132.3	0.03	46	762	894	8	48	8.5
8	630	63.5	20	10	0	0	6.5	0.19	119.7	0.03	40	665	995	15	80	15
9	630	63.5	20	10	0	0	6.5	0.20	126	0.03	40	665	995	8	45	8.5
10	630	63.5	20	10	0	0	6.5	0.22	138.6	0.03	40	665	995	6	35	5.5
11	630	73.5	20	0	0	0	6.5	0.21	132.3	0.03	40	665	995	12	70	12.5
12	630	68.5	20	0	5	0	6.5	0.21	132.3	0.03	40	665	995	9.5	57	9.5
13	630	68.5	20	5	0	0	6.5	0.21	132.3	0.03	40	665	995	9	53	9
14	630	63.5	20	0	10	0	6.5	0.21	132.3	0.03	40	665	995	8.5	45	8.5
15	630	58.5	20	15	0	0	6.5	0.21	132.3	0.03	40	665	995	7	40	7.5
16	630	63.5	20	7.5	0	2.5	6.5	0.21	132.3	0.03	40	665	995	7	41	8
17	630	63.5	20	5	0	5	6.5	0.21	132.3	0.03	40	665	995	6.5	35	7
18	630	63.5	20	2.5	0	7.5	6.5	0.21	132.3	0.03	40	665	995	5	31	6.5
19	630	63.5	20	0	0	10	6.5	0.21	132.3	0.03	40	665	995	4.5	26	5

According to the experimental theory in Part 2.3, relevant parameters and viscosity values of concrete specimens in table 3 are given in table 4.

Table 4 Tested viscosity values of grout and calculated viscosity values of UHSC

No.	η_{C1} (Pas)	ϕ_2	$f_2(\phi_2)$	η_{C2} (Pas)	ϕ_3	$f_3(\phi_3)$	η_{C3} (Pas)
1	1.60	0.4471	5.3988	8.64	0.3924	3.9747	34.33
2	1.60	0.4196	4.6000	7.36	0.3768	3.6733	27.04
3	1.60	0.3938	4.0030	6.40	0.3625	3.4265	21.95
4	1.60	0.3696	3.5453	5.67	0.3475	3.1940	18.12
5	1.60	0.4007	4.1506	6.64	0.3965	4.0604	26.96
6	1.60	0.4366	5.0710	8.11	0.3579	3.3518	27.20
7	1.60	0.4530	5.5986	8.96	0.3386	3.0661	27.47
8	3.55	0.4285	4.8372	17.17	0.3818	3.7650	64.65

9	2.45	0.4240	4.7153	11.55	0.3793	3.7194	42.97
10	1.35	0.4152	4.4908	6.04	0.3744	3.6294	21.92
11	2.90	0.4250	4.7418	13.75	0.3798	3.7293	51.28
12	2.40	0.4230	4.6891	11.25	0.3787	3.7081	41.73
13	2.20	0.4222	4.6691	10.27	0.3783	3.7011	38.02
14	1.80	0.4211	4.6373	8.35	0.3777	3.6886	30.79
15	1.40	0.4169	4.5325	6.35	0.3753	3.6472	23.14
16	1.50	0.4199	4.6078	6.91	0.3770	3.6774	25.42
17	1.35	0.4202	4.6156	6.23	0.3772	3.6802	22.93
18	1.15	0.4205	4.6235	5.32	0.3774	3.6830	19.58
19	1.05	0.4208	4.6314	4.86	0.3775	3.6872	17.93

3.2 Experimental results analysis of viscosity of UHSC with low viscosity

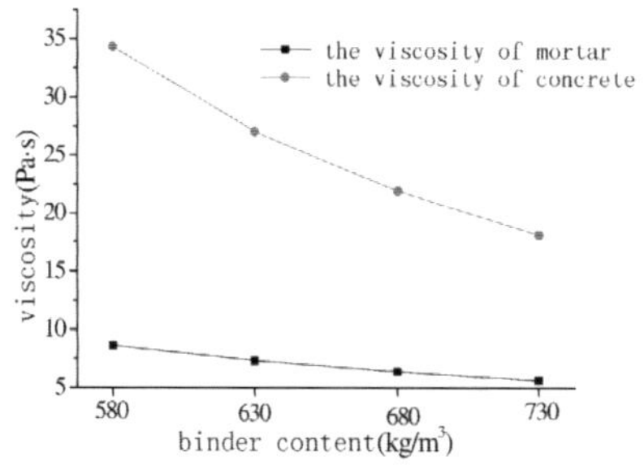

Figure.2. Effect of binder content on viscosity of mortar and concrete

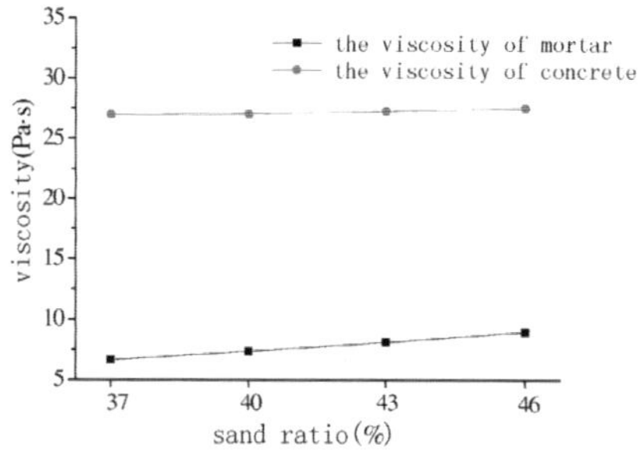

Figure.3. Effect of sand ratio on viscosity of mortar and concrete

3.2.1 Effect of binder content on viscosity of mortar and concrete.Based on the tested paste viscosity value (η_{C1}) of specimen 1, 2, 3 and 4, effects of different binder content on calculated mortar viscosity (η_{C2}) and concrete viscosity (η_{C3}) are given in figure.2 respectively. figure.2 shows that when binder content increased, both (η_{C2}) and (η_{C3}) decreased remarkably. The (η_{C2}) decreased from 8.64Pas to 5.67Pas, while the (η_{C3}) decreased from 34.33Pas to 18.12Pas. The decrease of concrete viscosity was far more than that of mortar viscosity. There are two reasons to explain this trend : the first one is that slurry quantity of mortar and fresh concrete increase with the increase of binder content, When both superplasticizer dosage and water-binder ratio are constant, the real superplasticizer dosage and real water consumption in unit volume fresh concrete will increase. The former increase sufficiently releases the water wrapped in flocculating constituent, while the latter increase makes the particles water film thicker, which result in declined viscosity. The other reason is that aggregate (sand and stone) volume in unit volume of fresh concrete will gradually decrease with the binder content increase, which decreased water absorption quantity and increased slurry quantity on the aggregate particles surfaces.

3.2.2 Effect of sand ratio on viscosity of mortar and concrete.Based on the tested paste viscosity values (η_{C1})of specimen 5, 2, 7 and 8, effect of sand ratio on calculated mortar viscosity (η_{C2}) and concrete viscosity (η_{C3}) are given in figure.3 respectively. figure.3 shows that (η_{C2}) and (η_{C3}) slightly increased with increase of sand ratio: (η_{C2}) increased from 6.64Pas to 8.96Pas, while (η_{C3}) increased from 26.97Pas to 27.5Pas. The increased (η_{C2}) is due to the fact that when the sand ratio increases, the sand quantity in unit volume concrete also increases while the binder ratio and dosage remain unchanged. Therefore, the slurry film covering on the aggregate particles surface was thinner, resulting in the mortar viscosity increase . For the increase of (η_{C3}), it is due to the fact that the sand water absorption is higher than that of coarse aggregate (basalt). With the increase of sand ratio, the fine sand quantity increases and the real water quantity in unit volume decreases, resulting in the increased viscosity. Meanwhile, compared with coarse aggregate, fine aggregate has a relatively higher Blaine specific surface, and all the aggregate surface are required to be covered or wrapped by slurry. Hence when the sand ratio increases, the required slurry quantity will increase. With the same slurry quantity in unit volume, the slurry film covering on the aggregate surface becomes thinner, resulting in the increased viscosity.

3.2.3 Effect of water-binder ratio on viscosity of mortar and concrete.Based on the tested paste viscosity value (η_{C1}) of specimen 8, 9, 2, and 10, effect of water-binder ratio on calculated mortar viscosity (η_{C2}) and concrete viscosity (η_{C3}) is given in figure.4 respectively. figure.4 shows that when the water-binder ratio increased, both (η_{C2}) and (η_{C3}) declined remarkably. The (η_{C2}) decreased from 17.17Pas to 6.04Pas, while the (η_{C3}) decreased from 64.63Pas to 21.92Pas. The variation trend is due to the fact that the solid particles themselves are not characterized by the viscosity, and the (η_{C2}) and (η_{C3}) depend on the particle surface's water film thickness, which then depends on the initial water consumption and the water quantity wrapped in the flocculating constituent. With the constant superplasticizer dosage , the water quantity wrapped in the flocculating constituent will maintain the same. Therefore, with increase of initial water consumption, the water volume in unit volume increases. Consequently, the water film on the particle surface in unit volume mortar/concrete becomes thicker, resulting in the decrease of (η_{C2}) and (η_{C3}).

3.2.4 Effect of class and dosage of FA on viscosity of mortar and concrete.Based on the tested paste viscosity value (η_{C1}) of specimen 11, 13, 2, and 15 mixed with FA (I) and that of specimen 11, 12 and 14 mixed with FA (II), effects of different FA class and dosage on calculated mortar viscosity (η_{C2}) and concrete viscosity (η_{C3}) are given in figure.5 respectively. figure.5 shows that when the FA dosage increased both (η_{C2}) and (η_{C3}) declined remarkably. With the same FA class, when FA dosage increased, both mortar viscosity and concrete viscosity decreased further. With the different FA class and the same FA dosage, the viscosity of mortar and concrete mixed with FA (I) decreased more than that mixed with FA (II). This could be attributed to the "ball-bearing" and water-reducing effects of FA Meanwhile, the FA density is less than that of cement, which resulted in an increased volume slurry and slurry film thickness on the particles surface when the cement is replaced by FA with equivalent weight of cement. The smaller density of FA (I) contribute to the better viscosity reducing effect. Meanwhile, the particle size of FA (I) is smaller than that of FA (II). Hence FA (I) can better fill in the space among the cement particles, which obtains denser packing. Due to the same reason, FA (I) can also replace more water from the space among relatively bigger particles. Therefore, the water film on the surface of slurry becomes thicker, resulting in the viscosity decrease.

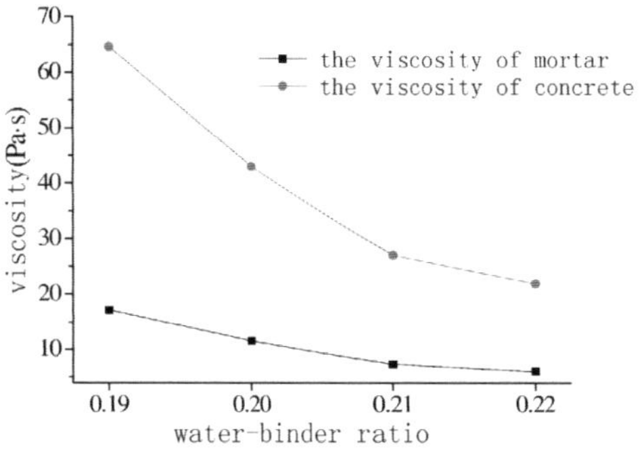

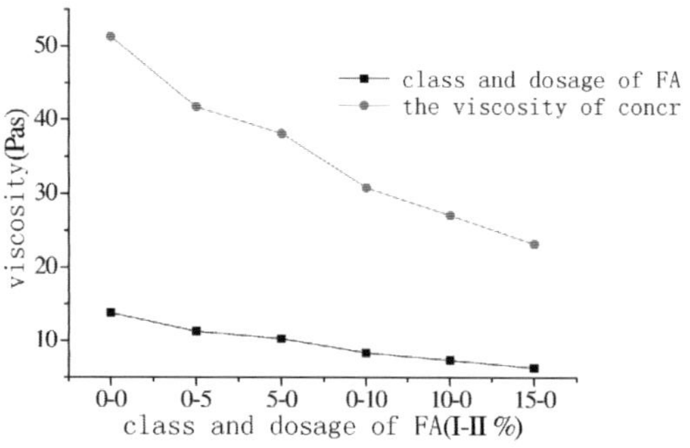

Figure.4. Effect of water-binder ratio on viscosity of mortar and concrete

Figure.5. Effect of class and dosage of FA on viscosity of mortar and concrete

3.2.5 Effect of microbead dosage on viscosity of mortar and concrete.Based on the tested paste

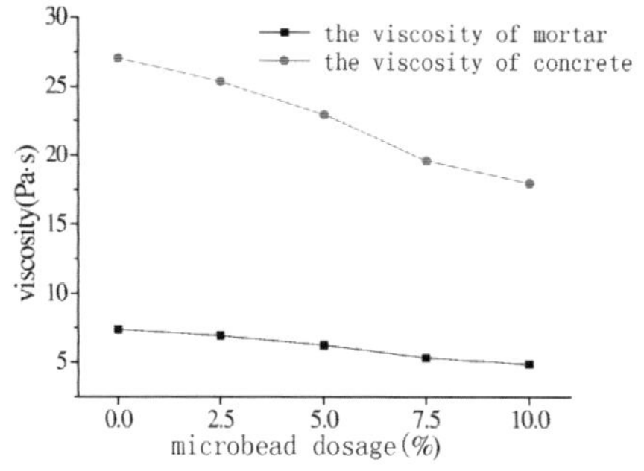

Figure.6. Effect of Microbead dosage on viscosity of mortar and concrete

viscosity value (η_{C1})of specimen 2, 16, 17, 18 and 19, effects of different microbead dosage on calculated mortar viscosity (η_{C2})and concrete viscosity(η_{C3})are given in figure.6 respectively. figure.6 shows that with the increase of microbeadreplacing percentage, both (η_{C2})and (η_{C3}) declined remarkably. The(η_{C2})decreased from 7.36Pas to 4.86Pas, while the (η_{C3}) decreased from 27.04Pas to 17.97Pas. Furthermore, the viscosity reducing effect of microbead is better than that of FA. One reason is due to their different particle morphology. Although most FA particles are spheres, all microbead particles are perfect spheres, with smooth surface. Therefore, the "ball-bearing" effect of microbead, filling in the space of cement particles, is better than that of FA. Another reason is due to the smaller particle size of continuously gradated microbead. Microbead can achieve better dense packing, making the space among particles lesser and replacing much more water from the space among relatively bigger particles. Hence, the water film on the surface of particles becomes thicker, resulting in the viscosity decrease.

3.3 The relationship between viscosity and workability of UHSC with low viscosity

Base on the discussion of rheological properties of fresh concrete, it can be seen that if the viscosity value of fresh concrete is controlled in a proper range, the plastic viscosity (η) is related to the flowing speed of fresh concrete under the same external force. And the T50, V-funnel and inverted slump time, characterizing the workability of fresh concrete, are also characteristic parameters of its flowing speed. Therefore, it may be reasonable to assume that there must be a kind of inherent relationship between viscosity and the characteristic parameters. Based on the experimental results in tables 3 and 4, the relationships between the fresh concrete viscosity and the three parameters, T50, V-funnel and inverted slump time were shown in figure.7, respectively.

figure.7 showed good linear relationships between fresh concrete viscosity and the three parameters With the increase of fresh concrete viscosity, the fresh concrete's T50, V-funnel and inverted slump time all increased. The obtained relationship between T50 and plastic viscosity is shown as y=0.1957x+2.012 (correlation coefficient R^2 value equals to 0.9334); the obtained relationship between V-funnel and plastic viscosity is shown as y=1.1222x+12.0631 (R^2 value equals to 0.9298); and the obtained relationship between inverted slump time and plastic viscosity is shown as y=0.2012x+2.4805 (R^2 value equals to 0.9078). These relationships show that viscosity of fresh concrete is well consistent with T50, V-funnel and inverted slump time respectively. Hence, the T50, V-funnel and inverted slump time can be used to intuitively reflect the fresh concrete viscosity.

However, in practical engineering, when fresh concrete viscosity is too low and segregation occurs, T50 becomes very short, but the V-funnel and inverted slump time become longer, because the sinking coarse aggregate blocks up the outlet of V-funnel or inverted slump. And when fresh concrete viscosity is relatively high, T50 becomes longer, with increase of both V-funnel and inverted slump time. This shows that it is necessary to simultaneously use the three characteristic parameters, namely T50, V-funnel and inverted slump time, to characterize the workability of UHSC with low viscosity.

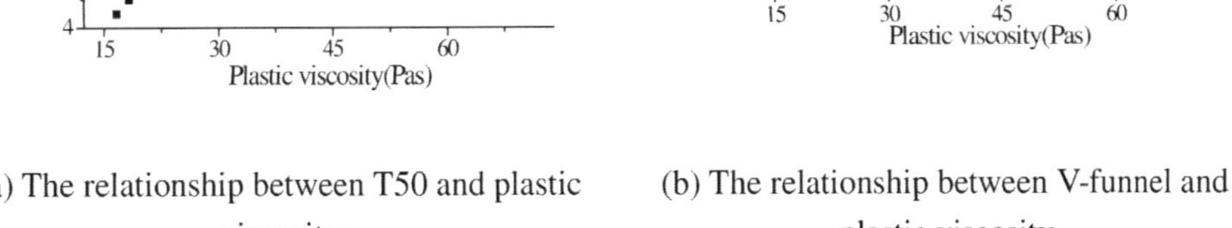

(a) The relationship between T50 and plastic viscosity

(b) The relationship between V-funnel and plastic viscosity

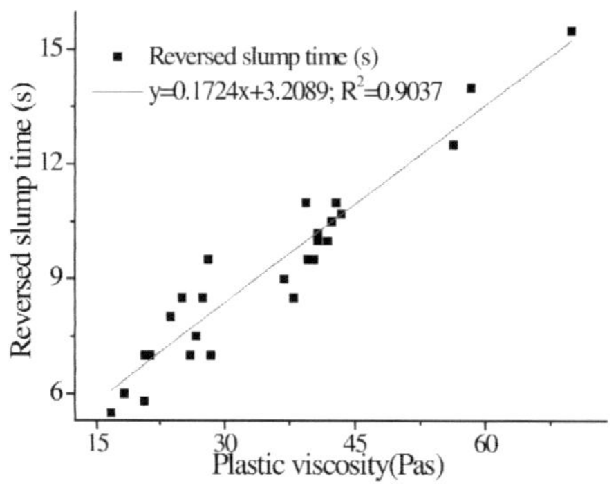

(c) The relationship between inverted slump time and plastic viscosity

figure.7. The relationship between: (a) T50 and plastic viscosity; (b) V-funnel and plastic viscosity; and (c) inverted slump time and plastic viscosity

4. Conclusions

Based on the results of this study, the following conclusions can be drawn:

(1) When binder content, microbead dosage, fly ash dosage or the water-binder ratio was increased or sand ratio was decreased, the fresh concrete viscosity would reduce correspondingly. However their effects were not the same.

(2) It was feasible to test the plastic viscosity of slurry with mineral fillers by BROOKFIELD R/S Plus Rotating Rheometer first, and then to calculate the plastic viscosity of mortar and concrete according to the micromechanics theory proposed by A.Ghanbari and B.L.Karihaloo. And the calculated results had high consistence with the objective reality.

(3) With viscosity increase of fresh UHSC, three workability characteristics of fresh UHSC, namely the T50, V-funnel and inverted slump time all increased. Good linear relationship between the viscosity and each characteristic was observed respectively. The relationships between T50 and fresh concrete viscosity, V-funnel and fresh concrete viscosity, and inverted slump time and fresh concrete viscosity can be represented as follows: $y=0.1957x+2.012(R^2=0.9334)$, $y=1.1222x+12.0631(R^2=0.9298)$ and $y=0.2012x+2.4805$ (R^2= 0.9078). These relationships show that fresh concrete viscosity is well consistent with T50, V-funnel and inverted slump time respectively. Therefore, by simultaneously using these three characteristic parameters, the fresh concrete viscosity can be intuitively reflected.

References

[1] Mikael Westerholm. 2006 *Rheology of the mortar phase of Concrete with Crushed Aggregate*[D]. Luleå University of Technology Department of Chemical Engineering and Geosciences Division of Mineral Processing.

[2] Sidney Mindess, J. Francis Yong, David Darwin. 2005 *Concrete* [M]. Beijing: Chemical Industry Press.

[3] Ferrars C.F., De Larrard F., Martys N.S.. 2001 *Fresh concrete rheology-recent developments*[J]. *Materials Science of Concrete VI. Proceedings. American Ceramic Society.* pp 215-241.

[4] Wallevik, O.H., Gjorv, O.E. 1988: *Rheology of Fresh Concrete*[J]. *Advances in Cement Manufacture and Use.Eng.* pp 133-134.

[5] Ghanbari A and Karihaloo BL. 2009, *Prediction of the plastic viscosity of self-compacting steel fibre reinforced concrete*[J]. *Cement and Concrete Research.* **39(12)**: pp 1209–1216.

[6] Deeb R, Ghanbari A and Karihaloo BL. 2012 Development of self-compacting high and ultra high performance concretes with and without steel fibres[J]. *Cement and Concrete Composites.***34(2):** pp 185–190.

[7] Karihaloo BL, Ghanbari A. 2012 Mix proportioning of self compacting high- and ultra-high-performance concretes with and without steel fibres[J].*Magazine of Concrete Research.* **64(12):** pp 1089-1100.

[8] Struble L, Sun G K 1995 *Viscosity of Portland cement paste as a function of concentration*[J]. *Advanced Cement Based Materials.* **2(2):** pp 62-69.

[9] Krieger IM, Dougherty TJ. 1959 *A mechanism for non-Newtonian flow in suspensions of rigid spheres*[J]. *Journal of Rheology.* **3(1):** pp 137-152.

Processing of waste material of radix physochlainae for preparation of fine chemicals after extraction

A He, A Yohannes, X T Feng and S Yao

School of Chemical Engineering, Sichuan University, Chengdu 610065, China

E-mail: cusack@scu.edu.cn

Abstract. Waste residues of Chinese traditional medicine radix physochlainae (Huashanshen) contain a large amount of hemicelluloses after extraction. After the removal of the cellulose and lignin, main components of the solution are different degree of hydrolysis products of hemicelluloses. In the degradation process, hemicelluloses firstly become pentose, and then pentose loses 3 molecules of water and turns into furfural. This study explored a series of conditions of the method; finally the yield of furfural can reach 8.5% (calculated with the weight of raw residues) under the condition of pH of 0.2-0.3, temperature of 104-106°C, hydrolysis duration for 10 minutes. Furfural can be further processed to be resin materials.

1. Introduction

In recent years, with the rapid development of traditional Chinese medicine and natural medicine industry, the use of herbal raw materials is increasing significantly. However, there exists the problem of the low utilization ratio of Chinese herbal medicines, and related enterprises produce a large amount of herbal residue every year, which is about 70% weight of the total amount of raw material. More than 90% manufacturers directly treat it as an industrial waste. The popular treatment mainly includes piling, burying or burning, which will not only pollute the environment, but also cause the waste of resources. According to the domestic and foreign experience, the treatment of organic solid waste should be based on resource utilization, which can realize the harmless reduction of waste material [2]. The residue always contains a large amount of crude fiber, lignin and hemicelluloses, and the degradation and transformation of application can reduce the environmental pollution and improve utilization ratio for renewable resource. As an example, radix physochlainae is a common alkaloids-rich herb used as traditional medicine, which has the clinical effects to warm lung phlegm, relieve cough and asthma, and tranquilize the mind [3]. Based on the waste residue after the extraction of bioactive ingredients, the study firstly removed cellulose and lignin effectively through appropriate pretreatment, and then the obtained hemicelluloses with different hydrolysis degree was hydrolyzed to prepare furfural with degradation process catalyzed by sulfuric acid. Furfural can be further processed to be resin materials. It can be a good way to overcome the shortcomings including low utilization of raw materials and high energy consumption.

2. Experimental

2.1. Materials, reagents and instrument

Radix physochlainae (Huashanshen) was purchased from Xindu herbal market, Chengdu. Sulfuric acid, furfural and chloroform were purchased from Chengdu Hangjia chemical reagent. Moreover, the electrical heating sleeve, DF-101D type constant temperature magnetic stirrer (Gongyi Yuhua Instrument Co. Ltd), MODEL-3 rotary evaporator (Shanghai Yarong biochemical instrument factory), SHZ-D (III) circulating water vacuum pump (Gongyi Yuhua instrument factory), TU1800 UV spectrophotometer (Beijing Puxi General Instrument Corp) and XH-MC-1 microwave reactor (Beijing Xianghu science and Technology Development Co Ltd) were used.

2.2. *Standard curve of furfural*

In the wavelength range of 200~600 nm, its maximum absorption wavelength of furfural was at 276 nm as shown in Fig.1(a). 0.0625 g authentic compound of furfural was weighed and dissolved in distilled water, and then transferred to a 100 mL volumetric flask with a pipette. After being diluted to 100 mL, the concentration of 6.25 μg/mL standard solution of furfural was obtained. 1 mL~10 mL of standard solution was further diluted to 10 mL, and 0.625 μg/mL, 1.250 μg/mL, 1.875 μg/mL, 2.500 μg/mL, 3.125 μg/mL and 3.750 μg/mL work solutions were prepared, respectively. Distilled water was used as the reference solution. According to the principle of least squares, the relationship of absorbance data (y) and concentration (x) of furfural was fitted and the standard curve is shown in Figure 1(b), the regression equation is $y=0.1626x - 0.0042$ and $R^2=0.9998$. The results showed that furfural solution showed good linear relationship in the concentration range of 0.625 μg/mL~3.750 μg/mL.

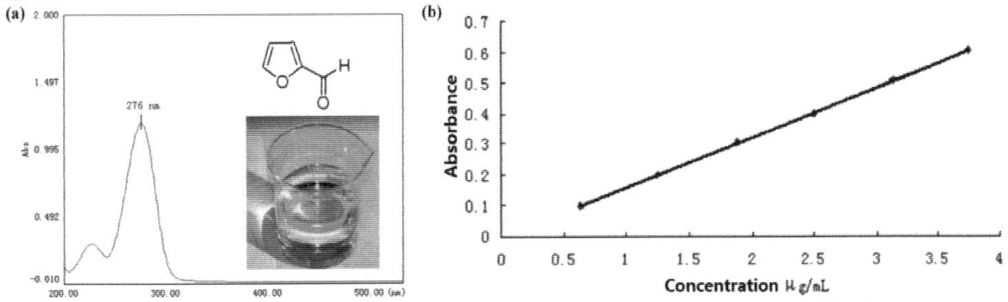

Fig. 1 UV full length scanning result (a) and standard curve (b) of furfural

2.3. *Processing procedures*

(1) Dry radix physochlainae (Huashanshen) was grinded to100 mesh, and then refluxed with 95% ethanol (ratio of mass to volume=10, g/mL) for 1 hour each time. The process was repeated 3 times to extract those useful bioactive compounds thoroughly. The waste materials were collected by filtration. (2) 10 g dry waste residue of radix physochlainae after pretreatment was weighed and placed in the flask, and then 8% NaOH solution with the 20:1 ratio of liquid to solid (mL/g) was added and reacted under stirring for 8 h under 60°C. After cooled to room temperature, the reaction system was filtered under vacuum and concentrated sulfuric acid was used to adjust the pH value of the filtrate to 2, the precipitation process was carried out on 30°C for one hour and the solid was removed by centrifugation with the speed of 500 rpm. (3) The supernatant was further mixed with 2 mol/L sulfuric acid to pH 0.2. The solution was transferred to the 1000 mL three neck flask and reacted under magnetic stirring and the increasing speed of temperature was controlled. The reaction system reached the refluxing temperature after a certain period of time. (4) The system was maintained under the refluxing temperature. After a certain period of time, furfural began to be generated in the system, which was collected through steam distillation in the end.

3. Results and discussion

3.1. *Effect of dehydration time*

Through two steps pentosan can be dehydrated to furfural; in the first step pentosan is dehydrated to become pentose with high yield and speed [4]. Compared with other studies [5,6], although this experiment employed one-pot method, the reaction substrate was no longer a pentosan-rich plant. After treatment of the first two steps in above processing procedure, most of the cellulose and lignin have been removed effectively, and the obtained solution was mainly composed of polypentose molecules with different degree of polymerization. So it can be considered to shorten the time of hydrolysis in the first step. On the other hand, the chemical stability of furfural is poor, and it is easy to be oxidized to produce bran acid; all of light, heat and air may cause the oxidation. The system of furfural production is acidic and the temperature is also high. If furfural is not immediately separated from the system, it will produce by-products, which can be found as the black material in the bottle wall. Therefore, it should be appropriate to shorten the time of dehydration. In order to determine the optimum conditions for the production of furfural, the time range of hydrolysis from 0 min to 40 min (from the start of the refluxing) was investigated. According to the experimental results, the dehydration time is suitable for 10 min, which can avoid the oxidation of furfural in the acidic environment. After the time more than 10 minutes, the yield did not continue to rise significantly, and with the extension of time more and more black jelly began to generate and adhere to the wall of the bottle, which is the by-product of oxidation of furfural. Therefore, it is advisable to keep the degradation process for 10 min after reaching the refluxing status (104~106°C).

3.2. Effect of distilment time

In the stage of steam distillation, on the one hand, the steam refluxing temperature is higher than the system temperature, which is conducive to the formation of furfural; on the other hand, steam will take furfural out of the system timely to avoid its oxidation in acidic condition under high temperature. In the study, the water vapor condensed liquid was collected; as the result, a total of 10 equivalent fractions were obtained and each fraction was 250 mL (about 30 min). According to the results of quantitation, furfural content and cumulative furfural production were shown in Figure 2(a) and 2(b). From the experimental results, it can be seen that the yield of furfural in individual fractions appears a sharp decline after No.7, and in the tenth fraction of furfural content already becomes very low. The whole trend is to increase first and then decrease. For the economic consideration, 9×250 mL of distillation liquid is more reasonable for the efficient collection of furfural. The average yield of furfural was 8.5%.

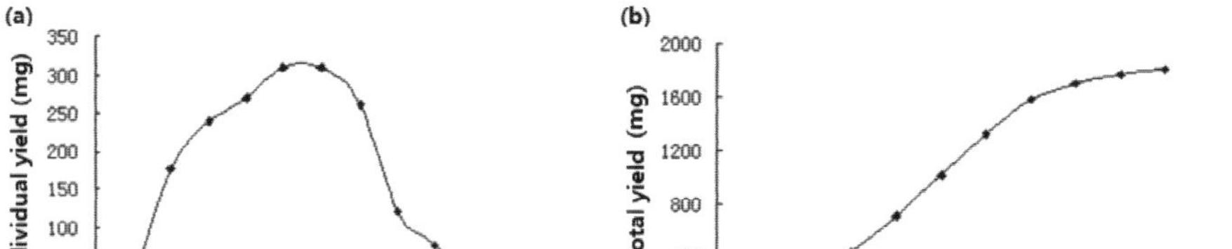

Fig. 2 Furfural content in each fraction (a) and accumulative total of furfural in steam distillation (b)

3.3. Separation of furfural product

The distillation liquid of furfural solution was collected and combined. On the basis of the great difference of furfural solubility in aqueous solution under different temperature, the furfural product can be separated. Especially, for the existence of the large amount of salt produced in the front steps of pH adjustment, the aqueous solution is almost saturated, which can further reduce the solubility of furfural in aqueous solution after cooling. As the result, furfural exists in the upper layer, and it can be obtained by separating funnel. Residual furfural in aqueous solution was detected by UV, and the recovery rate was calculated as 68.44%. Although furfural extraction from aqueous solution cannot be completely, more than half of the furfural can be obtained through the liquid; meanwhile it reduces the amount of furfural in aqueous solution and the throughput of distillation.

3.4. Microwave assistance
Microwave radiation can pass through the microwave-transparent solvents freely and then arrive at the inner parts of plant materials, resulting in a sudden rise of temperature in related system. Continuous high temperature will make internal pressure exceed the expansion capacity of cell wall, which can greatly accelerate the degradation rate and enhance degradation efficiency obviously. On the basis of previous research, this study investigated the effects of different power of microwave on the yield of furfural by using 100 ~ 900 W adjustable microwave reactor. It was found that the yield of 8%~9% was also achieved within 6~7 min when 500 W assisted microwave was used to help catalytic degradation. However, if the microwave power is too great, the local temperature will be over high, which can easily result in severe carbonation of raw materials and the significant increase of production of byproducts. The technology is proved to be an important method for the degradation of herbal residues, and it can be also taken into consideration in the process of extraction.

3.5. Further processing for resin material
In order to evaluate the applicability of this kind of furfural product prepared from herbal materials, resin material was selected as an object in the section. Furfural resin is a kind of polymer which can be obtained by condensation polymerization or copolymerization with furfural as the main raw material. It has high temperature resistance, good chemical resistance and electrical insulation, and can be used in the production of tube plates, machinery parts and equipment lining coating, etc. We firstly used furfural to form a linear polymer under acid catalysis, and then double bond polymerization continues to occur in the presence of acid. Finally, the insoluble and non-fused cross-linked polymer can be obtained. The resin product is a black sticky liquid, which can become solid after the hardening treatment with benzene sulfochloride or *p*-toluene sulfonic acid. The result indicated the self-made furfural from waster residue had good property in above processing process.

4. Summary
In the extraction process, herbal medicine has experienced a series of grinding, swelling and breaking of cell processes (especially in the use of novel solvents just like ionic liquids in recent years, which have strong influence on the herbs skeleton), which is very conducive to the further processing. The corresponding residues always contain a large amount of crude fiber, lignin and hemicelluloses, and the degradation and transformation can be applied to prepare high value-added products and reduce environmental pollution effectively. This paper developed an easy and feasible way to exploit the roots of the herbal material, meanwhile provided a meaningful reference for similar studies.

References
[1] Shi L C, Ye C and Li X 2012 Treatment method and comprehensive utilization of drug residues in traditional Chinese medicine production enterprises *Guide of China Medicine* **10** 385-386.
[2] Zhen G M and Huang G H 2006 Biology and control of compost environment, *Science Press*, Beijing
[3] National Pharmacopoeia Commission, 2010 China Pharmacopoeia (2010 edition), *China Medical Science and Technology Publishing House,* Beijing
[4] Xu J, Xiao C H and Hui Y H 2010 Production technology of furfural and its prospect *Journal of Puyang Vocational and Technical College* **23** 150-152.
[5] Li Z S and Yi W G 2010 Study on furfural preparation from corn cob, *Fine Chemical Intermediates* **404** 53-55.
[6] Xu G F, Lin W and Cai L X 2010 Studies on furfural production with wheat stalk. *Journal of Putian University* **17** 75-77.

Oxidation characteristics of MgF_2 in air at high temperature

H K Chen, Y Y Jie and L Chang

Department of Chemistry and Chemical Engineering, Baoji University of Arts and Science, Baoji 721013, PR China

E-mail:hk7115@yahoo.com

Abstract. High temperature oxidation properties of MgF_2 in air were studied. The changes of phase composition, macro surface morphology, weight and elemental composition of MgF_2 samples with temperature were investigated by using XRD, EDS and gravimetric analyses. The results show that the oxidation reaction of MgF_2 converted to MgO occurred at high temperature, and the reaction was accelerated by the increase of temperature and the presence of impurities. This result clarifies the understanding of the high temperature oxidation behavior of MgF_2 in air, and provides a theoretical basis for the reasonable application of MgF_2 in optical coating materials, electronic ceramic materials and magnesium melt protection.

1. Introduction

MgF_2 is an excellent optical coating material. Because of its outstanding properties such as wide transparent band, low refractive index, wide energy gap, high mechanical strength and laser damage threshold, it is considered to be the first choice for the preparation of the films with high reflectivity and high damage threshold [1]. MgF_2 is also an important electronic ceramic material. It has been attracting extensive attention in the development of metal ceramic materials due to its good insulating performance. In addition, in the process of smelting of magnesium and its alloys, MgF_2, as a major component of composite protective film, plays a key role in the protection of magnesium melt against its combustion by SF_6 and other fluorine containing gases. The content of MgF_2 determines the protective effect of SF_6 and other fluorine containing gases for magnesium melt [2~5].

It is generally believed that MgF_2 is a stable ionic compound, which does not decompose under normal circumstances, and it will not be converted into other compounds. However, recent studies have found that MgF_2 become less stable at high temperatures when impurities participate and it has a tendency to change to MgO. In the experiment of MgF_2/Al composite films deposited by electron beam evaporation in vacuum, Weimer et al. [6] found that there was about 1% MgO was present in the MgF_2/Al composite film. They believe that the emergence of MgO may be due to the raw materials containing 2% of the impurities Na that leads to the oxidation of MgF_2. Xu Zhiyuan and co-workers [7] researched the preparation process of a 30%Ag-MgF_2 cermet material. It was found that when the 30%Ag-MgF_2 cermet material was prepared in high oxygen partial pressure at temperature higher than 600 °C, MgF_2 component in the Ag-MgF_2 cermet will change to MgO by oxidation. But pure MgF_2 in the atmosphere, even though at 1000 °C, is rarely oxidized, while Ag participates in, MgF_2 will be severely oxidized into MgO under this condition. Sun Daming et al. [8] investigated the oxidation behavior of MgF_2 in Ag-MgF_2 cermets. They also found that when there is Ag participation, MgF_2 at high temperature and high oxygen partial pressure atmosphere (such as air) can be oxidized to produce MgO; otherwise, MgF_2 is rarely oxidized or not oxidized at all.

However, others have also come to the other results. Liu Meiyan et al [9] studied the oxidation of MgF_2 in air by X ray diffractometer and found that with the increase of temperature above 800 °C, whether or not it is doped with other elements, MgF_2 will be converted to MgO. The higher the temperature, the greater the conversion rate. Obviously, these results are different from those of Xu Zhiyuan and Sun Daming et al.

Therefore, whether MgF_2 can be oxidized in the air at high temperatures, and whether the participation of impurities has an effect on the oxidation of MgF_2 are still controversial issues. Because of the oxidation of MgF_2 decreases the quality of MgF_2, which diminishes its practical performance badly, it is necessary to further study the high temperature oxidation behavior of MgF_2 in the air. In this paper, the oxidation properties of MgF_2 in the air at high temperatures were studied. The purpose of the paper is to clarify the conditions and rules of the oxidation reaction of MgF_2 in the air, which will provide a theoretical basis for the reasonable processing and application of MgF_2 as optical coating materials and electronic ceramic materials and for the optimization of protection conditions of fluorine-containing gases for magnesium melt.

2. Experimental

The raw materials used for the present study were high purity MgF_2 (HP MgF_2) and analytical reagent MgF_2 (AR MgF_2) powder. Their chemical composition is shown in Table 1.

Table 1. Chemical composition of HP MgF_2 and AR MgF_2 (wt%)

Specimen	Na	Fe	Si	Pb	Ca	SO_4^{2+}	H_2O	MgF_2
HP MgF_2	0.001	0.001	0.002	<0.001	0.002	0.002	0.001	99.99
AR MgF_2	0.05	0.3	0.2	0.05	0.3	0.6	1.0	97.5

The oxidation tests of MgF_2 in the air were carried out in a SX2-10-12 box-type resistance furnace. Approximately 1g (accurate to 0.0001g) of MgF_2 powder, which was ground to 200 meshes in an agate mortar before the test, was weighed by using an electronic balance and put into a 50mm×28mm×10mm alumina crucible. Then the sample and crucible were moved together into the resistance furnace, heated to the set temperature at a rate of 5-7 °C/min, and held at the temperature for 2h. After that, the sample was cooled down to room temperature, weighed and stored in a dryer for the X ray diffraction (XRD) analysis, macroscopic morphology and energy dispersive spectrometer (EDS) analyses. The selected oxidation temperatures were 600, 700, 750, 800, 850, 900 and 1000 °C, respectively.

XRD analysis of the oxidized MgF_2 samples was carried out on a Rigaku Ultima IV X-ray diffractometer with a Cu-Kα source operated at 40kV and 40mA. The scanning speed is 4°/min, the step width is 0.02°, and the diffraction angle is 20° to 80°. The elemental composition of the oxidized MgF_2 samples was investigated by an EDAX Genesis APEX energy dispersive spectroscopy assembled in Quanta FEG 250 field emission scanning electron microscope.

3. Results

3.1. XRD analysis

The X-ray diffraction analysis results of HP MgF_2 samples after oxidation at different temperatures in the air for 2h are shown in figure 1. It can be seen that there was obvious MgO appeared in HP MgF_2 at 800 °C. From 800 °C to 1000 °C, with the increase of temperature, the content of MgO increased gradually and MgF_2 content decreased gradually. This result shows that MgF_2 was oxidized and converted to MgO at high temperature in the air, and the degree of the oxidation gradually increased with the increase of temperature. These results are consistent with those of Liu Yanmei et al [9], but different from those of Xu Zhiyuan and Sun Daming et al [6~8].

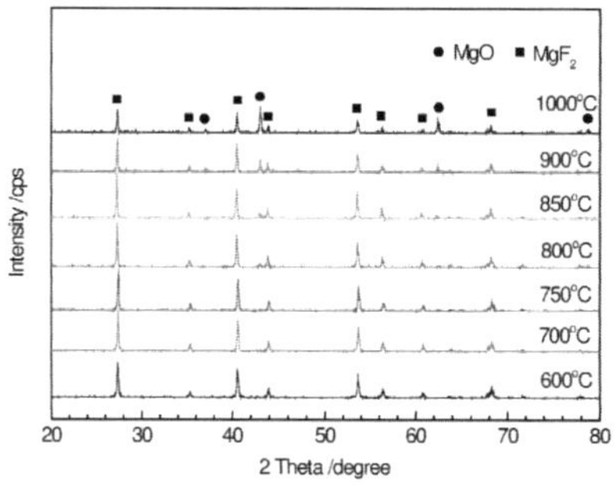

Figure 1. XRD results of HP MgF_2 samples after oxidized at different temperatures for 2h.

In order to investigate the influence of impurities on the oxidation process of MgF_2, HP MgF_2 and AR MgF_2 samples were put into the resistance furnace together and oxidation experiments on them was carried out. Figure 2 shows the results of X ray diffraction analysis of the samples after oxidized at different temperature for 2h. As can be seen, in this case, HP MgF_2 and AR MgF_2 also began to produce MgO at 800 °C. From 800 °C onwards, as the temperature increased, MgO content increased gradually and MgF_2 content decreased gradually. Up to 1000 °C, the main phase of AR MgF_2 has become MgO. In addition, it can be seen that at the same temperature, the content of MgO in AR MgF_2 was greater than that of MgO in HP MgF_2. These results indicate that not only the increase of temperature accelerates the process of oxidation reaction of MgF_2, but also the presence of impurities accelerates the oxidation process of MgF_2. The results confirm the conclusion of Weimer et al [6], Xu Zhiyuan and Sun Daming et al [7,8] that the impurities accelerate the oxidation of MgF_2, but the results are inconsistent with those of Liu Yanmei et al [9].

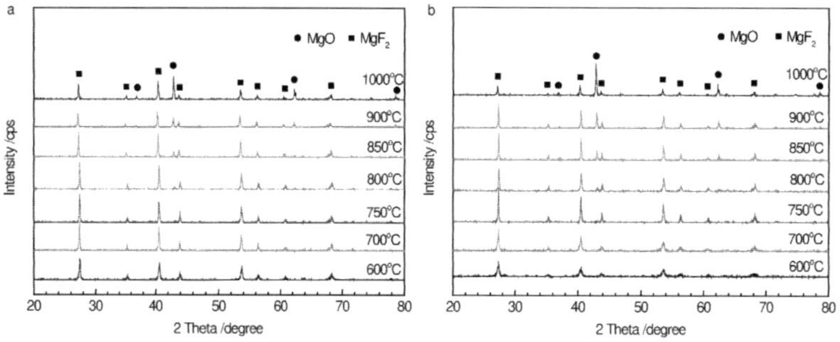

Figure 2. XRD results of MgF_2 samples after simultaneously oxidized at different temperatures for 2h: (a) HP MgF_2; (b) AR MgF_2.

Compared with figure 1 and figure 2 (a), we can see that at the same temperature, the content of MgO produced by the individual oxidation of HP MgF_2 is less than that produced by HP MgF_2 that was simultaneously oxidized with AR MgF_2. This result indicates that the rate of individual oxidation of HP MgF_2 is less than that of HP MgF_2 that was simultaneously oxidized with AR MgF_2 at the same temperature. The reason for this result may be that the high temperature evaporation or decomposition products of impurities in AR MgF_2 promoted the process of MgF_2 oxidation. The problem needs to be further studied, but the result can also confirm that the participation of impurities can accelerate the oxidation reaction of MgF_2.

3.2. Macroscopic morphology

Figure 3 shows the macro-morphology photos of HP MgF_2 and AR MgF_2 samples after oxidation at the same time at different temperatures in the air for 2h. As is seen from figure 3, the macro-morphology of HP MgF_2 samples showed no significant difference with the increase of temperature, while the macro-morphology of AR MgF_2 samples changed significantly with the increase of temperature. At 750 °C, the color of AR MgF_2 sample has been turned from white to pale red. As the temperature increased to 800 °C, there were cracks appeared in the sample. From 800 °C onwards, with the increase of temperature, the cracks gradually increased and was accompanied by obvious shrinkage. Combined with the results of XRD analysis, it can be considered that the macro-morphology changes of HP MgF_2 and AR MgF_2 samples are mainly caused by the oxidation of MgF_2. Therefore, from the macro-morphology changes in the samples, it can be seen that the oxidation rate of AR MgF_2 was higher than that of HP MgF_2, and the rate increased with the increase of the temperature. These results also show that the participation of impurities and the increase of temperature accelerate the rate of MgF_2 oxidation.

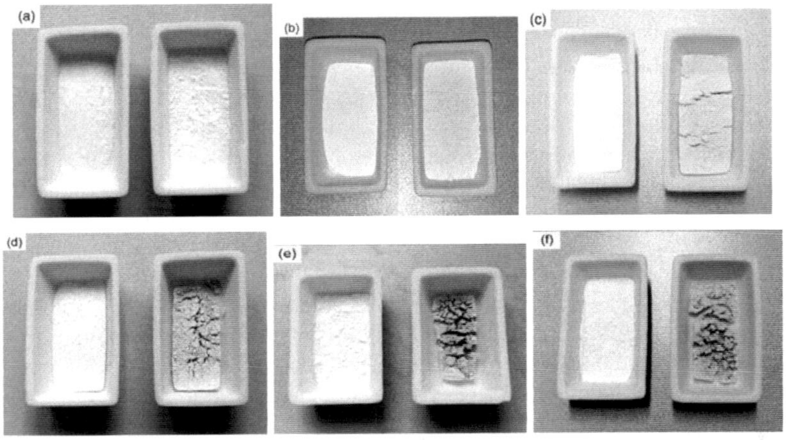

Figure 3. Macro surface morphologies of MgF_2 samples after simultaneously oxidized at different temperatures in the air for 2h: (a) before oxidation; (b) 750 °C; (c) 800 °C; (d) 850 °C; (e) 900 °C; (f) 1000 °C (The left is HP MgF_2 and the right is AR MgF_2 in each figure).

3.3. Weight loss

Figure 4 shows the weight loss of HP MgF_2 and AR MgF_2 samples after individual and simultaneous oxidation in the air at different temperatures for 2h. As shown in figure 4, from 800 °C onwards, the weight loss rate of HP MgF_2 and AR MgF_2, whether they were oxidized individually or simultaneously, showed a significant increase with the increasing of temperature. The weight loss rate of AR MgF_2 was the highest, the weight loss rate of HP MgF_2 that was simultaneously oxidized with AR MgF_2 was in the middle, and the weight loss rate of HP MgF_2 that was individually oxidized was the least. Since MgF_2 does not decompose and evaporate at atmospheric pressure within 1000 °C [10], and the oxidation process of MgF_2 into MgO will result in weight loss (the molar mass of MgF_2 is 62.3g/mol^{-1}, MgO is 40.3g/mol^{-1}, and theoretically the weight loss rate of 1mol MgF_2 completely oxidized and converted to MgO is 35.31%), it can be considered that the weight loss of MgF_2 samples means the occurrence of MgF_2 oxidation, and the greater the weight loss rate, the greater the degree of the oxidation. Therefore, from the weight changes of MgF_2 samples, we can see that the oxidation reaction of MgF_2 converted to MgO occurred in the air at high temperature, and the increase of temperature and the addition of impurities accelerated the rate of the reaction. In addition, it can be seen from figure 4 that the weight loss rate of AR MgF_2 after oxidation at 1000 °C is close to the theoretical value, which shows that the majority of AR MgF_2 has changed to MgO at the temperature.

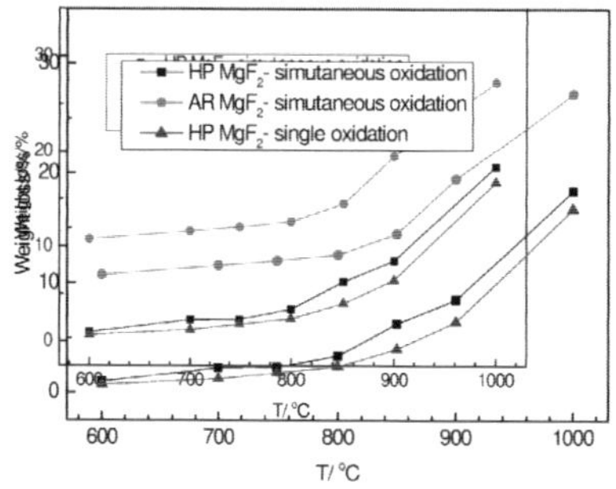

Figure 4. Weight loss of MgF_2 samples after oxidation in the air at different temperatures for 2h.

3.4. EDS analysis

The EDS analysis results of the elemental composition of HP MgF_2 and AR MgF_2 samples after individual or simultaneous oxidation in the air at different temperatures for 2h are shown in table 2. According to table 2, from 800 °C onwards, the content of O element in HP MgF_2 and AR MgF_2, whether they were oxidized individually or simultaneously, was significantly increased with the increase of temperature. The increase of O element content in AR MgF_2 with the temperature was the highest, the increase in HP MgF_2 that was simultaneously oxidized with AR MgF_2 was in the middle, and the increase in HP MgF_2 that was individually oxidized was the least. This shows that from 800 °C onwards, HP MgF_2 and AR MgF_2 had an obvious oxidation and the degree of oxidation was increased with the temperature. The oxidation rate of AR MgF_2 was higher than that of HP MgF_2 oxidized simultaneously with AR MgF_2, and the oxidation rate of HP MgF_2 oxidized individually was the least. The results also indicate that MgF_2 in the air at high temperatures does undergo the oxidation reaction of conversion to MgO, and the presence of impurities accelerates the oxidation reaction of MgF_2. These results are consistent with the results of XRD analysis above.

Table 2. Elemental composition of MgF_2 samples after oxidizing at different temperatures for 2h (wt%)

Sample	Element	600°C	700 °C	750 °C	800 °C	850 °C	900 °C	1000 °C
HP MgF_2	O	1.15	1.40	1.53	2.31	3.98	6.86	16.84
(Single	F	56.75	56.12	55.96	54.85	51.90	47.67	29.68
oxidation)	Mg	42.10	42.48	42.51	42.84	44.12	45.47	53.48
HP MgF_2	O	1.19	1.51	1.69	2.74	4.28	9.58	18.79
(Simultaneous	F	48.37	48.78	50.36	50.02	48.06	41.04	22.50
oxidation)	Mg	50.44	49.71	47.95	47.24	47.66	49.38	58.71
AR MgF_2	O	1.79	2.62	2.98	4.88	7.55	19.34	28.43
(Simultaneous	F	50.80	50.93	50.43	48.28	45.64	25.88	6.29
oxidation)	Mg	47.41	46.45	46.59	46.84	46.81	54.78	65.28

4. Discussion

MgF_2 and MgO are ionic compounds. Because of the electronegativity of fluorine element (3.90) is greater than that of oxygen (3.50) and the change in standard Gibbs free energy for the reaction of MgF_2 and O_2 to produce MgO and F_2 is highly positive, which is 502 $kJ\cdot mol^{-1}$ at 25 °C, the chemical stability of MgF_2 is greater than that of MgO, and the oxidation reaction of MgF_2 converted into MgO will not occur at normal temperature and pressure in the air. Can the oxidation reaction of MgF_2 converted to MgO occur at high temperatures? There are still different views on this issue [7~9]. And there are also different views on the problem whether the participation of impurities can accelerate the reaction of MgF_2 transformation into MgO at high temperature [7~9].

Our experiments confirmed that when the temperature was higher than 800 °C, the reaction of MgF_2 transformed into MgO in the air did take place, and with the further increase in temperature, the rate of the reaction increased gradually. The impurities in MgF_2 accelerated the rate of the reaction of conversion of MgF_2 to MgO.

According to the literature [9], there is an abnormal expansion zone in MgF_2 at about 700 °C, and the two F atoms of MgF_2 have a tendency to rotate around the Mg atom with the increase of temperature. Therefore, we hypothesized that the mechanism of oxidation and conversion of MgF_2 into MgO at high temperature might be as follows: At high temperature, F^- ions in MgF_2 have very high thermal motion energy. They are easy to break away from the shackles of Mg^{2+} under the collision of O_2 molecules in the air, give electrons to Mg^{2+}, and leave in the form of F atoms. Then the F atoms combine to form F_2 molecules and escape. Since O element has a very high electronegativity (3.50) next only to F^-, when F^- ions leave, the O_2 immediately combines with the atomic Mg to produce MgO. The reason for the presence of impurities to accelerate the rate of MgF_2 oxidation may be that the impurities play an intermediary role in the process of MgF_2 oxidation. It is involved in the process of MgF_2 oxidation and there may be a coupling effect between the reaction of impurities and the reaction of MgF_2 oxidation reaction, which reduces the free energy and activation energy of MgF_2 oxidation reaction and makes the reaction rate of MgF_2 oxidation increase. The detailed mechanism needs to be further investigated.

Our experimental result that MgF_2 in air at high temperatures can be converted to MgO is in agreement with that of Liu Yanmei et al [9], but different from that of Xu Zhiyuan and Sun Daming et al [7, 8]. The reason may be that Xu Zhiyuan and Sun Daming's experiments were carried out in an open high temperature furnace, due to the presence of air convection, the temperature of O_2 contacted with MgF_2 was relatively low, which caused MgF_2 not to be oxidized or slightly oxidized. The result of our experiment that the presence of impurity in MgF_2 accelerated the oxidation reaction process of MgF_2 confirms the conclusion by Weimer, Xu Zhiyuan and Sun Da Ming that impurities accelerate the oxidation reaction of MgF_2, but the result is not consistent with that of Liu Yanmei et al. The reason may be that in the experiment of Yan Mei Liu et al, the time of MgF_2 oxidation reaction is too short (only 30 minutes) so that the role of impurities has not come into play, which causes the result that impurities have no effect on the oxidation of MgF_2.

5. Conclusions

The oxidation properties of MgF_2 in the air at 600 °C~1000 °C were studied. The results show that the oxidation reaction of MgF_2 transformation into MgO in the air actually did take place at high temperature. The rise of temperature and the impurity added to MgF_2 accelerated the reaction process of MgF_2 oxidation. These results clarified the understanding of the high temperature oxidation behavior of MgF_2 in air and provided a theoretical basis for the reasonable processing and application of MgF_2 as coating materials and electronic ceramic materials, and for optimizing the protection condition of fluorine-containing gases on magnesium melt.

Acknowledgements

This work was financially supported by the National Natural Science Foundation of China (51471003) and the fund of the State Key Laboratory of Solidification Processing in NWPU (SKLSP201311).

References

[1] Li S, Liao H, An J, Wang J, Fu R, Xiao X, Bai H and Xie M 2013 *J. Yunnan Normal Univ.* (Nat. Sci. Ed.) (in Chinese), **33** 34
[2] Pettersen G, Øvrelid E, Tranell G, Fenstad J and Gjestland H 2002 *Mater. Sci. Eng.* A, **332** 285
[3] Cashion S, Ricketts N and Hayes P 2002 *J. Light Met.* **2** 37
[4] Chen H, Liu J and Huang W 2010 *Corros. Sci.* **52** 3639
[5] Chen H, Liu J and Huang W 2007 *Mater. Charact.* **58** 51

[6] Weimer J, Kim J, Zukic M and Torr D 1995 *J. Vac. Sci. Technol.* A, **13** 1008
[7] Xu Z, Sun D, Li A and Sun Z 2002 *J. Chin. Ceram. Soc.* **30** 505
[8] Sun D, Sun Z, Li A and Xu Z 1999 *Vacuum*, **55** 39
[9] Liu Y, Zhao Z, Li A, Yang K, Han J and Sun Z 2002 *J. Anhui Univ.* (Nat. Sci. Ed.) (in Chinese), **26** 46
[10] Nie S, Xiong S and Liu B 2006 *Mater. Sci. Eng.* A, **422** 346

SDS-assisted hydrothermal synthesis of porous $CdIn_2S_4$ microspheres

X F Bai [1,2], J S Li [2], W Wu [2]

[1] Institute of Petrochemistry, Heilongjiang Academy of Sciences, Harbin 150040, China

[2] School of Chemistry and Material Sciences, Heilongjiang University, Harbin 150080, China

E-mail: tommybai@126.com

Abstract: The porous $CdIn_2S_4$ microspheres were synthesized via a sodium dodecyl sulfate (SDS)-assisted hydrothermal technology. The as-prepared $CdIn_2S_4$ products were characterized by X-ray diffraction, field emission scanning electron microscopy and UV-Vis diffusive reflectance spectroscopy. The results showed that hydrothermal time and the surfactant addition had great effect on the structure, morphology and optical property of $CdIn_2S_4$ products.

1. Introduction

$CdIn_2S_4$ has been studied as an important semiconducting photocatalytic material due to its chemical stability, photocatalytic activity [1] and optoelectronic property [2-3], especially its applications in photovoltaic conversion, photocatalytic hydrogen production and light-emitting diodes [4]. $CdIn_2S_4$ with cubic spinel structure has narrow band gap and strong absorption in the visible light region, which largely improve the efficiency of light utilization. Kale synthesized $CdIn_2S_4$ nanotubes and "marigold-like" structure by controlling hydrothermal conditions [5]. Zhou prepared $CdIn_2S_4$ nanosheets with cubic spinel phase and high crystallinity in the mercaptoacetic acid-assisted hydrothermal process without the help of templates or surfactants. The nanosheets obtained a quasi-rounded morphology with the diameter of about 80 nm [6]. Guo synthesized $CdIn_2S_4$ hollow spheres with typical opening hollow architecture via a facile hydrothermal process without any template or surfactant [7].

In this study, a series of porous $CdIn_2S_4$ microspheres were prepared by hydrothermal method using sodium dodecyl sulfate (SDS) as template. The influences of hydrothermal time, concentration of SDS on the morphology and photocatalytic properties of $CdIn_2S_4$ were studied in detail.

2. Experimental section

2.1. Preparation of materials

All chemical materals were analytical grade without further purification. In a typical procedure, $Cd(NO_3)_2 \cdot 6H_2O$ (2 mmol), $In(NO_3)_3 \cdot 4H_2O$ (4 mmol), thioacetamide (TAA,16 mmol) and different amounts of SDS were added and dissolved in 75 ml of distilled water. The as-prepared solution was then transferred into a Teflon-lined autoclave(100 ml). The autoclave was sealed and controlled at 160°C for a certain time, and then naturally lowered to room temperature. The $CdIn_2S_4$ product was obtained by centrifugation then washed with ethanol and deionized water for several times.

2.2. Characterization

The structure, morphology and optical property of as-prepared porous $CdIn_2S_4$ microspheres were characterized by XRD (Bruker D8 Advance X-ray diffractometer) using Cu irradiation(λ=0.15418 nm) with 0.5°/s scanning speed from 10° to 70°(2θ), SEM (HITACHI S-4800) with accelerating voltage was 30 kV and UV-Vis-DRS(Shimadzu UV-2450 spectrophotometer).

3. Results and discussion

3.1. Effect of the hydrothermal time

A series of porous $CdIn_2S_4$ microspheres were prepared in aqueous solution with the same amount of SDS at 160℃ for different hydrothermal reaction time. The structure and optical property of above-prepared products were characterized by XRD and UV-Vis-DRS. The results showed in Figure 1 (a-e) and Figure 2

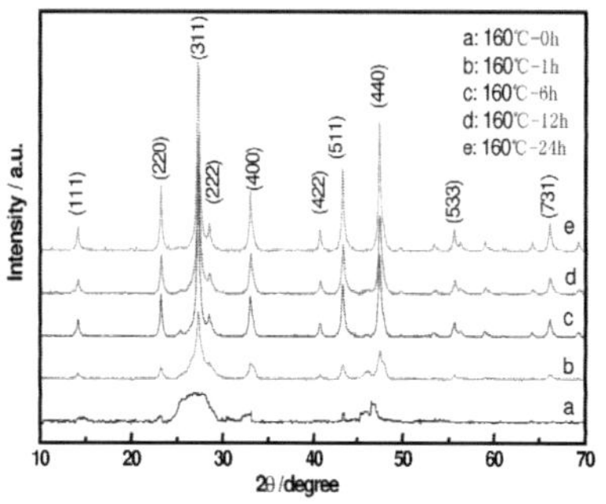

Figure 1. XRD patterns of $CdIn_2S_4$ prepared for 0 h (a); 1 h(b); 6 h(c); 12 h(d); 24 h(e)

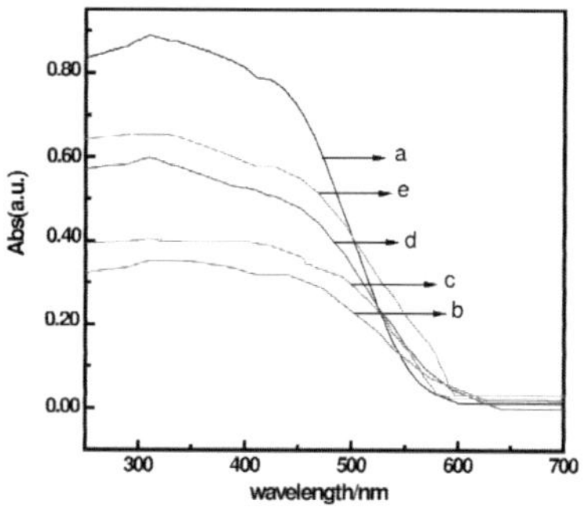

Figure 2. UV-Vis-DRS of $CdIn_2S_4$ prepared for 0 h (a); 1 h(b); 6 h(c); 12 h(d); 24 h(e)

It was revealed from Figure 1 that there are ten characteristic diffraction peaks at 2θ-values of 14.13°, 23.18°, 27.25°, 28.48°, 33.00°, 40.74°, 43.32°, 47.41°, 55.51°, 66.12°, which corresponded to (111), (220), (311), (222), (400), (422), (511), (440), (533), (731) crystal faces of pure cubic spinel phase of $CdIn_2S_4$ (ICSD-JCPDS card No.27-0060). It could be seen that hydrothermal time affected the crystalline of the products. With the hydrothermal time prolonging, the diffraction peak intensities were enhanced. However, the apparent broadening peaks of the precipitate without hydrothermal treatment indicated the absence of products with highly crystalline. As shown in Figure 2, the maximum absorption edge of $CdIn_2S_4$ prepared for 1 h was located at 560 nm corresponding to about the band gap of 2.21 eV. With the hydrothermal time prolonging, the maximum absorption edges of $CdIn_2S_4$ samples showed red shift to 610 nm corresponding to 2.03 eV.

In order to describe the producing process of the $CdIn_2S_4$ microspheres, the morphologies of the products prepared for different hydrothermal treating-time were respectively determined by SEM measurement (Figure 3(a-e)).

Figure 3. SEM images of $CdIn_2S_4$ prepared for 0 h (a); 1 h(b); 6 h(c); 12 h(d); 24 h(e)

As shown in Figure 3, unique flower-like $CdIn_2S_4$ microspheres with the average diameter of about 5 μm could be obtained. The formation of microspheres during hydrothermal process could be divided into several obvious growing stages. It was found from Figure 3(a) that mass ruleless dollops without pore structure were produced without hydrothermal treatment, and obvious aggregation of particles could be observed, which indicated that the self-assembly growth of crystallites was educed and accelerated by hydrothermal reaction. However, as shown in Figure 3(b), abundant floss-like spheres with rough surface were formed under this condition, and the size of the loose spheres was about 1-3 μm. With prolonging hydrothermal time to 6 h, the microspheres became tight and developed swelled in size. It could be seen from Figure 3(c) that flower-like $CdIn_2S_4$ microspheres composed of numerous flakes were produced through self-assembly growth process. Many slit-like pores were formed among these curved nanosheets and the inside of petals because of the flakes interconnected with each other. Figure 3 (d) showed the morphology of the product treated for 12 h, the microspheres were dispersed with good monodispersion and uniform diameter, and the size of microsphere remained unchanged. Surprisingly, some opening hollow microspheres were obtained during this stage. When hydrothermal time was 24 h, mass microspheres with particular surface consisted of pyramids were produced, as shown in Figure 3(e).

3.2. Effect of SDS assistance

A series of porous $CdIn_2S_4$ microspheres were prepared in aqueous solution with the different adding amount of SDS at 160℃ for 12 h. The structure and optical property of above-prepared products were characterized by XRD and UV-Vis-DRS. The results showed in Figure 4(a-d) and Figure 5.

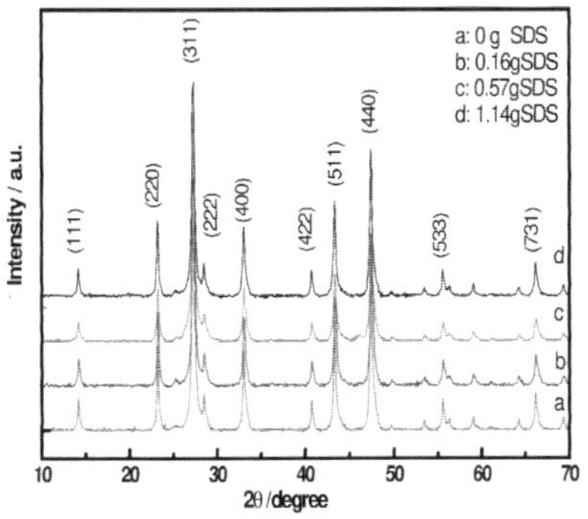

Figure 4. XRD patterns of $CdIn_2S_4$ prepared with different adding amount of SDS
0 g(a); 0.16 g(b); 0.57 g (c); 1.14 g (d)

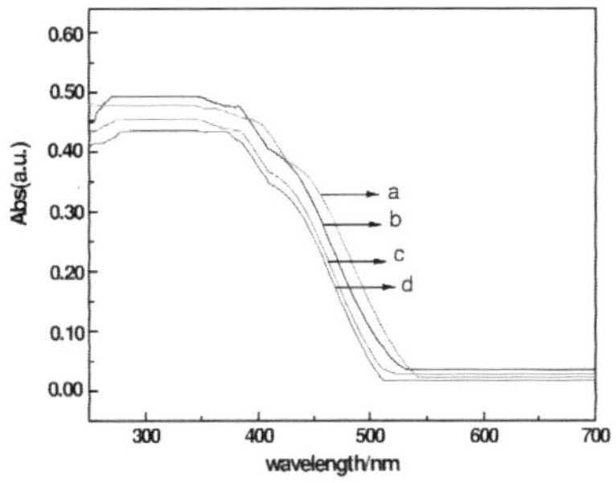

Figure 5. UV-Vis-DRS of $CdIn_2S_4$ prepared with different adding amount of SDS
0 g(a); 0.16 g(b); 0.57 g (c); 1.14 g (d)

IOP Conf. Series: Materials Science and Engineering **170** (2017) 012036 doi:10.1088/1757-899X/170/1/012036

As shown in Figure 4, with the amount of SDS increasing, the crystallinity of products remained similar, and all the diffraction peaks revealed pure cubic spinel phase of $CdIn_2S_4$. The results indicated that the concentration of SDS hardly affected the crystal structure of $CdIn_2S_4$ products. As shown in Figure 5, the UV-Vis-DRS showed that the absorption edges of samples displayed blue shift from visible region with the increasing of SDS concentration. This phenomenon was attributed to the difference on structure and morphology of SDS-assisted $CdIn_2S_4$ products. The addition of SDS could lead to the formation of small crystallitic particles, which made the absorption edges blue shift. The maximum absorption edge of $CdIn_2S_4$ prepared without SDS was located at 550 nm corresponding to band gap of about 2.25 eV; while the absorption edges of SDS-assisted hydrothermal products were located from 540 nm to 520 nm corresponding to band gaps from 2.30 to 2.38 eV.

In order to describe the producing process of the $CdIn_2S_4$ microspheres, the morphologies of the products prepared for different adding amount of SDS at 160℃ for 12 h were respectively determined by SEM measurement Figure 6(a-d).

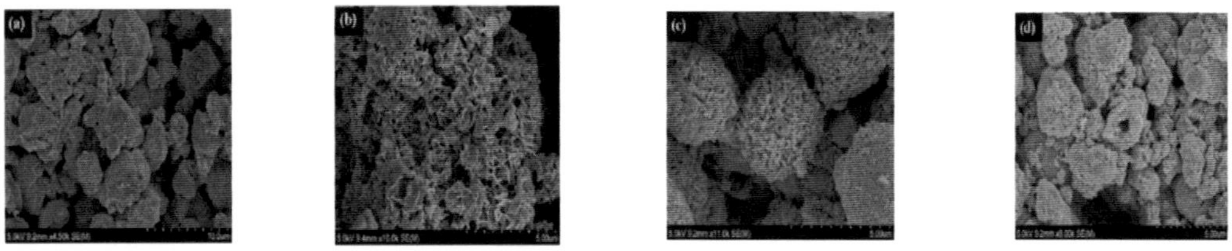

Figure 6. SEM images of $CdIn_2S_4$ prepared with different adding amount of SDS
0 g(a); 0.16 g(b); 0.57 g (c); 1.14 g (d)

As shown in Figure 6(a), a large amount of variform floss-like spheres with rough surface were obtained under hydrothermal condition without any template. With the amount of SDS increasing, the aggregation of flakes tended to self-organized into regular flower-like microspheres with an average diameter of 3-5 μm. Moreover, the gaps between petals became smaller, and many compact and dense microspheres were formed. It revealed that the SDS-assisted $CdIn_2S_4$ nanosheets were prone to self-organized into flower-like microspheres. The adding amount of SDS played a key role on the morphology by the formed different micelle shape. When the adding amount of SDS in aqueous solution was 0.57 g (2.8×10-2 mol/L), the spherical micelle could be helpful for the formation of the microspheres with regular slit pore. With the amount of SDS increasing, spherical micelle further transformed into claval micelle, which hindered the formation of even granular microspheres.

4. Conclusions

In summary, porous $CdIn_2S_4$ microspheres had been prepared through SDS-assisted hydrothermal technology. The results showed that hydrothermal time and the surfactant addition had great effect on the structure, morphology and optical property of $CdIn_2S_4$ products, respectively. The possible mechanism for the formation process of $CdIn_2S_4$ porous microspheres had been related with the self-assemble growth of $CdIn_2S_4$ crystals.

Acknowledgements
This work was financially supported by the Projects of Heilongjiang Academy of Sciences.

References

[1] Gou X L, Cheng F Y, Shi Y H 2006 *J Am Chem Soc.* **128** 7222
[2] Yu Y G, Chen G, Wang G 2013 *Int J of Hydrogen Energy* **38** 1278
[3] Sawant R R, Rajpure K Y, Bhosale C H 2007 *J Physica B* **393** 249
[4] Radautsan S I, Molodyan I P, Bulyarskii S V 2006 J Phys Stat Sol A **21** 617
[5] Kale B B, Baeg J O, Lee S M 2006 *J Adv Funct Mater.* **16** 1349
[6] Zhou Q,Zheng Z,Zhang Y 2014 *Mater.Lett.* **124** 61
[7] Fan L, Guo R 2008 *J Phys Chem C.* **112** 10700

The Effects of Cylinder Head Gasket Opening on Engine Temperature Distribution for a Water-Cooled Engine

J Y Jang [1,a] , G X Chi[2,b]

[1,2]Department of Mechanical Engineering, National Cheng-Kung University, Tainan, Taiwan 70101

E-mail:[a] jangjim@mail.ncku.edu.tw , [b]tom70155@gmail.com

Abstract. In a liquid-cooled engine, coolant is pumped throughout the water jacket of the engine, drawing heat from the cylinder head, pistons, combustion chambers, cylinder walls, and valves, etc. If the engine temperature is too high or too low, various problems will occur. These include overheating of the lubricating oil and engine parts, excessive stresses between engine parts, loss of power, incomplete burning of fuel, etc. Thus, the engine should be maintained at the proper operating temperature. This study investigated the effects of different cylinder head gasket opening on the engine temperature distributions in a water-cooled motorcycle engine. The numerical predictions for the temperature distribution are in good agreement with the experimental data within 20%.

1.Introduction

In an engine, 33% of combustion energy must be removed by the cooling system. If the engine temperature is too high or too low, various problems will occur. Therefore, the design of the water jacket is very important for engine design. Many studies of the engine have been analyzed by CFD simulations. Deger et al. [1] used CFD simulation to analysis a diesel engine exhaust manifold and got the temperature distribution of manifold. Mulemane and Soman [2] analyzed the cooling system of a four-cylinder water-cooled engine. It is shown that nucleate boiling tends to occur in the regions where the flow is not regular and the temperature is high. Jafarabadi et al. [3] analyzed a 12-cylinder diesel engine. They found that the heat transfer performance is enhanced by the boiling. Fontanesi and Giacopini [4] optimized the cooling water jacket of a six-cylinder diesel engine. Agarwal et al. [5] analyzed the heat transfer performance of the water jacket in a diesel engine. This study uses 3-D computational fluid dynamics finite difference method to investigate the effects of different cylinder head gasket openings on the engine temperature distributions in a water- cooled motorcycle engine.

2.Mathematical Analysis

A 3D model that is a 400c.c four strokes single cylinder water-cooled motorcycle engine was established, as shown in Fig. 1. Four different gasket openings, as shown in Fig. 2, are examined. Case1 (holes 1 and 2 are open, other 6 holes are closed) is the original design from the locomotive manufacturer. Case 2: holes 2 and 3 are open, others 6 holes are closed. Case 3: holes 2, 3 and 4 are open, other 5 holes are closed. Case 4: holes 1, 2, 3 and 4 are open, other 4 holes are closed. The fluid is considered 3-D incompressible turbulent flow with k- ε flow model and the flow is assumed to be steady with no viscous dissipation.

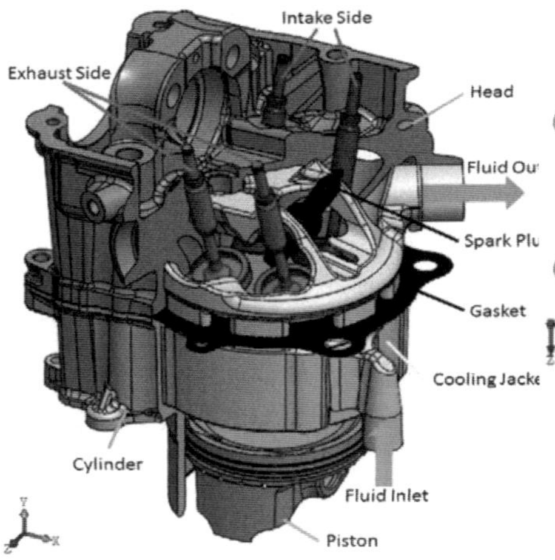

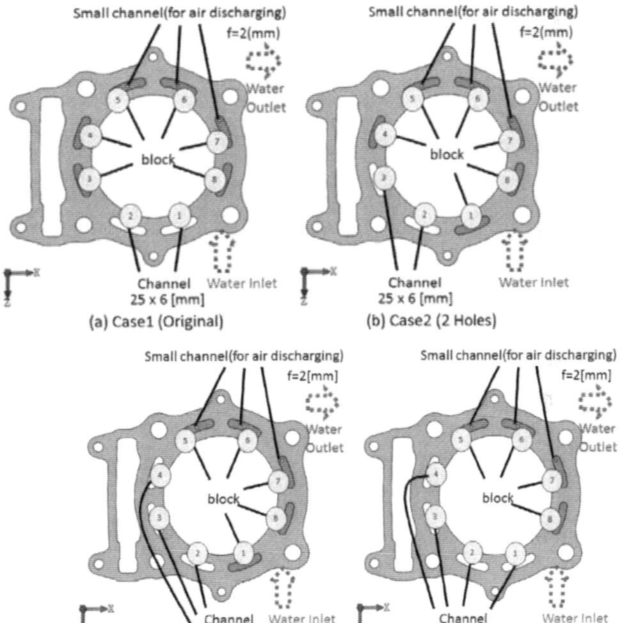

Fig.1 Physical model

Fig.2 Four different Gasket openings

. Each component of motorcycle engine is subjected to hot gases with gas temperature and heat transfer coefficient by Newton's law of cooling. To investigate the effects of different gasket openings on the uniformity of the valve seat temperature, coefficient of variation is defines as:

$$V = \frac{\sqrt{\frac{\Sigma\left(T_i - \bar{T}\right)^2}{n}}}{\bar{T}} \times 100(\%) \quad (1)$$

In this study, the governing equations are solved numerically using a control volume based finite difference formulation, ANSYS FLUENT [6]. The convergence criterion is satisfied when the residuals of all variables are less than 1.0 × 10^{-7}. Computations were performed in parallel calculation, and the computer consummation time is around 2 hours.

3.Results and Discussion

The experimental data of Case 1 are provided by the locomotive manufacturer. It is found that the calculated temperature data are in good agreement with experimental data within 20%. This verifies our theoretical model and numerical accuracy. The water volume flow rate and flow field for different gasket openings are shown in Table 1 and Fig 3, respectively. It is seen that as the number of the hole on the gasket is increased from 2 to 4, the flow rate is increased up to 18.30 %. After the water flows into the water jacket, fluid is divided into two parts. For Case 1, about 50% of the water flow rate enters the cylinder head through hole 1 in the clockwise direction, while about 37% enters the cylinder head through hole 2 in the counterclockwise direction. There is vortex between the hole 1 and hole 2. For Case 2, the flow rate is increased by 3.57% compared to the original design. About 50% of the water flow rate enters the cylinder head through hole 1, while about 46% enters the cylinder head through hole 2.

The heat transfer coefficient distribution and average heat transfer coefficient for different gasket openings are shown in Fig 4 and Table 2, respectively. It is interesting to note that for cases 3 and, 4, both water flow rates are increased due to the gasket opening holes are increased compared to case1, while the corresponding average heat transfer coefficients are reduced around 3%. For case 2, the average heat transfer coefficient is increased by around 6 %. For the original design of Case1, there is low h value under the gas exhaust side because there is vortex between hole 1 and hole 2. One can observe that the local heat transfer coefficient around the gas exhaust side is more uniform for case 2 compared to Case1. For cases 3 and 4, the local heat transfer coefficients are increased only around the gas intake side.

The temperate distributions of the cylinder head for different gasket openings are shown in Fig. 5. The highest temperature region appeares between the two exhaust valve seats in a combustion chamber. It is also seen that , case 2 has the lowest low and most uniform temperature distribution around the region of exhaust valve seats. Table 3 shows the exhaust valve seat mean temperature and coefficient of variation for different gasket openings. It is seen that for case 2, its has the minimum exhaust mean temaperaure and coefficient of variation.

Table 1 water flow rate for case1 to 4

	$\dot{Q}$ (L/min)	V_{in} (m/s)	$\frac{\dot{Q}-\dot{Q}_{case1}}{\dot{Q}_{case1}}$ (%)
Case 1	28.0	2.04	-
Case 2	29.0	2.11	3.57
Case 3	32.5	2.37	16.07
Case 4	33.1	2.41	18.30

Table 2. heat transfer coefficient for case1 to 4

	$\bar{h}$ (W/m²-K)	$\frac{\bar{h}-\bar{h}_{case1}}{\bar{h}_{case1}}$ (%)
Case1	8129.31	-
Case2	8667.56	6.62
Case3	7836.69	-3.61
Case4	7878.85	-3.08

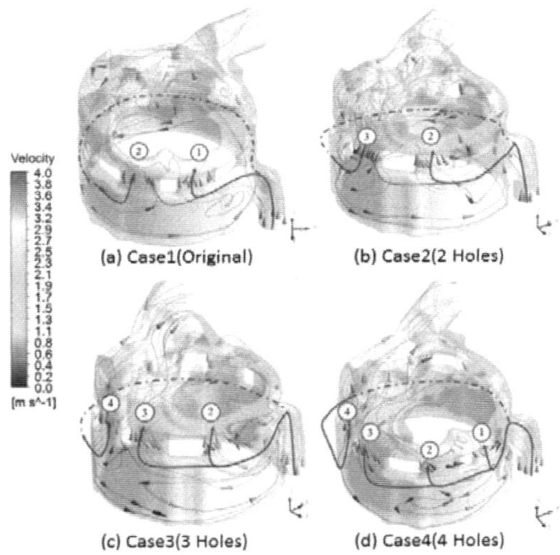

Fig.3 Flow flied for cases 1 to 4

Table 3 Exhaust valve seat mean temperature and coefficient of variation for case1 to 4

	$\bar{T}_{plug}$(°C)	$\bar{T}_{chain}$(°C)	V_{plug}(%)	V_{chain}(%)
Case1	312.71	313.67	13.41	11.78
Case2	305.15	312.20	12.12	11.57
Case3	317.32	334.78	12.57	11.87
Case4	319.71	317.48	13.16	11.18

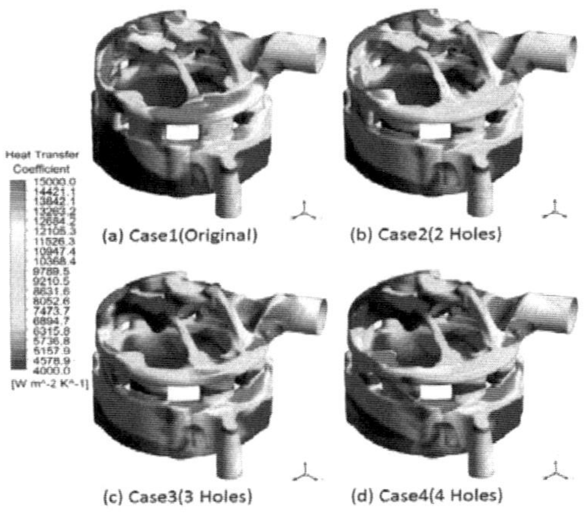

Fig.4 Heat transfer coefficient for cases 1 to 4

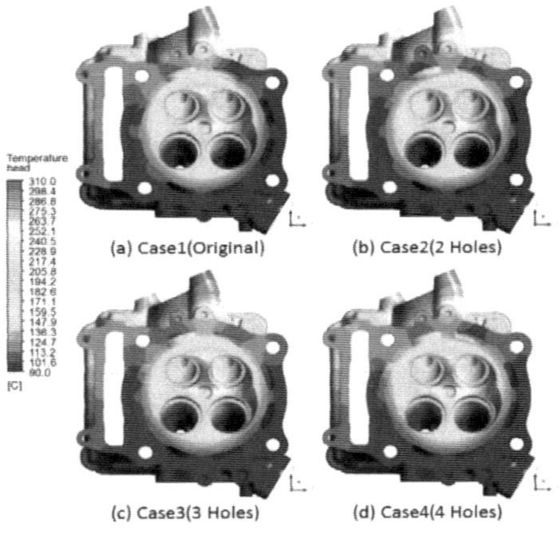

Fig.5 cylinder head temperature for cases 1 to 4

4.Conclusion

This study investigated the effects of different cylinder head gasket opening on the engine temperature distributions in a water-cooled_motorcycle engine. Compared to the original design of case 1, the water volume flow rate for case 2 is increased by 3.57 %, and the average heat transfer coefficient is increased by around 6 %. For cases 3 and 4, the water flow rates are increased by 16 to 18 %, while, the average heat transfer coefficients are reduced by around 3 %. Case 2 has the lowest and most uniform temperature distribution around the region of exhaust valve seats.

5.Acknowledgements

The financial support of this work is provided by Ministry of Science and Technology, Taiwan, under the contract number MOST 105-2622-8-006-001.

6.References

[1] Deger Y, Simperl B and Jimenez L. P 2004 Coupled CFD-FE-Analysis for the Exhaust Manifold of a Diesel Engine *ABAQUS Users 'Conference*, pp.199-208

[2] Mulemane K. and Soman R.2007 CFD Based Complete Engine Cooling Jacket Development and Analysis *SAE Technical Paper* **2007-01-4129**, pp.1-10

[3] Jafarabadi M., Chamani H., Malakizadi A. and Jazayeri S. A. 2008 A Fast Coupled Cfd-Thermal Analysis Of A Heavy Duty Diesel Engine Water Cooling System *2008 ASME International Mechanical Engineering Congress and Exposition*, pp.1-10

[4] Fontanesi S. and Giacopini M.2013 Multiphase CFD–CHT optimization of the cooling jacket and FEM analysis of the engine head of a V6 diesel engine, *Applied Thermal Engineering,* **Vol.52, No.2**, pp.293-303, 2013

[5] Agarwal D., Basu A. and Nandgaonkar M. R. 2015 Heat Transfer Analysis Of Medium Duty DI Diesel Engine, *International Journal of Informative & Futuristic Research*, **Vol.2, No.10**, pp.3627-3637, 2015

[6] ANSYS Fluent, a Release 15.0 Documentation, ANSYS, Inc.2013.

IOP Conf. Series: Materials Science and Engineering **170** (2017) 012038 doi:10.1088/1757-899X/170/1/012038

Preparation and properties of regenerated cellulose hydrogels

L Zhou[1], F Xie[1], H Li[1], W Li[1], W J Li[1,2,*] Y W Li[3]

[1]Key Laboratory of Display Materials & Photoelectric Devices, School of Materials Science and Engineering, School of Chemical and Tianjin University of Technology, Tianjin 300384, PR China.

[2]Zhejiang-California International nanosystems Institute, Zhejiang University, Hangzhou, 310012, PR China.

[3]Shandong Yingli Industrial CO.LTD, Shandong 262700, PR China.

E-mail:liwj@tjut.edu.cn

ABSTRACT. The regenerated cellulose (RCE) hydrogels were successfully prepared via an easy and green environmental method in N-methylmorpholine-N-oxide (NMMO) aqueous solution. The effect of cellulose content on the thermostability properties, swelling behavior and retention rate of hydrogels was investigated. The thermostability of RCE hydrogels was slightly enhanced with the addition of 8 wt% cellulose, the highest decomposition temperature rose from 335 ℃ to 352 ℃, and the least heat loss is about 75.60%. The equilibrium swelling ratio increases from 394.12% for 3% cellulose hydrogels to 619.46% for 8% cellulose hydrogels. The retention ratio increases from 1.13% to 28.46%.

1. Introduction

In the past decade, the hydrogels as new biomaterials prepared from natural polysaccharide have been extensively studied. It has a three-dimensional cross-linked network, only swelling but not soluble in solvents. Hydrogels have attracted wide applications in water treatment, food, cosmetics, sensors, contact lenses, tissue engineering and drug delivery systems due to their excellent hydrophilic properties, biocompatibilities, permeability and high swelling ratio [1,2]. As the ideal green environmental material, the cotton pulp with the high cellulose content is nontoxic and cheap. Macromolecular cellulose composed of glucose molecules and hydrophilic hydroxyl groups is a kind of regularity of the three-dimensional linear polymer, in which the main anhydrous glucose residues through β (1-4) glycosidic bond connect with D - pyran glucose [3,4]. Thus, the cellulose hydrogel and a polymer network system can be formed by the hydrophilic carboxyl groups combined with water molecules. Normally, the soft and white cellulose hydrogel can absorb a lot of water and maintain a certain shape. The mechanism of the hydrogel is the state of the swelling cellulose molecular chain with hydroxyl hydrogen bonding interaction connecting, gathering, and twining to form a hydrogel [5]. N-Methylmorpholine-N-oxide monohydrate (NMMO) is an effective "green" solvent for cellulose in industrial fiber-making. The NMMO shows low toxicity, high thermal stability, and the possibility of being recovered and reused. It affects hydrogen bonds and weakens van der Waals forces between the cellulose chain molecules, thus changing the cellulose structure [6].

2. Materials and methods

2.1. Preparation of RCE hydrogels

About 1g of CE was mixed with 0.02g of ascorbic acid. Then about 33.30g of an aqueous solution of NMMO (33.3g, 25g, 19g, 15.7g, 11.48g, respectively) was added to the mixture in a beaker. After that, the beaker was placed in a vacuum oven set at a temperature of 110oC and the vacuum of -0.1MPa. The heating was carried out until the mixture was transformed into a uniform transparent amber solution.

Finally, the deionized water was introduced into the beaker, and the deionized water was changed every 8h for the formation of the opalescent hydrogel containing about 3% of RCE.

The procedure was repeated by using various amounts of an aqueous solution of NMMO (25.00g, 19.00g, 15.70g, and 11.48g) to produce hydrogel containing about 4, 5, 6, and 8% of RCE, respectively.

2.2. Swelling ratio

The dried hydrogels were immersed in distilled water at room temperature. The weight of the ternary hydrogels was repeatedly measured every hour after the excess water on the surfaces was wiped with filter paper. Swelling ratios (SR) was calculated as follows:

$$\mathrm{SR}(\%)=(Wt - W0)/W0 \times 100\%. \quad (1)$$

where the Wt is the weight of the swollen hydrogel at a designated time t during swelling and the W_0 is the weight of the dried hydrogel.

2.3. Retention test

The Albumin Bovine V (BSA) powder (0.15g) was dissolved in the buffer solution (PH=7.4) for preparing the basic protein buffer (1.5 g/l). The buffer solution was composed by monosodium orthophosphate and disodium hydrogen phosphate. The concentration of BSA before and after filtration was determined by recording the protein buffer solution absorption wavelengths at 280nm, using the ultraviolet-visible spectrophotometer (TU-1901, Beijing Purkinje General Instrument Co. Ltd). The retention rate (R) was calculated as follows:

$$\mathrm{R}(\%)=(C - C_0)/C\times 100\%. \quad (2)$$

where the C and C_0 are the concentration of initial and filtered protein buffer solution.

3. Results and Discussion

The SEM image of the regenerated cellulose (RCE) hydrogels is shown in Fig. 1. It can be seen that the cellulose fibers with one-dimension are interfingering with each other and forming a three-dimensional network. Many porous structures with varying sizes in the hydrogels can be observed. After the water evaporates, lots of holes can be preserved in dried hydrogels due to the freeze-dried method, which keeps the hydrogels dry in the original state. Therefore, the numerous water molecules could easily diffuse into hydrogels to form the large pores, leading to the higher swelling ratio.

The FTIR spectrum of the regenerated cellulose (RCE) hydrogel is shown in Fig. 2. The broad absorption bands of the hydrogels at 3000–3500 cm^{-1} are assigned to stretching of a large number of hydroxyl groups on the cellulose. The stretching vibration of CH 2 is evidenced by the appearance of the absorption peaks at 2920 cm^{-1}. The band observed at 1625 cm^{-1} and 1375 cm^{-1} in the cellulose hydrogels can be attributed to COO^- stretching and bending, respectively. The results indicate that the carboxyl groups and hydroxyl groups existed in the hydrogels.

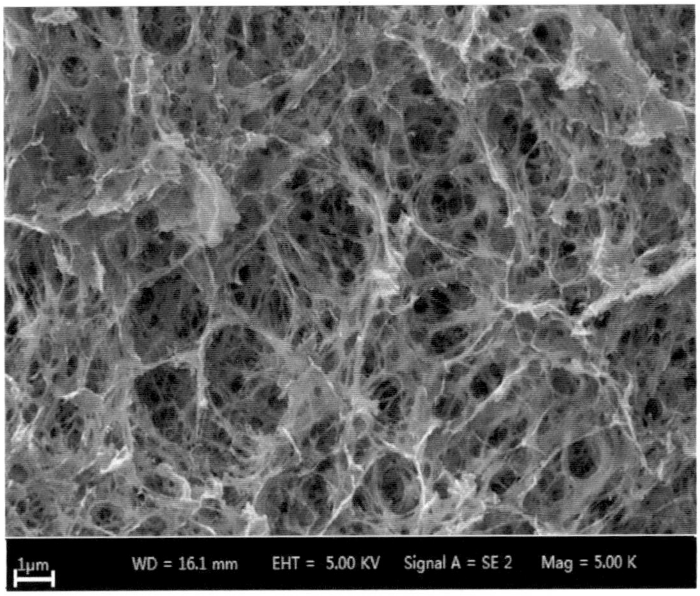

Figure 1. SEM image of the cellulose (RCE) hydrogels.

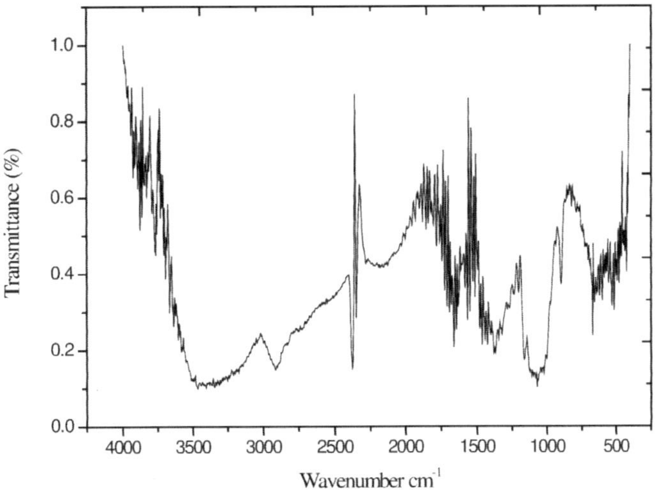

Figure 2. The FTIR spectrum of regenerated cellulose (RCE) hydrogels.

Fig. 3 shows the regenerated cellulose (RCE) hydrogels' TG curves and DTG curves, respectively. From the figure, we can see that the thermostability of RCE hydrogels are getting better with the cellulose content increasing. The least thermal degradation is 65% with 8% RCE hydrogels. The highest decomposition temperature is rising from 342°C to 352 °C.The increase in the thermal stability may be due to the increase of cellulose content in the hydrogel. As the cellulose content is increased, the number of OH groups also increase. Since the OH groups are responsible for the formation of hydrogen bonding between the cellulose molecules, increase the amount of OH groups increases the number of hydrogen bonds between the molecules and enhanced the intermolecular force so that the thermostability of RCE hydrogels is improved.

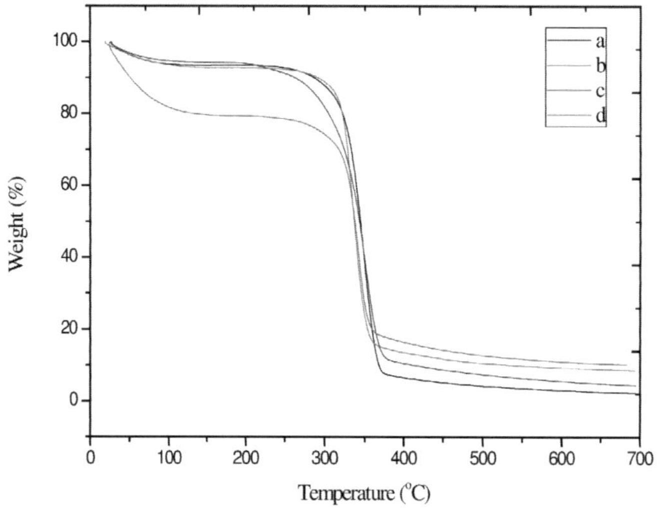

(1)

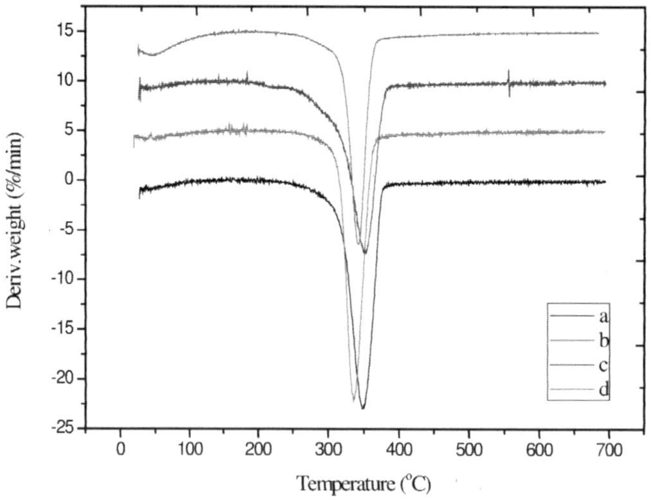

(2)

Figure 3. The regenerated cellulose (RCE) hydrogels' TG curves (1) and DTG curves (2): 4% RCE (a); 5% RCE (b); 6% RCE (c); 8% RCE (d).

Fig. 4 shows the typical swelling curves of the hydrogels with various contents of cellulose in distilled water. It is evident that the water uptake rate and equilibrium swelling ratio of hydrogels gradually increase with the increase of cellulose. This phenomenon is ascribed to the increase in the number of hydrophilic hydroxyl groups and the three-dimensional cross-linked network structure of hydrogels. For cellulose hydrogels, the rapid initial absorption occurs in the first 10 h and then gradually increases to reach equilibrium. The absorption rate is slightly increased with the increasing cellulose contents due to the fact that the water molecules are easily inhaled in the loose hydrogel at the commencement of absorption. When the absorbed water reached a certain extent, the original molecular chains are gradually stretched. The pressure is increased as the inhaled water increase, which prevents the additional water from being absorbed and penetrated through into matrix. Moreover, the equilibrium swelling ratio increases from 394.12% for 3% cellulose hydrogels to 619.46% for 8% cellulose hydrogels.The retention test of 3%, 4%, 5%, 6% cellulose hydrogels are 1.13%, 6.17%, 16.79% and 28.46%, respectively. It is obvious that the retention rate of hydrogels gradually increases with the increase of cellulose.

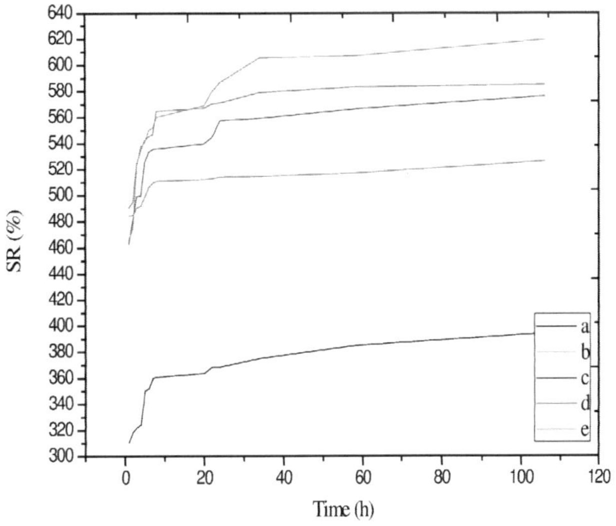

Figure 4. The typical swelling curves of the hydrogels with various contents of cellulose distilled water: 3% RCE (a); 4% RCE (b); 5% RCE (c); 6% RCE (d); 8% RCE (e).

4. Conclusions

In summary, regenerated cellulose (RCE) hydrogels were successfully prepared via a straightforward and green environmental method in N-methylmorpholine-N-oxide (NMMO) aqueous solution. The thermostability, the equilibrium swelling and the retention ratios of RCE hydrogels increased with the cellulose content.

Acknowledge

This work was financially supported by 321 talent project of Nanjing (Grant No.631783), National Natural Science Foundation of China (Grant No.51371126) and Nanjing Green and Clean Biological Technology Co.LTd

References

[1]. Kwon S.S., Kong B.J. and Park S.N.,2015 *Eur J. Pharm Biopharm.* **92**, 146–154
[2]. Xue B., Kozlovskaya V., Liu F., Chen J., Williams J.F., Campos-Gomez J., Saeed M. and Kharlampieva E., 2015 *ACS Appl Mater Inter.* **7**, 13633–13644
[3]. Klemm D., Heublein B., Fink H.P. and Bohn A. 2005 *Angew. Chem. Int. Ed.* **44**, 3358-3393
[4]. Dogan H., Hilmioglu N. D.2009 *Carbohyd Polym.* **75**, 90-94
[5]. Wang Q.Y., Cai J. and Zhang L.N. 2013 *J. Mater. Chem.* **A1**, 6678-6686
[6]. Zheng Y., Zhao J., Xu F. and Li Y. 2014 *Prog. Energy Combust.* **42**, 35–53